Ecohydrology of Water-Controlled Ecosystems

Soil Moisture and Plant Dynamics

T0211199

Understanding of the systems that control the circulation of water between atmosphere, soil, and plant life is clearly important for a general understanding of the hydrologic cycle and the way that ecosystems operate and survive. *Ecohydrology of Water-Controlled Ecosystems: Soil Moisture and Plant Dynamics* addresses the connections between the hydrologic cycle and plant ecosystems, with special emphasis on arid and semi-arid climates.

This book presents a quantitative understanding of the impacts of soil moisture on ecosystem dynamics using a probabilistic framework. It investigates the vegetation response to water stress (drought), the hydrologic control of cycles of soil nutrients, and the dynamics of plant competition for water. The book also offers insights into processes closely related to soil moisture dynamics, such as soil–atmosphere interaction and soil gas emissions. This modern and important topic is treated by building suitable mathematical models of the physics involved and then applying them to study the ecosystem structure and its response to rainfall and climate forcing in different parts of the world, including savannas, grasslands, and forests.

The book will appeal to advanced students and researchers from a large range of disciplines, including environmental science, hydrology, ecology, earth science, civil and environmental engineering, agriculture, and atmospheric science.

IGNACIO RODRÍGUEZ-ITURBE is Theodora Shelton Pitney Professor of Environmental Sciences and Professor of Civil and Environmental Engineering at Princeton University. Professor Rodríguez-Iturbe is the author of over 200 research papers and several books, including *Fractal River Basins: Chance and Self-Organization* with Andrea Rinaldo (Cambridge, 1997). A member of the US National Academy of Engineers, and many other academies throughout the world, he is the winner of numerous national and international awards including the Stockholm Water Prize, the Horton and Macelwane Medals of the American

Geophysical Union, the Langbein Lecture (also AGU), the Huber Prize and V. T. Chow Award (American Society of Civil Engineering), the Horton Lecture (American Meteorological Society), the Premio Mexico, and the Premio Nacional de Ciencias de Venezuela.

AMILCARE PORPORATO is Associate Professor of Civil and Environmental Engineering at Duke University. He is the author of more than 50 research papers. He received the Arturo Parisatti International Prize, awarded by the Istituto Veneto di Scienze, Lettere e Arti. He has been a research associate in the Civil Engineering, Environmental and Water Resources Division at Texas A&M University and the Department of Civil Engineering at Princeton University. He was chairman of the last two Ecohydrology sessions of the American Geophysical Union Spring Meeting.

Ecohydrology of Water-Controlled Ecosystems

Soil Moisture and Plant Dynamics

Ignacio Rodríguez-Iturbe

Princeton University, New Jersey, USA

Amilcare Porporato

Duke University, North Carolina, USA

CAMBRIDGE
UNIVERSITY PRESS

CAMBRIDGE UNIVERSITY PRESS
Cambridge, New York, Melbourne, Madrid, Cape Town, Singapore, São Paulo

Cambridge University Press
The Edinburgh Building, Cambridge CB2 2RU, UK

Published in the United States of America by Cambridge University Press, New York

www.cambridge.org
Information on this title: www.cambridge.org/9780521819435

© I. Rodríguez-Iturbe and A. Porporato 2004

This publication is in copyright. Subject to statutory exception
and to the provisions of relevant collective licensing agreements,
no reproduction of any part may take place without
the written permission of Cambridge University Press.

First published 2004
This digitally printed first paperback version, with corrections, 2006

A catalogue record for this publication is available from the British Library

Library of Congress Cataloguing in Publication data

Rodríguez-Iturbe, Ignacio.
Ecohydrology of water-controlled ecosystems: soil moisture and plant dynamics/Ignacio
Rodríguez-Iturbe, Amilcare Porporato.
p.   cm.
Includes bibliographical references (p.).
ISBN 0 521 81943 1
1. Ecohydrology.   2. Plant–water relationships.   3. Plant–soil relationships.
I. Porporato, Amilcare.   II. Title.
QH541.15.E19R64 2005
581.7 – dc22   2004046566

ISBN-13 978-0-521-81943-5 hardback
ISBN-10 0-521-81943-1 hardback

ISBN-13 978-0-521-03674-0 paperback
ISBN-10 0-521-03674-7 paperback

The publisher has used its best endeavors to ensure that the URLs for external websites referred to in this
publication are correct and active at the time of going to press. However, the publisher has no responsibility
for the websites and can make no guarantee that a site will remain live or that the content is or will
remain appropriate.

The colour plates referred to within this publication have been removed for this digital
reprinting. At the time of going to press the original images were available in colour
for download from http://www.cambridge.org/9780521036740

Como siempre, para Mercedes
I. R.-I.

A Sandra
A. P.

Contents

The colour plates referred to within this publication have been removed for this digital reprinting. At the time of going to press the original images were available in colour for download from http://www.cambridge.org/9780521036740

Foreword

Merging beauty with importance!

Among the numerous and diverse subjects within the geosciences, hydrologic science is arguably the fastest evolving discipline. Born from chapters and appendices of standard hydraulics and agricultural science textbooks, hydrology went through its first metamorphism 40 years ago to become a prominent science dealing with the physical laws that govern water movement within watersheds. A second metamorphism occurred in the last 30 years with the realization that water is among the primary controlling factors of the Earth's climate, and thus, the coexistence of all three physical states and the cycling among them became a research priority. Over the past ten years, however, a third metamorphism has started to develop and is primarily motivated by the recognition that the water cycle strongly influences element cycling such as nitrogen and carbon. Within the terrestrial biosphere, water availability regulates the growth of plants and controls the rate of nitrogen uptake and carbon assimilation. Hence, the interaction between the hydrologic cycle and vegetation received simultaneous attention within the climate, hydrologic, and ecological communities. *Ecohydrology* is the emerging discipline that concentrates on the cycling of water and other elements within the context of the Earth's biological productivity, the subject of this book.

From its birth, ecohydrology bifurcated early on into a phase that is primarily observational and focused on a plant's response to its microclimate (primarily spear-headed by ecologists) and a phase that focused on detailed water flow models combined with plant models of varying complexity (primarily spear-headed by hydrologists). The trajectory of ecohydrology following this bifurcation was brief and predictable – increases in observational cataloging and increases in model complexity with little intersections amongst these two trajectories. Enter Ignacio Rodríguez-Iturbe and

Amilcare Porporato – who promoted low-dimensional models that offer simplicity in interpreting physical and biological processes within the context of field experiments yet retain much of the inherent system nonlinearity. As such, this text draws upon the ripening fruits of nonlinear science, the wealth of stochastic precipitation models that formed the initial growth phase of hydrologic sciences, and the probabilistic treatment that shaped molecular physics some 80 years ago – now re-introduced into the hydrologic consciousness.

The text, composed of 11 chapters, addresses six themes central to water-controlled ecosystems. By no means can 11 chapters cover the entire depth and breadth of ecohydrology and give justice to all the literature on the topic. The authors chose to focus primarily on the propagation of stochastic rainfall patterns within the nonlinear component of the soil–plant–atmosphere continuum. The thrust is primarily devoted to how stochasticity in precipitation produces different modes of variability and how these modes affect ecosystem structure and function. The first five chapters deal with the stochastic treatment of soil moisture and its impact on plant water stress (*theme 1*). The sixth chapter introduces the coupled water and carbon uptake by plants and their interaction with the atmospheric boundary layer (*theme 2*). The seventh chapter is a preliminary treatise on plant strategies and water use – using a binary classification of extensive and intensive plant users of soil moisture (*theme 3*). Chapters 8 and 9 revisit the soil moisture dynamics with emphasis on scaling in space and time (*theme 4*). Longer-term soil moisture dynamics ranging from seasonal to interannual and spatial scaling from point-processes to hill-slope are developed. Chapter 10 introduces the connection between cycles of soil organic matter and nutrients, with particular emphasis on carbon and nitrogen (*theme 5*). Chapter 11 charts a new direction to the connection between spatio-temporal patterns of precipitation and vegetation structure. The emergence of spatially organized patterns in vegetation is explored via the methods developed in the previous chapters in the context of cellular automaton (*theme 6*) among others. The book includes numerous examples and draws upon published datasets from a wide range of water-controlled ecosystems in three continents.

Ignacio Rodríguez-Iturbe's hallmark statement (using a charming Venezuelan accent) in seminars is "we don't study . . . because it is *important* but because it is *beautiful*." Ironically, he ended up drafting a book with Amilcare Porporato whose impact on ecohydrology is probably comparable to the impact of Robert May's landmark papers on population dynamics some 25 years ago. Judging this statement must wait for the future. For now, it is a privilege to commend the following pages to individual readers and to the broad scientific community.

Gabriel Katul, Professor of Hydrology and Environmental Fluid Mechanics, Nicholas School of the Environment and Earth Sciences, Duke University

Preface

The last decade has seen a reformulation of the disciplinary basis of hydrology, which will be even more accentuated during the years ahead, with a dramatic increase of the intimate links between hydrology and the life sciences. We are convinced that the role of the hydrologic cycle throughout a wide range of temporal and spatial scales will be seen as a keystone in some of the most crucial areas related to biocomplexity, biodiversity, and the nature of the environment.

This book deals with the spatial and temporal linkages between hydrologic and ecological dynamics. It is a book on ecohydrology, which we define as the science that seeks to describe the hydrologic mechanisms that underlie eco-logical patterns and processes. The interplay between climate, soil, and vegeta-tion is central to hydrology itself and it is crucially influenced by the scale at which the phenomena are studied as well as by the physiological characteristics of the vegetation, the pedology of the soil, and the type of climate. Ecohydrology is a key component of what are loosely called biogeosciences, in reference to the interrelationship among the biological, geophysical, and geochemical approaches to understand the earth system. Hydrologic phenom-ena play a commanding role in this field and hydrologically oriented research has much to contribute towards what surely will be one of the most exciting scientific frontiers of the first part of the twenty-first century.

As Peter S. Eagleson has eloquently said:

We need to get away from a view of hydrology as a purely physical science. Life on earth also has to be a self-evident part of the discipline. In particular, I am thinking of vegetation and its powerful interactive relationship with the atmosphere, at both a local and a global level. In attempting to get the full picture, we must not be afraid to express the role of plants in our mathematical equations.
(Interview with Peter Hanneberg, in Our Struggle for Water, Stockholm: Stockholm International Water Institute, 2000.)

This statement describes well the core question that this book attempts to study; namely, how can we develop a geographically and temporally broad understanding of the dynamic coupling between biological and hydrological processes in natural systems? Within this framework we then focus on water-controlled ecosystems and the relationship between their hydrologic and vegetation dynamics. Thus, for example, one will intuitively accept that plant biomass production depends not only on the total rainfall during the growing season but also on the intermittency and magnitude of the rainfall events. Nevertheless, to quantitatively model the linkage between soil moisture dynamics and carbon assimilation accounting for the stochasticity of rainfall is a challenging problem which requires extensive simplification of very complex processes of a physical and biological nature. The dynamics of soil moisture, transpiration, and carbon assimilation takes place at the hourly time scale with soil, plant, and atmospheric boundary layer characteristics affecting the diurnal course of photosynthesis and transpiration. The linkage of this complex dynamics with climate fluctuations and/or changes in aspects such as ecosystem structure requires temporally upscaling the hourly processes. Any attempt to do this necessarily needs a simplified analytical description of the dynamics involved.

The disciplinary reformulation of hydrology which the above implies is inserted in a new intellectual frontier for the environmental sciences. Lars Hedin et al. (2002) have defined this frontier as "the natural convergence of the historically distinct disciplines of biology and physical science. This disciplinary convergence will over the next several decades transform our understanding of basic processes that control the stability and sustainability of natural environmental systems. The ensuing findings will have extraordinary implications for our abilities to predict and manage how humans impact the health of ecosystems across local, regional, and global scales. Such knowledge is a critical component of a safe, sustainable, and prosperous future."

As the title of the book implies, soil moisture is considered as the crucial link between hydrologic and biogeochemical processes. This key role is described throughout the book in terms of its controlling influence on transpiration, runoff generation, carbon assimilation, and nutrient absorption by plants among many other phenomena. Through its impact on these and other processes, soil moisture is the central hydrologic variable synthesizing the interaction between climate, soil, and vegetation. Its temporal and spatial dynamics are at the heart of ecohydrology.

The challenges that we have mentioned are part of what Roger Newton (1993) so beautifully describes in his book *What Makes Nature Tick?*: "Science at the most fundamental level is very far from being merely an efficient enumeration of experimental facts and empirical rules, nor is its structure simply determined by induction from observations. To think of it only as an orderly collection of intriguing and useful bits of information is to misunderstand its cultural value and its fascination altogether. Science is, in fact, an intricate edifice erected from complex, imaginative designs in which esthetics is a more powerful incentive than utility. Beauty, finally, comprises its greatest intellectual appeal." The simplifying assumptions made in different parts of the analysis can obviously be criticized as naive and/or incomplete from many different points of view. This is always the case when attempting to model nature and even more so with the type of dynamics that is the subject of this book. Nevertheless, we are convinced that it is through necessarily simplistic models with a strong esthetic incentive that general principles and an illuminating quantitative framework of analysis are laid out to facilitate progress in enormously complex problems.

The subject and results of this book are by no means exhaustive or conclusive since ecohydrology has been experiencing major scientific advances in recent years, and this will undoubtedly continue, even more so, in the years ahead. Nevertheless, we believe it is an appropriate moment to present some of the results with a unifying perspective and through a coherent framework, hopefully facilitating in this manner future work on these topics.

We owe recognition to institutions and individuals. Princeton University, Politecnico di Torino and, more recently for one of us, Duke University have provided a most supportive environment that has enabled our close collaboration, even when we were continents apart. To these institutions, our grateful thanks. Peter S. Eagleson from MIT has been a source of friendship and guidance for I. R.-I. for 30 years. More recently, his path-breaking work on ecohydrology has been an inspiration for us in attempting to work in this area. Rafael L. Bras, from MIT, Andrea Rinaldo, from the University of Padova, and Juan Valdes, from the University of Arizona have been, for many years, very special friends and close companions in science for I. R.-I. Luca Ridolfi, from Politecnico di Torino, has been a source of invaluable support and close friendship for A. P. from his days as a student, and more recently a friend and coworker for I. R.-I. The trust and support of Luigi Butera and Sebastiano T. Sordo and the friendly help of Roberto Revelli, from Politecnico di Torino, were important to A. P. during his stay at Princeton University. David R. Cox (Oxford) and Valerie Isham (University College, London) have also been a source of personal and professional support for which we are very grateful.

Paolo D'Odorico (University of Virginia), Francesco Laio (Politecnico di Torino), Paolo Perona (ETH, Zurich), Davide Poggi (Duke University), Andrew Guswa (Smith College), Coral Fernandez-Illescas (Harvard University), Mark van Wijk (University of Wageningen), Edoardo Daly (Politecnico di Torino), Eduardo Zea (Princeton University), and Kelly Caylor (Princeton University) were close research collaborators as graduate students and postdocs. We are extremely grateful for their generous and enthusiastic contributions to this book. We could not have dreamt to have more creative and hard-working colleagues.

Simon A. Levin and Steve W. Pacala, from Princeton University, were always ready to give us guidance and support, trying to remedy the enormous voids we present in our ecological knowledge (or lack of it!). We have benefited more than we can express from our close contact with them.

Dennis Baldocchi and Nancy Kiang, from University of California, Berkeley, Robert J. Scholes, from CSIR, Pretoria, Steve Archer, from University of Arizona, Tucson, Hank Shugart, from University of Virginia, Charlottesville, Philip Fay, from University of Minnesota, Duluth, and Eric Small, from University of Colorado, Boulder, kindly provided data, reviews, photographs, and unpublished results. It is a pleasure to acknowledge their help and generosity.

Gabriel Katul and John Albertson, from Duke University, have always offered support and many useful suggestions. Their expertise is gratefully acknowledged. They have also been a source of precious and friendly assistance for A. P. in his move to Duke University.

Michael A. Celia has been a close friend for I. R.-I. for years. More recently he has also been a close research collaborator who has contributed enormously to the efforts of this book. Special thanks for all this and also to him and Eric F. Wood for the close daily companionship during our academic life in Princeton.

Over the years the US National Science Foundation has funded our research in the topic of this book, most recently through the grants on Biocomplexity and the National Center of Earth Surface Dynamics. We gratefully acknowledge this support. A. P. also acknowledges the Office of Science, Biological and Environmental Research Program (BER), US Department of Energy, through the Great Plains Regional Center of the National Institute for Global Environmental Change (NIGEC).

1

Introduction

Soil moisture and plants are the two main subjects of this book, the former being at the center of the hydrologic cycle and the latter ones representing the primary component of terrestrial ecosystems. The analysis of their interrelationships points at the very heart of ecohydrology, the science that studies the mutual interaction between the hydrologic cycle and the ecosystems (Rodríguez-Iturbe, 2000; Porporato and Rodríguez-Iturbe, 2002).

The interaction between water balance and plants is responsible for some of the fundamental differences among various biomes (e.g., forests, grasslands, savannas) and for the developments of their space–time patterns. The first objective of ecohydrology is thus to understand the intertwined characteristics of climate, soil, and vegetation that make a biome what it is, and to relate hydrologic dynamics to the space–time response of vegetation in a region. Throughout the book, we will concentrate on water-controlled (or water-stressed) ecosystems, where water may be a limiting factor not only because of its scarcity but also because of its intermittent and unpredictable appearance.

Understanding what is the relative importance of the interactions between soil moisture and plants and how this importance changes from one ecosystem to another will be the guide to our modeling effort, which in many ways is inspired by the principle that "the purpose of models is not to fit the data but to sharpen the questions" (Karlin, 1983). We will use simplified analytical models to describe the various mechanisms responsible for the dynamics of soil moisture, from the most basic ones at a point to the more complicated cases involving different spatial and temporal scales. The necessary assumptions will be stated clearly to warn the reader about the possible limitations of each analysis, always striving for analytical tractability and results of some general validity.

In this introductory chapter we give an overview of the problems that will be analyzed in the following. The discussion of different examples to explain the philosophy underlying the developments of the analysis mostly follows the general papers by Rodríguez-Iturbe et al. (2001a) and Porporato and Rodríguez-Iturbe (2002).

1.1 Ecohydrology of water-controlled ecosystems

Water-controlled ecosystems are complex, evolving structures whose characteristics and dynamic properties depend on many interrelated links between climate, soil, and vegetation (Figure 1.1). On one hand, climate and soil control vegetation dynamics (e.g., Lange et al., 1976; Boyer, 1982; Jones, 1992; Kramer and Boyer, 1995; Larcher, 1995), on the other hand vegetation exerts important control on the entire water balance and is responsible for many feedbacks to the atmosphere (e.g., Schlesinger et al., 1990; Kutzbach et al., 1996; Zeng et al., 1999). Many important issues depend on the quantitative understanding of the ecohydrology of water-controlled ecosystems, including environmental preservation and proper management of resources (e.g., Noy-Meir, 1973; Shmida et al., 1985; Archer et al., 1988; Scholes and Walker, 1993; Archer, 1994).

Soil moisture is the key variable synthesizing the action of climate, soil, and vegetation on the water balance and the dynamic impact of the latter on plants (e.g., Noy-Meir, 1973; Lange et al., 1976; MacMahon and Schimpf, 1981; Tinley, 1982; Stephenson, 1990; Neilson, 1995; Laio et al., 2001a; Porporato et al., 2001; Rodríguez-Iturbe et al., 2001a; Porporato and Rodríguez-Iturbe, 2002). If rainfall, at a first analysis, may be regarded as a gross surrogate of soil moisture to determine plant ecosystem structure at the continental scale (e.g., Figure 1.2), the actual assessment of plant conditions depends on soil moisture dynamics *in situ*. This is clear in Figure 1.3, which shows the importance of the variability of soil moisture, as affected by both soil texture and interannual rainfall fluctuations, in the recruitment of *Bouteloua gracilis* (blue grama), a small dominant grass in the Colorado steppe (Laurenroth et al., 1994).

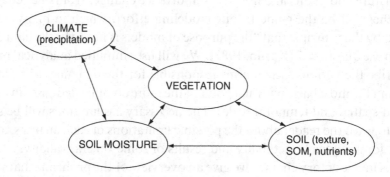

Figure 1.1 Schematic representation of the climate, soil, and vegetation system (SOM = Soil Organic Matter).

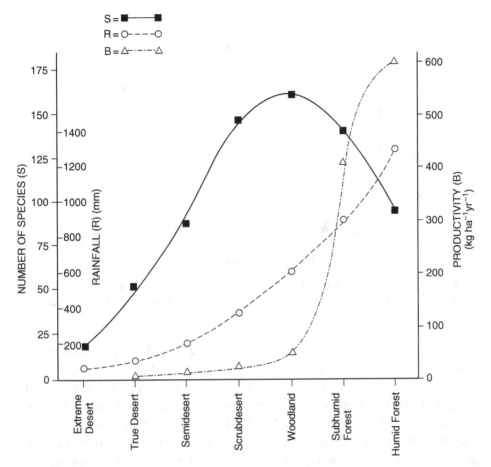

Figure 1.2 General link between precipitation, biomass, and biodiversity in water-controlled ecosystems. After Shmida and Burgess (1988).

The idea that soil moisture dynamics is at the core of water-controlled ecosystems is not new. It permeated some of the pioneering and seminal works in the field, such as those by Gardner (1960), Cowan (1965), Noy-Meir (1973), and Eagleson (1978a, 1978b, 1982) among others. As stated by Noy-Meir (1973), the soil is the store and regulator in the water flow system of the ecosystems, both as a temporary store for the precipitation input allowing its use by organisms and as a regulator controlling the partition of this input between the major outflows: runoff, evapotranspiration redistribution, and the flow between the different organisms.

Plants play a special role in water-controlled ecosystems, having an active role in water use that heavily conditions the soil water balance and at the same time being impacted by the arid conditions they contribute to produce.

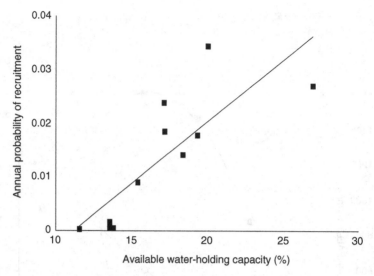

Figure 1.3 Soil-moisture dependence of recruitment of new plants of *Bouteloua gracilis* for a site in Colorado. After Laurenroth et al. (1994); see also Section 5.3.

Differences in soil moisture dynamics are among the principal reasons for the existence of particular functional vegetation types and ecosystem structures. Special adaptation to water stress and intra/inter-species interactions are likely connected to the dynamics of the climate–soil–vegetation system and both the coexistence of different functional vegetation types and the development of temporal and spatial vegetation patterns depend on the emergence of specific temporal niches of soil water and nutrient availability (Cody, 1986; Scholes and Archer, 1997). Chapters 2, 3, 4, and 6 will set the stage to model the basic interaction between soil moisture and plants.

1.2 Simplifying assumptions

There are two characteristics that make the quantitative analysis of the problem especially daunting (Rodríguez-Iturbe et al., 2001a): (i) the very large number of different processes and phenomena that make up the dynamics, and (ii) the extremely large degree of variability in time and space that the phenomena present. The first of the above characteristics obviously calls for simplifying assumptions in the modeling scheme while still preserving the most important features of the dynamics, while the second one calls for a stochastic description of some of the processes controlling the overall dynamics.

In the attempt to capture the essential dynamics of ecosystems, plants responding in a similar way to a syndrome of environmental factors are often grouped together referring to plant functional types (Gitay and Noble, 1997). This is extremely useful for providing results that are applicable to different ecosystems throughout the world, so that the analysis may be generally referred to typical average conditions rather than to particular places or species. It does not mean, however, that the specific adaptation mechanisms to water stress and intra/inter-species interactions (e.g., competition for water and reproduction) are not essential for the dynamics of a particular biome. As shown in Figure 1.4, the impact of soil moisture on water stress may be quite different among species (as well as each individual) and this may in turn drive the emergence of specific temporal niches of soil water and nutrient availability and patterns of coexistence (e.g., Fernandez-Illescas et al., 2001; Porporato et al., 2001). As noted by Tilman (1996), biodiversity seems to stabilize ecosystem properties: while interspecific competition magnifies the effect of a perturbation on the abundances of individual species, the competitive release experienced by disturbance-resistant species acts to stabilize total biomass in species-rich communities.

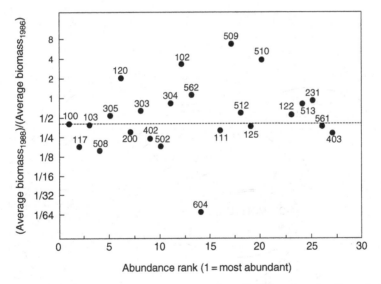

Figure 1.4 Change in the average biomass of 24 abundant species, from before the drought (1986) to the peak of the drought (1988) in a long-term study of 207 grassland plots in Minnesota, graphed against their ranked abundance. The dashed line shows the response of community biomass with each dot representing a species identified by the number. Notice the quite diversified response to drought and how many species performed better during drought than did plant community biomass. After Tilman (1996).

1.3 Levels of description

As it is typical of complex systems, the dynamics of the climate–soil–vegetation system presents different levels of complexity according to the scale of interest (Porporato and Rodríguez-Iturbe, 2002): the importance of the various hydrologic processes may be different whether one considers daily, seasonal, or interannual fluctuations or point (i.e., plot), regional (i.e., hill-slope to catchment), or continental scales. Such a distinction of scales is important as it naturally suggests different levels of analysis in which only the main interactions may be retained (Figure 1.5).

A first basic level of analysis of the climate–soil–vegetation system is at the spatial scale of a few meters (e.g., plot scale) and at the temporal scale of the growing season. At such scales, soil moisture dynamics directly impacts vegetation through plant water stress, while the rainfall input and the soil characteristics may be considered as external components (Figure 1.6). The first part of the book deals with this level (Chapters 2–6).

A second and more complex level of analysis involves the links between soil moisture, soil nutrient cycles, and the related evolution of soil properties. Such factors are all interrelated with vegetation dynamics and represent an area relatively less studied by both ecologists and hydrologists. In Chapter 10 we

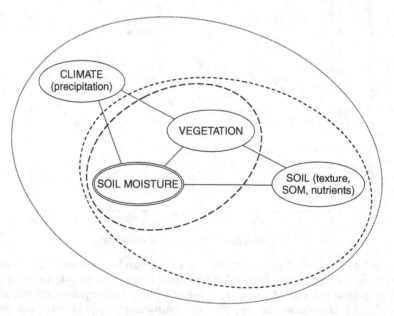

Figure 1.5 Levels of description of the climate–soil–vegetation system having at the center soil moisture (SOM = Soil Organic Matter). After Porporato and Rodríguez-Iturbe (2002).

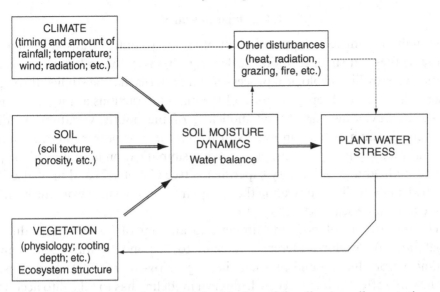

Figure 1.6 First and main level of description of the climate–soil–vegetation system. Climate and soil characteristics may be considered as external forcing components while soil moisture dynamics pivots the mutual links between the vegetation and water stress. As modified by Porporato and Rodríguez-Iturbe (2002) after Rodríguez-Iturbe et al. (2001a).

will concentrate on the direct action of soil moisture on carbon and nitrogen cycles with a short discussion of the role of feedbacks from plant dynamics.

Finally, a more general level of description is introduced when the system is analyzed at large spatial scales (e.g., continental). Large-scale patterns of vegetation types (and also water stress) induce corresponding patterns of albedo and transpiration characteristics that heavily condition the dynamics of the atmospheric boundary layer and the formation of convective precipitation. Thus the climatic/atmospheric component becomes influenced by the feedbacks induced by the soil–plant system. The climate component is no longer an external forcing and becomes an essential part of the dynamics (e.g., Segal et al., 1988; Eltahir, 1996; Sellers et al., 1997; Porporato et al., 2000; Pielke, 2001). Such a problem, briefly discussed in Chapter 6, connects ecohydrology to hydrometeorology, another important discipline from the family of the hydrologic sciences. The implications of soil–plant–atmosphere interaction involve not only the exchanges of water between the components of the system, but also CO_2 fluxes between ecosystems and the atmosphere and the many feedbacks on the nitrogen cycle (e.g., Siqueira et al., 2000; Dickinson et al., 2002).

1.4 Temporal scales

Although micrometeorology and plant physiology control transpiration and photosynthesis at small temporal scales (e.g., hourly), the soil moisture fluctuations controlling ecological processes and patterns may be studied through a suitable functional representation of the main interactions at the daily time scale, thus avoiding the explicit modeling of the hourly variations in the different parameters (e.g., internal storm structure, evapotranspiration diurnal variations, photosynthesis, etc.). Such a simplification is empirically supported by field and laboratory experiments (see Chapter 2) and may also be justified analytically by upscaling the soil–plant water balance from the hourly to the daily time scale (see Chapter 6).

The uncertainty of both the timing and amount of rainfall has induced vegetation to develop different strategies to respond to water stress and optimize reproduction and productivity (e.g., Noy-Meir, 1973). As a consequence, any effort towards ecohydrological modeling has to take into account the stochastic character of soil moisture dynamics. Accordingly, precipitation input is modeled as a stochastic process interpreted at the daily time scale. For the sake of simplicity, at the beginning (Chapters 2–6), only cases with negligible seasonal and interannual rainfall components are considered. The growing season is thus assumed to be statistically homogeneous: the effects of the initial soil moisture condition are considered to last for a relatively short time and the probability distribution of the soil water content to settle in a state independent of time.

However, steady conditions are not always representative, especially at the beginning of the growing season. For example, the soil moisture initial condition is very different between the savannas of Nylsvley in South Africa and the forests in northwestern United States. At Nylsvley, the growing season from September to April receives 98% of the annual rainfall and is preceded by a warm and dry winter season, which makes initial soil moisture conditions at the start of the growing season practically irrelevant. On the contrary, in northwestern United States a relatively dry growing season is preceded by a wet and cold winter season: these factors, combined with a deep active soil layer, make the initial value of soil moisture storage a commanding factor in water use by plants (see Chapters 3 and 7). As a consequence, the case of Nylsvley is likely to be well represented by the statistically steady-state properties of the soil moisture process, while for the forests of northwestern United States the transient properties of soil moisture dynamics are crucial for an adequate statistical representation of the conditions during the growing season.

In other cases, seasonal components in transpiration and precipitation may have dramatic consequences for plants. For example, in many subtropical climates, rainfall and transpiration are in phase and have highest rates during the growing season, while in Mediterranean climates rainfall, being mostly concentrated during the winter season, is in counterphase with the growing season transpiration. This results in very different seasonal patterns of soil water availability and thus, even if the total annual precipitation and average temperature are the same, in very different vegetation structures. Figure 1.7

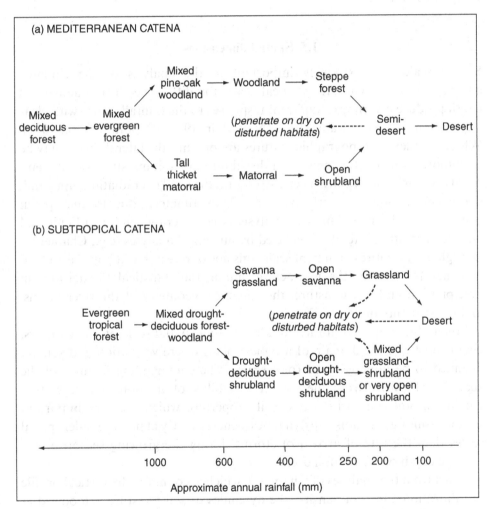

Figure 1.7 Different vegetation catenae (from an arboreal temperate formation to a contracted desert) as an example of specific ecosystem structures resulting from different seasonal patterns of water availability. (a) Mediterranean catena; (b) subtropical catena. As modified by Porporato and Rodríguez-Iturbe (2002) after Shmida and Burgess (1988).

reports the typical vegetation catenae (from an arboreal temperate formation to a contracted desert) for Mediterranean and subtropical climates: in the region between 400 and 250 mm of annual rainfall, for instance, one may find steppe forests and open shrublands in Mediterranean climates, or open savannas and drought-deciduous shrublands in ecosystems with subtropical climates. At longer time scales, interannual fluctuations also become important, superimposing themselves on the daily and seasonal components. Seasonal and interannual hydrologic fluctuations will be the subject of Chapters 8 and 11.

1.5 Spatial dimensions

Spatial scales are extremely important in the analysis of the climate–soil–vegetation system. At the local scale, the differences in the active soil depth produce a variety of different responses to the rainfall input, with clear implications for plant water stress (Noy-Meir, 1973; Porporato et al., 2001). Wherever relevant topographic features are present, the lateral fluxes may be an important factor for the spatial distribution of soil moisture and its temporal evolution; slope and aspect also control the local net radiation input and, consequently, soil moisture dynamics. Unfortunately, due to the spatial components, the mathematical analysis is considerably more difficult and analytical solutions are often replaced by numerical analyses (e.g., Chapter 9). Brought to the entire catchment scale, this line of research will hopefully blend in a natural manner the interaction among the statistical fluctuations in precipitation and soil moisture, the statistical geometry of the river basins, and the structure of vegetation.

Much of the analysis in this book refers to a spatial scale of a few meters, characteristic of the domain of a typical plant where vegetation and soil are assumed to be homogeneous (plot scale). The commanding features of the spatial scales arise from the spatial variability of the soil and vegetation condition. Soil composition and soil properties, which are very important for the resulting soil moisture dynamics, generally vary at much smaller spatial scales than the rate of arrival of storms during the growing season or the average depth of the rainfall events.

Apart from the analyses of Chapter 9, which account for the vertical profile of infiltration and root density, the dynamics of soil moisture is modeled at a point for vertically averaged conditions and the vertical dimension is all embedded in the dependence of soil moisture dynamics on the rooting depth. The effect on soil moisture dynamics of lateral fluxes resulting from topographic features is also considered in Chapter 9. As is often the case in arid and semi-arid ecosystems, all the analyses of this book assume no interaction between the saturated zone and the active soil layer. For practical purposes, this means that our applications do not cover regions or sites where, because of humid climate, adjacent water courses, topographic converging features, or any combination of the above, the water table may have an active role in determining plant conditions.

When averaging over large spatial scales, on the order of kilometers, the impact of spatial variability in soil moisture is drastically reduced and the temporal scale at which the processes are effectively correlated is very much increased. As said before, these larger scales are very useful for climatologically oriented studies and require the analysis of the climate–soil–vegetation system at the most general level (see Figure 1.5). However, they are not very illuminating for the study of vegetation response and the accompanying hydrologic dynamics, and will only be dealt with here in developing the atmospheric boundary layer growth model in Chapter 6.

1.6 Soil moisture and the cycles of soil nutrients

In many regions of the world, soil moisture dynamics is also the key factor determining the duration of the periods in which primary production and nutrient mineralization can occur. The production of plant residues is directly dependent on the growth of vegetation through its photosynthetic capacity, which in turn depends on water availability. Consider, for example, the difference in the photosynthesis rates of evergreen and drought-deciduous species even when operating under the same climatic conditions. Drought-deciduous species have higher photosynthetic rates than evergreen species because they have larger amounts of nitrogen in their leaves on a dry-mass basis. "Where nitrogen, and hence growth-rate, is limited, leaf turnover is reduced and, in fact, there is a selection toward increased leaf duration. The longest leaf durations generally occur in plants occupying the most nutrient-deficient habitat" (Mooney, 1983). Thus evergreen plants tend to exist in soils with lower nutrient status and where nutrients are more slowly recycled than in the case of drought-deciduous species.

Soil moisture is also a controlling factor of mineralization and uptake. Figure 1.8 presents two interesting examples of this dependence and shows how the equilibrium nitrogen cycling rate for a site is directly related to its water availability (Aber et al., 1991). Such hydrologic controls on the soil nutrient cycles are part of the second level of analysis of the soil–climate–vegetation system (Figure 1.5), which will be the focus of Chapter 10.

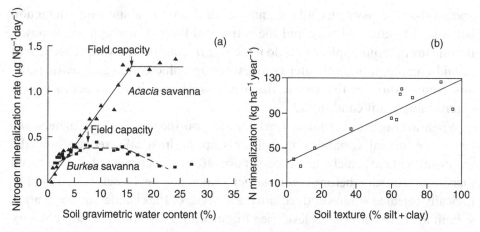

Figure 1.8 (a) Dependence of nitrogen mineralization rate on soil water content in the broad-leafed and fine-leafed savanna at Nylsvley. After Scholes and Walker (1993). (b) Measured nitrogen mineralization rates for several stands in Wisconsin in relation to soil texture, which is closely related to soil water availability. Modified after Aber et al. (1991).

1.7 Soil moisture dynamics and ecosystem structure

The dependence of water stress on soil moisture and soil nutrient dynamics drives the growth, reproduction, and competitive abilities of plants. Since the different species (as well as each individual) have a specific response to soil moisture dynamics (see Figure 1.4), the hydrologic fluctuations continually (and randomly) shift the habitat preference in favor of different species and plant functional types.

The time scales of the hydrological forcing (especially the interannual fluctuations) interact with those of plant growth at various levels. In the case of trees, for example, these may easily span from the day to the growing season and the decadal time scale (Figure 1.9). For this reason, the understanding of such dynamics requires the combination of hydrologic models with a suitable quantitative description of the growth and death of plants as well as their competition and colonization properties.

The dynamics is further complicated by the spatial interactions that are due to both soil variability and plant competition. Temporal and spatial dynamics are thus highly intertwined and give rise to regular and irregular spatial patterns in continuous evolution. Figure 1.10 is an example of a savanna in southern Texas where the complex spatial and temporal patterns of tree–grass coexistence are driven by interannual rainfall fluctuations (Archer et al., 1988). In some extreme cases, the interaction of vegetation with the site water balance and nutrient cycles may even lead to dramatic changes in the equilibrium

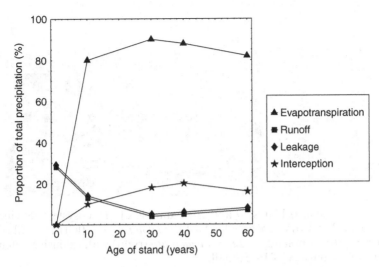

Figure 1.9 Hydrological budget for an oak stand in the eastern European forest steppe. Redrawn after Molchanov (1971), as quoted in Larcher (1995).

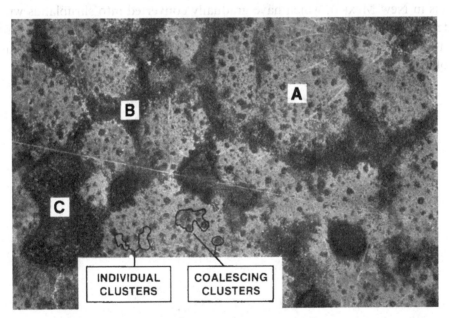

Figure 1.10 Aerial view of the savanna of La Copita (southern Texas, USA). The two-phase patterns of discrete clumps of woody vegetation are scattered throughout a grassy matrix (A). Bordering the two-phase portions of the landscape are monophasic woodlands associated with more mesic drainages (B) and low-center polygons of playas (C). The clusters, organized about mesquite (*Prosopis glandulosa*), represent chrono sequences whose species compositions at latter stages of development are similar to that of the closed canopy woodlands in region B. The largest clusters in the two-phase zone represent a mosaic of coalesced clusters. After Archer et al. (1988).

Figure 1.11 (see also Plates 19 and 20) Example of an ecosystem (Sevilleta LTER, New Mexico) where grasslands of *Bouteloua eripoda* (left) are gradually converted into shrublands of *Larrea tridentata* (right) with extensive patches of bare soil. Courtesy of Eric Small.

conditions of ecosystems of entire regions. This is the case of some arid grasslands in New Mexico, which have gradually converted into shrublands with extensive patches of bare soil (Figure 1.11 and Plates 19 and 20). The last part of this book will present an attempt to tackle these intriguing and difficult problems (Chapter 11).

2

Stochastic soil moisture dynamics and water balance

This chapter presents a probabilistic model for the temporal dynamics of soil moisture, with the goal of providing a simplified yet realistic description amenable to analytical solutions. Such an analysis is the necessary starting point for the quantitative understanding of the impacts of soil moisture on ecosystems' dynamics, such as the vegetation response to water stress, the hydrologic control on cycles of soil nutrients, and the dynamics of plant competition for water. The same approach may also be useful to obtain insights into other processes closely related to soil moisture dynamics, such as soil–atmosphere interactions, soil production, and soil gas emissions.

The probabilistic soil moisture model that we describe in this chapter was originally proposed by Rodríguez-Iturbe et al. (1999a) and improved by Laio et al. (2001a). Here all the various physical processes involved in the soil moisture dynamics, the simplifying assumptions, and their related implications are discussed in detail. Particular attention is devoted to the role of soil properties and plant transpiration characteristics.

The solution of the problem in probabilistic terms, which is made necessary by the introduction of the stochastic representation of rainfall, is carried out for statistically steady-state conditions during a growing season. The results allow assessment of the roles of climate, soil, and vegetation on the soil moisture probability density function (pdf) and on the average long-term water balance. The chapter closes with a brief discussion of possible simplifications of the soil moisture dynamics that are of interest when considering the large-scale water balance.

2.1 Soil water balance at a point

The soil water balance is the mass conservation of soil water as a function of time. In general, it is a very complex problem that depends on many interactions and processes. We will consider the water balance vertically averaged over the root zone, focusing on the most important components as sketched in Figure 2.1.

The state variable regulating the water balance is the relative soil moisture, s, which represents the fraction of pore volume containing water. The total volume of soil is given by the sum of the volumes of air, water, and mineral components, i.e., $V_s = V_a + V_w + V_m$. The porosity is defined as

$$n = \frac{V_a + V_w}{V_s},$$ (2.1)

and the volumetric water content, θ, is the ratio of water volume to soil volume, so that the relative soil moisture is

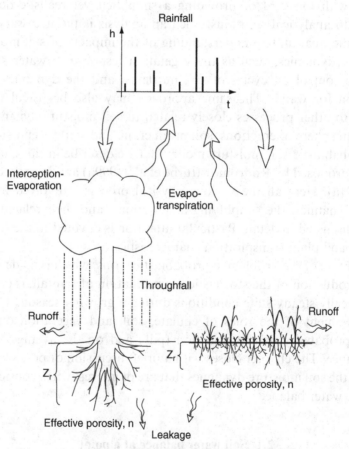

Figure 2.1 Schematic representation of the various mechanisms of the soil water balance with emphasis on the role of different functional vegetation types. After Laio et al. (2001a).

$$s = \frac{V_w}{V_a + V_w} = \frac{\theta}{n}, \tag{2.2}$$

$0 \le s \le 1$. Denoting the rooting depth by Z_r, snZ_r represents the volume of water contained in the root zone per unit area of ground.

With the above definitions and under the simplifying assumption that the lateral contributions can be neglected (i.e., negligible topographic effects over the area under consideration), the vertically averaged soil moisture balance at a point may be expressed as

$$nZ_r \frac{ds(t)}{dt} = \varphi[s(t), t] - \chi[s(t)], \tag{2.3}$$

where t is time, $\varphi[s(t), t]$ is the rate of infiltration from rainfall, and $\chi[s(t)]$ is the rate of soil moisture losses from the root zone. The terms on the r.h.s. of Eq. (2.3) represent water fluxes, i.e ., volumes of water per unit area of ground and per unit of time (e.g., mm day^{-1}).

The infiltration from rainfall, $\varphi[s(t), t]$, is the stochastic component of the balance. It represents the part of rainfall that actually reaches the soil column, i.e.,

$$\varphi[s(t), t] = R(t) - I(t) - Q[s(t), t], \tag{2.4}$$

where $R(t)$ is the rainfall rate, $I(t)$ is the amount of rainfall lost through canopy interception, and $Q[s(t), t]$ is the rate of runoff.

The water losses from the soil are from two different mechanisms

$$\chi[s(t)] = E[s(t)] + L[s(t)], \tag{2.5}$$

namely, $E[s(t)]$ and $L[s(t)]$, the rates of evapotranspiration and leakage, respectively.

Equation (2.3) is a stochastic, ordinary differential equation for the state variable $s(t)$. It is of fundamental hydrologic importance and underlies most of the dynamics studied in this book. For simplicity of notation, in what follows we will dispense with the indication of the time dependence of soil moisture whenever it is not necessary.

2.1.1 Rainfall modeling

At small spatial scales, where the contribution of local soil-moisture recycling to rainfall is negligible, the rainfall input can be treated as an external random forcing, independent of the soil moisture state. The inclusion of the stochastic nature of rainfall is essential when attempting proper modeling of soil moisture

dynamics, since the response of vegetation (i.e., resistance to drought, productivity, reproduction and germination, etc.) in water-controlled ecosystems
is strongly linked to the intermittent and unpredictable character of rainfall
availability (e.g., Noy-Meir, 1973).

Since both the occurrence and amount of rainfall can be considered to be
stochastic, the occurrence of rainfall is idealized as a series of point events in
continuous time, arising according to a Poisson process of rate λ and each
carrying a random amount of rainfall extracted from a given distribution. The
temporal structure within each rain event is ignored and the marked Poisson
process representing precipitation is physically interpreted at a daily time
scale, where the pulses of rainfall correspond to daily precipitation assumed
to be concentrated at an instant in time.

With these assumptions, the distribution of the times τ between precipitation events is exponential with mean $1/\lambda$ (e.g., Cox and Miller, 1965), i.e.,

$$f_T(\tau) = \lambda e^{-\lambda\tau}, \text{ for } \tau \geq 0, \tag{2.6}$$

while the depth of rainfall events is assumed to be an independent random
variable h, described by an exponential probability density function

$$f_H(h) = \frac{1}{\alpha} e^{-\frac{1}{\alpha}h}, \text{ for } h \geq 0, \tag{2.7}$$

where α is the mean depth of rainfall events. Since the model is interpreted at
the daily time scale, α may be estimated as the mean daily rainfall in days when
precipitation occurs. In the following we will often refer to the value of the
mean rainfall depth normalized by the active soil depth, i.e.,

$$\frac{1}{\gamma} = \frac{\alpha}{nZ_r}. \tag{2.8}$$

Both the Poisson process and the exponential distribution are of common use
in simplified models of rainfall at a daily time scale. The exponential distribution fits daily rainfall data well and, at the same time, allows analytical
tractability (Benjamin and Cornell, 1970; Eagleson, 1978a, 1978b).

Notice that the rainfall rate $R(t)$ in Eq. (2.4) can be linked to the probability
distributions Eqs. (2.6) and (2.7) if one expresses the marked Poisson process
as a temporal sequence, i.e.,

$$R(t) = \sum_i h_i \delta(t - t_i) \tag{2.9}$$

where $\delta(\cdot)$ is the Dirac delta function, $\{h_i, i = 1, 2, 3, \ldots\}$ is the sequence of random rainfall depths with distribution given by Eq. (2.7), and $\{\tau_i = t_i - t_{i-1}, i = 1, 2, 3, \ldots\}$ is the interarrival time sequence of a stationary Poisson process with rate λ.

For the moment, the values of α and λ are assumed to be time-invariant quantities, representative of a typical growing season. In cases where there are significant changes in the parameters throughout the season, it may be useful to consider the early growing and the late growing seasons separately, each one with its own set of parameters. Chapter 8 will deal with seasonal and inter-annual variability of α and λ.

2.1.2 Interception

Part of the rainfall is intercepted by the aerial part of vegetation, while the remaining directly reaches the soil as throughfall (Figure 2.1). Especially in arid and semi-arid climates, where rainfall events are generally short and evaporation demand is high, a sizeable fraction of the intercepted rainfall is lost directly through evaporation, while the rest reaches the soil as stem flow (e.g., Dingman, 1994; Scholes and Archer, 1997; Waring and Running, 1998). The precise mechanisms of interception are quite complicated to model and depend on vegetation type as well as on the intensity and duration of rainfall. Because of its strong dependence on the type of plant and more specifically on its leaf area index, the amount lost by interception can indeed be quite different for, say, trees and grasses. As an example, well-vegetated trees in South African savannas can intercept up to over 0.2 cm of rainfall per storm event (Scholes and Walker, 1993), while grasses usually intercept much less. At small spatial scales interception is also known to alter the spatial distribution of the water reaching the soil (e.g., through stem flow and crown shading; Scholes and Archer, 1997), but these effects are not considered here.

In order to keep analytical tractability, interception is incorporated in the stochastic model by simply assuming that, depending on the kind of vegeta-tion, a given amount of water can be potentially intercepted from each rainfall event (Rodríguez-Iturbe et al., 1999a). This implies fixing a threshold for rainfall depth, Δ, below which no water reaches the ground. If the depth of a given rainfall event, h_i, is higher than Δ, then the actual rainfall depth reaching the soil, h', is assumed to be $h_i - \Delta$, and Δ is lost by interception. The amount of water intercepted can be related to the type of vegetation by simply using different values for Δ. This model of interception is sketched in Figure 2.2a. The effect of fluctuations in wind and air temperature on

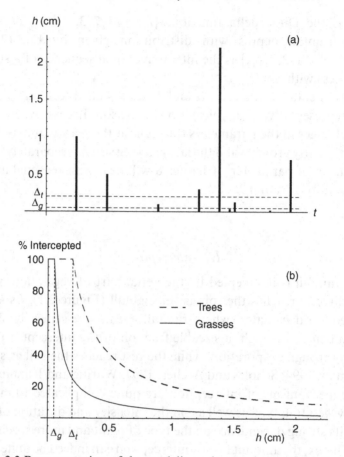

Figure 2.2 Representation of the modeling scheme adopted for interception. (a) Temporal sequence of rainfall events (h is the rainfall depth) along with the thresholds of interception, $\Delta_t = 0.2$ cm and $\Delta_g = 0.05$ cm, typical for trees and grasses in some savannas. (b) Percentage of intercepted rainfall as a function of the total rainfall per event. After Laio et al. (2001a).

interception losses is assumed to be negligible compared to the role of differences in canopy coverage.

Although admittedly being a simplistic model for interception, the previous scheme provides a representation of the process that is in good agreement with the experimental evidence. As shown in Figure 2.2b, the percentage of rainfall intercepted as a function of total rainfall reproduces quite well the one found in field experiments (e.g., Feddes, 1971; Lai and Katul, 2000).

From a mathematical viewpoint the consideration of a threshold on the rainfall Poisson process does not complicate its analytical tractability. The rainfall process is in fact transformed into a new marked-Poisson process, called a censored process, where the frequency of rainfall events is now

$$\lambda' = \lambda \int_{\Delta}^{\infty} f_H(h)\, dh = \lambda e^{-\Delta/\alpha} \tag{2.10}$$

and the depths h' have the same distribution as h, given by Eq. (2.7). Thus, one can simply write

$$R(t) - I(t) = \sum_i h'_i \delta(t - t'_i) \tag{2.11}$$

where $\{\tau'_i = t'_i - t'_{i-1}, i = 1, 2, 3, \ldots\}$ is the interarrival time sequence of a stationary Poisson process with frequency λ'.

2.1.3 Infiltration and runoff

When the soil has enough available storage to accommodate all the incoming water of the rainfall event, the increment in water storage is equal to the rainfall depth of the event; whenever the rainfall depth exceeds the available storage, the excess is converted into surface runoff. Since it depends on both rainfall and soil moisture content, infiltration from rainfall is a stochastic, state-dependent component, whose magnitude and temporal occurrence are controlled by the entire soil moisture dynamics. Because of the vertically lumped representation of soil moisture dynamics, the temporal propagation of the wetting front into the soil is not considered. However, this is not judged to be too restrictive when the soil moisture dynamics is considered at the daily time scale. Chapter 9 addresses the conditions of validity and the degree of approximation of such a hypothesis.

The probability distribution of the infiltration component may be easily written in terms of the exponential rainfall-depth distribution of Eq. (2.7) and the soil moisture state s. Referring to its dimensionless counterpart y (i.e., the infiltrated depth of water normalized by nZ_r) one can write

$$f_Y(y, s) = \gamma e^{-\gamma y} + \delta(y - 1 + s) \int_{1-s}^{\infty} \gamma e^{-\gamma u}\, du, \quad \text{for } 0 \le y \le 1 - s, \tag{2.12}$$

where γ is defined in Eq. (2.8). Equation (2.12) is thus the probability distribution of having a jump in soil moisture equal to y, starting from a level s. The mass at $(1 - s)$ represents the probability that a storm will produce saturation when the soil has moisture s (Figure 2.3). This sets the upper

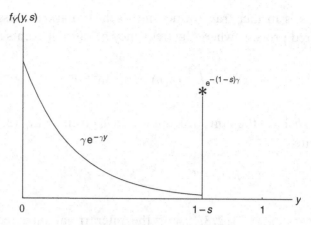

Figure 2.3 Sketch of the probability density function describing infiltration from rainfall y. The asterisk represents the atom of probability corresponding to soil saturation. After Rodriguez-Iturbe et al. (1999a).

bound of the process at $s = 1$, making the soil moisture balance evolution, e.g., Eq. (2.3), a bounded shot noise process. Although the bounded character of the process complicates its mathematical analysis, it will be seen that the Markovian nature of the process allows for a complete analytical solution in the case of steady-state conditions.

Similarly to Eq. (2.11), infiltration from rainfall can be written in Eq. (2.3) as

$$\varphi[s(t), t] = nZ_r \sum_i y_i \delta(t - t_i') \tag{2.13}$$

where $\{y_i, i = 1, 2, 3, \ldots\}$ is the sequence of random infiltration events having a distribution as in Eq. (2.12).

The probabilistic analysis of the temporal distribution of runoff production requires the study of the crossing properties of the level $s = 1$, using the framework discussed in the next chapter. It is only noted here that the runoff process by itself is not Markovian and that the temporal distribution of its time of occurrence is not exponential. It is also worth noting that runoff production may be interpreted in a different but equivalent manner by considering an alternative process where runoff occurs from deterministic losses that take place at an infinite rate above $s = 1$. In this case the state variable s is not required to be formally bounded at $s = 1$, as the instantaneous decay above $s = 1$ makes the process completely equivalent to the bounded one for $s < 1$. This interpretation will turn out to be useful for a more intuitive interpretation of some of the results that will be presented later.

2.1.4 Active soil depth

The soil is modeled as a horizontal layer of depth Z_r with homogeneous characteristics. As previously defined, the product of soil depth and porosity gives the active soil depth, nZ_r, which is the height (or volume per unit surface area) available for water storage. From a physical viewpoint, the active soil depth in Eq. (2.3) controls the response time of the system. As will be seen in detail later, different values of nZ_r greatly affect the interaction of soil and vegetation with the climate forcing. It is interesting to point out how the intermittent nature of the forcing, which makes the system undergo continuous fluctuations, leads to the aforementioned role of the active soil depth, even in statistically steady conditions. From a mathematical viewpoint, this important interplay between temporal forcing and soil depth can be understood from Eq. (2.3), in that if the infiltration from rainfall were constant in time instead of intermittent, the steady-state condition would be independent of soil depth.

The values of porosity and soil depth are generally influenced by many factors. Porosity usually shows some dependence on soil texture; plant roots and the action of small animals may also alter the value of the so-called macroporosity and produce preferential directions for water movement. However, since the effective importance of these latter variations of porosity is difficult to quantify without making the modeling scheme restrictively specific, we will refer only to variations due to different soil textures (see Table 2.1). The actual soil depth involved in the water balance is a difficult parameter to estimate and may show a large range of spatial variation, depending on soil pedology and, especially, vertical root distribution (e.g., Canadell et al., 1996; Jackson et al., 1996; Schulze et al., 1996). As a general rule, trees generally have deeper roots than grasses, and tend to be accompanied by a slightly higher effective porosity.

A number of links and feedbacks between vegetation, climate, and soil will not be considered, being of secondary importance at the time scale of concern. However, for analyses focusing on the long-term dynamics of vegetation, the variations of porosity, soil aging, and soil depth, which are linked to soil production and in turn also to soil moisture dynamics, should be taken into account. Such variations generally take place at much longer time scales than the ones considered here, but can become faster in tropical ecosystems (Larcher, 1995; see also Chapter 10). In those cases, nZ_r and the other soil characteristics may not be considered static parameters, but evolving variables.

The relevance of the active soil depth in regard to its role in large-scale soil moisture dynamics and its relevant ecological consequences will be evident

Table 2.1 *Parameters describing the soil characteristics used in the model for five different soil textures. After Laio et al. (2001a).*

	$\overline{\Psi}_s$ (MPa)	b	c	K_s (cm day^{-1})	n	β	s_h	s_w	s^*	s_{fc}
Sand	$-0.34 \cdot 10^{-3}$	4.05	11.1	>200	0.35	12.1	0.08	0.11	0.33	0.35
Loamy sand	$-0.17 \cdot 10^{-3}$	4.38	11.7	≈ 100	0.42	12.7	0.08	0.11	0.31	0.52
Sandy loam	$-0.70 \cdot 10^{-3}$	4.90	12.8	≈ 80	0.43	13.8	0.14	0.18	0.46	0.56
Loam	$-1.43 \cdot 10^{-3}$	5.39	13.8	≈ 20	0.45	14.8	0.19	0.24	0.57	0.65
Clay	$-1.82 \cdot 10^{-3}$	11.4	25.8	≈ 20	0.5	26.8	0.47	0.52	0.78	≈ 1

Data from Harr (1962); De Wiest (1969); Clapp and Hornberger (1978); Cosby et al. (1984); Dingman (1994).
$\overline{\Psi}_s$ is the geometric mean of different measurements (Clapp and Hornberger, 1978).
$s_h, s_w,$ and s^* correspond to Ψ_s of $-10, -3, -0.03$ MPa, respectively.
s_{fc} is the s level at which leakage becomes negligible compared to evapotranspiration (see text for details).

throughout the entire book. Milly and Dunne (1993) and Milly (1997) have also studied the effect of soil depth on the global water balance and its response to climate change. Nepstad et al. (1994), Canadell et al. (1996), and Scholes and Archer (1997) have dealt with the impact of soil depth on plant water stress and on the possible selection of temporal niches of soil moisture availability by different vegetation types.

2.1.5 Evapotranspiration

Since interception has already been subtracted (e.g., Eq. (2.11)), the term $E[s, (t)]$ in Eq. (2.5) represents the sum of the losses resulting from plant transpiration and evaporation from the soil. Although these are governed by different mechanisms, we will consider them together for the moment.

When soil moisture is high, the evapotranspiration rate depends mainly on the type of plant and climatic conditions (e.g., leaf area index, wind speed, air temperature and humidity, etc.). As long as soil moisture content is sufficient to permit the normal course of the plant physiological processes, evapotranspiration is assumed to occur at a maximum rate E_{max}, which is independent of s. When soil moisture content falls below a given point s^*, which depends on both vegetation and soil characteristics, plant transpiration (at the daily time scale) is reduced by stomatal closure to prevent internal water losses and soil water availability becomes a key factor in determining the actual evapotranspiration rate (e.g., Schulze, 1986; Hale and Orcutt, 1987; Smith and Griffith, 1993; Larcher, 1995; Nilsen and Orcutt, 1998; see also Chapters 4 and 6). Transpiration and root water uptake continue at a reduced rate until soil moisture reaches the so-called wilting point s_w.[1] At this level, suction to extract water from soil is so high that it damages the plant tissues (e.g., Lange et al., 1976; Schulze, 1986; Nilsen and Orcutt, 1998). Small water losses from the plant continue via cuticular transpiration and, if soil water is not replenished by rainfall, wilting and irreversible plant damage begin to appear (plant water stress is discussed in detail in Chapter 4). Below wilting point, s_w, soil water is further depleted only by evaporation at a very low rate up to the so-called hygroscopic point, s_h.

It will be seen in Chapters 4 and 6 that below s^* the rate of transpiration is reduced by a series of complex mechanisms and feedbacks, exerted at different

---

[1] We note that s^* and s_w are two very important thresholds used to describe in a practical manner the behavior of transpiration as a function of soil moisture at the daily time scale. Apart from that, however, they do not have a specific physical meaning and the fact that they are often referred to as the "point of incipient stomatal closure" and "the wilting point" respectively should not be misleading in this respect. The climate, soil, and vegetation characteristics controlling their values will be discussed in Chapter 6.

levels in the soil–plant–atmosphere continuum. The resulting relationship
between the transpiration and soil moisture content in the range between s^*
and s_w is well approximated by a linear decrease (e.g., Schulze, 1986; Hale and
Orcutt, 1987). Figure 2.4 shows an example of the relationship between leaf
(or stomatal) conductance and soil moisture content measured during a controlled
experiment. The strongly nonlinear, threshold-like dependence on soil

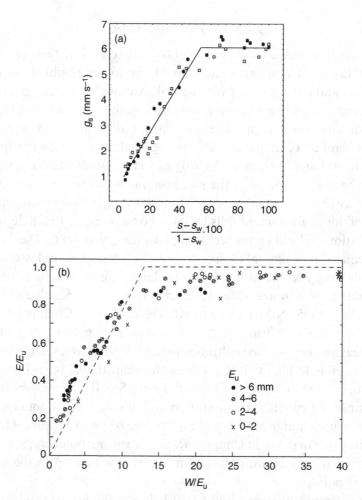

Figure 2.4 (a) Leaf stomatal conductance, g_s, versus soil moisture level for
Nerium oleander, measured maintaining two different levels of vapor pressure
deficit (solid circles and open squares, respectively). Adapted from Schulze
(1986) and after Gollan et al. (1985). (b) Normalized daily transpiration as a
function of available soil moisture from simulation and data of stressed
yellow birch. E is the daily transpiration (mm), E_u is the unstressed daily
transpiration (mm), and W is the available water in the root zone (mm). After
Federer (1979); see also Federer and Gee (1976).

moisture is clearly evident. Notice also from Figure 2.4 that the soil moisture level at which stomatal closure begins is not very sensitive to changes in vapor pressure deficit. Since leaf conductance exerts the most important control on transpiration (transpiration rate is practically proportional to leaf conductance; see Chapter 6), the same kind of dependence on soil moisture is also true for transpiration (e.g., Jones, 1992; Nobel, 1999), as clearly shown in the example of Figure 2.4b. For most plants in temperate and semi-arid regions, this typical behavior at the daily time scale, can be found in the literature at the level both of the single plant and of the entire stand (e.g., Gardner and Ehlig, 1963; Federer, 1979; Spittlehouse and Black, 1981; Federer, 1982; Schulze, 1993; Paruelo and Sala, 1995; Williams and Albertson, 2004). The origin of this dependence on soil moisture will be investigated in detail in Chapter 6, where the hourly transpiration course will be integrated to the daily time scale. Due to different and specialized photosynthetic pathways (e.g., the CAM pathway), notable exceptions to this behavior in extremely arid regions have been reported (e.g., MacMahon and Schimpf, 1981; Larcher, 1995), so that suitable modifications should be considered for the analogous modeling of transpiration-vs-soil moisture in particular species of desert flora.

From the above arguments, daily evapotranspiration losses will be assumed to happen at a constant rate E_{max} for $s^* < s < 1$, and then to linearly decrease with s, from E_{max} to a value E_w at s_w. Below s_w, only evaporation from the soil is present and the loss rate is assumed to decrease linearly from E_w to zero at s_h. The dependence of evapotranspiration losses on soil moisture is thus summarized in the following expression

$$E(s) = \begin{cases} 0 & 0 < s \leq s_h \\ E_w \frac{s - s_h}{s_w - s_h} & s_h < s \leq s_w \\ E_w + (E_{max} - E_w) \frac{s - s_w}{s^* - s_w} & s_w < s \leq s^* \\ E_{max} & s^* < s \leq 1, \end{cases} \tag{2.14}$$

whose behavior is shown in Figure 2.5.

As the smallest time scale of interest here is the daily scale, E_{max} can be interpreted as the average daily evapotranspiration of a unitary surface uniformly covered with vegetation under well-watered conditions during the growing season. Possible estimates of E_{max} values may be obtained using physically based expressions, such as the Penman–Monteith equation (see Chapter 6). However, in the following we prefer to refer to measured data, as is done, for example, by Paruelo and Sala (1995). For regions with hot growing seasons, values of E_{max} are typically around 0.5 cm day^{-1} for grasses

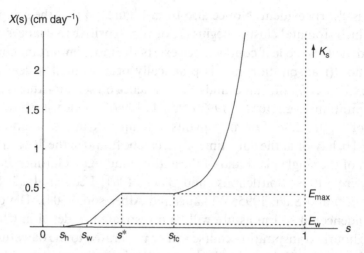

Figure 2.5 Soil water losses (evapotranspiration and leakage), $\chi(s)$, as a function of relative soil moisture for typical climate, soil, and vegetation characteristics in semi-arid ecosystems. After Laio et al. (2001a).

and 10% less for trees under well-watered conditions, although there is considerable variability among different plants and regions (e.g., data from Scholes and Walker, 1993). Of this total value approximately $0.1\,\mathrm{cm\ day}^{-1}$ can be attributed to evaporation from the soil (e.g., Nobel, 1999, page 360). The higher maximum transpiration rate of grasses in these regions is partly explained by the more efficient C_4 pathway of photosynthesis that many grasses of semi-arid ecosystems have with respect to the more common and less efficient C_3 pathway of trees.[2]

Chapter 6 will clarify how both the value of soil moisture, s^*, at which transpiration starts being reduced and the wilting point, s_w, depend on the type of vegetation and soil properties. An indication of their possible values can be obtained from the soil–plant–atmosphere continuum scheme, which describes the transfer of water by plants from soil to the atmosphere in terms of the water potential, Ψ, defined as the difference in free energy per unit volume evaluated with respect to that of pure water at zero reference level and under standard temperature and pressure conditions (e.g., Slatyer, 1967; Oertli, 1976; Larcher,

---

[2] In the so-called C_3 plants the primary product of the photosynthesis is a three-carbon sugar, while in C_4 species it is a four-carbon compound (see Chapter 6). As opposed to C_3 plants, in C_4 plants carbon dioxide and oxygen do not compete for the same enzyme, so that the higher rate of CO_2 fixing leads to concentrations of carbon dioxide inside the stomata that are much lower than those of C_3 species. This is generally associated with higher photosynthetic rates and water-use efficiencies for C_4 plants (Campbell and Norman, 1998). The biochemical and physiological differences between C_3 and C_4 plants have important consequences for ecological and hydrological processes. Regions with warm season rainfall tend to have greater C_4 abundance than do regions with cool season precipitation (Larcher, 1995; Lambers et al., 1998).

1995; Nobel, 1999). Assuming that any given plant is characterized by typical levels of plant water potential below which evapotranspiration is reduced and finally stopped, and considering that the daily averaged plant water potential is related to soil matrix potential (see Chapters 4 and 6), for each type of plant the typical values of soil matrix potential, Ψ_{s,s^*} and Ψ_{s,s_w}, at which stomatal closure begins and is completed, can be determined.

In general, wilting points for plants in water-controlled ecosystems are considerably lower than the value of -1.5 MPa commonly assumed for temperate crops. For plants in semi-arid environments, typical values of soil matrix potential at the wilting point, Ψ_{s,s_w}, can be as low as -3 MPa or even -5 MPa (e.g., Ludlow, 1976; Richter, 1976; MacMahon and Schimpf, 1981; Scholes and Walker, 1993; Larcher, 1995). There is considerable variation among different plants, but the permanent wilting point is frequently lower for grasses than for trees, even though grasses reach water stress conditions before trees do. There are fewer available data for Ψ_{s,s^*}, which in some water-restricted ecosystems is estimated to be around -0.03 MPa, with values somewhat larger for grasses than for trees (e.g., Scholes and Walker, 1993).

Soil matrix potential can be related to relative soil moisture using the soil-water retention curves (e.g., Clapp and Hornberger, 1978; Dingman, 1994; Hillel, 1998). Such curves depend on the type of soil and provide the dependence of s^* and s_w on soil texture. Following Clapp and Hornberger (1978) one can use empirically determined soil-water retention curves of the form

$$\Psi_s = \overline{\Psi}_s s^{-b} \tag{2.15}$$

where $\overline{\Psi}_s$ and b are experimentally determined parameters (see Table 2.1). Typical curves for loamy sand and loam are shown in Figure 2.6. It is worth noting that in order to obtain values of s_w and s^* in reasonable agreement with measured field data, one should use $\overline{\Psi}_s$ as given by the geometric mean of the measured values (reported by Clapp and Hornberger, 1978, page 603) instead of its arithmetic mean, which would yield unrealistically high values of soil moisture. As can be seen from Figure 2.6, the role of soil texture is very important. Different types of soil may yield very different levels of relative soil moisture corresponding to the same value of soil matrix potential. Typical values of s_w are found around 0.1 for loamy sand, while for loam they are near 0.25. Similarly, s^* is found to be around 0.3 for loamy sand and moves to above 0.55 in the case of loam (see Table 2.1). The values of s_w and s^* are crucially important both in the water balance and in the study of plant water stress. The above discussion clearly shows how the control of transpiration by vegetation,

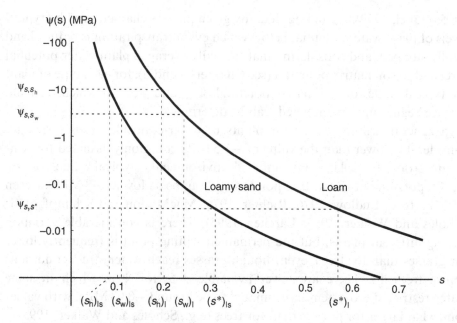

Figure 2.6 Soil moisture retention curves used to derive values of the soil water levels s_h, s_w, and s^* from the corresponding values of soil matrix potential, $\Psi_{s,s_h} = -10\,\text{MPa}$, $\Psi_{s,s_w} = -3\,\text{MPa}$, $\Psi_{s,s^*} = -0.03\,\text{MPa}$. Two different soil types, a loamy sand (ls) and a loam (l), are considered, using the parameters reported in Table 2.1. After Laio et al. (2001a).

besides depending on physiological plant characteristics, relies directly on soil properties.

Thanks to the relatively small losses involved at low soil moisture levels, a precise evaluation of the values of E_w and s_h is not very important for the temporal evolution of the soil moisture process and the different components of water balance, as long as an approximately correct order of magnitude is used in their description. A reference value of $-10\,\text{MPa}$ will be used for Ψ_{s,s_h} and $0.01\,\text{cm day}^{-1}$ for E_w. The value of s_h can be obtained from Ψ_{s,s_h} using the soil-water retention curves in the same way as for s_w and s^* (see Figure 2.6).

A final comment is in order concerning the temporal interplay of rainfall and evapotranspiration. The fact that evapotranspiration is considerably reduced during precipitation events would be a source of further complication if one considered the specific temporal dynamics within the rainfall event. This, however, is not relevant in the present model since rainfall is modeled as a sequence of instantaneous pulses, while evapotranspiration takes place continuously. Thus in our mathematical framework the instantaneous contribution of evapotranspiration is infinitesimal compared to that of rainfall in

Eq. (2.3) whenever there is a precipitation event. This might lead to a slight overestimation of the total evapotranspiration.

2.1.6 Leakage losses

Leakage losses are assumed to happen by gravity at the lowest boundary of the soil layer, neglecting possible differences in matrix potential between the soil layer under consideration and the one immediately below. The loss rate is assumed to be at its maximum when soil is saturated and then rapidly decays as the soil dries out, following the decrease of the hydraulic conductivity $K(s)$. The decay of the hydraulic conductivity is usually modeled using empirical relationships of different forms (e.g., Sisson et al., 1988; Hillel, 1998). Here the hydraulic conductivity is assumed to decay exponentially from a value equal to the saturated hydraulic conductivity K_s at $s = 1$, to a value of zero at a field capacity, s_{fc}. The exponential form, chosen for reasons of mathematical tractability, has been commonly employed in the literature (e.g., Davidson et al., 1963; Cowan, 1965; Sisson et al., 1988) as an alternative to the more customary power law. The assumed behavior of leakage losses is thus expressed as

$$L(s) = K(s) = \frac{K_s}{e^{\beta(1-s_{fc})} - 1} \left[e^{\beta(s-s_{fc})} - 1 \right], \quad s_{fc} < s \leq 1, \qquad (2.16)$$

where β is a coefficient that is used to fit the above expression to the power law (e.g., Clapp and Hornberger, 1978; Dingman, 1994; Hillel, 1998)

$$K(s) = K_s s^c, \qquad (2.17)$$

with $c = 2b + 3$ and b defined in Eq. (2.15). A possible criterion to minimize the discrepancies between the two expressions (2.16) and (2.17) is to impose the condition that they subtend the same area between s_{fc} and $s = 1$. This provides $\beta = 2b + 4$ so that the value of β depends on the type of soil, varying from $\simeq 12$ for sand to $\simeq 26$ for clay. Using the typical values reported in Table 2.1, the two expressions have very similar behavior. The behavior of leakage losses is represented in Figure 2.5. The strong nonlinear behavior of these losses as a function of soil moisture content is evident in $\chi(s)$ for $s_{fc} \leq s \leq 1$.

The field capacity s_{fc} is the soil moisture content at which drainage by gravity practically ceases to occur (for a discussion of its physical and practical relevance see Hillel, 1998; see also footnote 1 of Section 2.1.5). Here it is operationally defined as the value of soil moisture at which the hydraulic conductivity according to Eq. (2.17) becomes negligible (10%) compared to the maximum daily evapotranspiration losses, $E_{\max}$ (for this purpose fixed at $0.5\,\mathrm{cm\ day^{-1}}$).

From a physical viewpoint, the modeling of leakage driven by a unit vertical gradient due to gravity implies no interaction with the underlying soil layers and the water table. Such simplification may pose some restrictions in the use of the model, even though in most water-controlled ecosystems the contribution due to capillary rise from the water table or deeper soil layers, if any, is generally of secondary importance.

2.1.7 Soil-drying process

During interstorm periods the model Eq. (2.3) describes deterministic decays of soil moisture starting from initial values that depend on the previous history of the entire process. Upon normalization with respect to the active soil depth, the complete form of the losses described before is

$$
\rho(s) = \frac{\chi(s)}{nZ_r} = \frac{E(s) + L(s)}{nZ_r} =
\begin{cases}
0 & 0 < s \leq s_h \\[2mm]
\eta_w \frac{s - s_h}{s_w - s_h} & s_h < s \leq s_w \\[2mm]
\eta_w + (\eta - \eta_w)\frac{s - s_w}{s^* - s_w} & s_w < s \leq s^* \\[2mm]
\eta & s^* < s \leq s_{fc} \\[2mm]
\eta + m\left[e^{\beta(s - s_{fc})} - 1\right] & s_{fc} < s \leq 1,
\end{cases}
\tag{2.18}
$$

where $\rho(s)$ stands for the normalized loss function to which we will refer hereafter, and

$$
\eta_w = \frac{E_w}{nZ_r},
\tag{2.19}
$$

$$
\eta = \frac{E_{\max}}{nZ_r},
\tag{2.20}
$$

$$
m = \frac{K_s}{nZ_r\left[e^{\beta(1 - s_{fc})} - 1\right]}.
\tag{2.21}
$$

Before studying the probabilistic structure of the overall process, it is useful to analyze the behavior of the system when undergoing a prolonged drought following a rainy period. The analytical expression for the soil moisture decay from an initial condition $s_0 \geq s_{fc}$ in the absence of rainfall events is

$$s(t) = \begin{cases} s_0 - \frac{1}{\beta}\ln\left\{\dfrac{\left[\eta-m+me^{\beta(s_0-s_{fc})}\right]e^{\beta(\eta-m)t}-me^{\beta(s_0-s_{fc})}}{\eta-m}\right\} & 0 \leq t < t_{s_{fc}} \\ s_{fc} - \eta(t - t_{s_{fc}}) & t_{s_{fc}} \leq t < t_{s^*} \\ s_w + (s^* - s_w)\left[\dfrac{\eta}{\eta-\eta_w}e^{-\frac{\eta-\eta_w}{s^*-s_w}(t-t_{s^*})} - \dfrac{\eta_w}{\eta-\eta_w}\right] & t_{s^*} \leq t < t_{s_w} \\ s_h + (s_w - s_h)e^{-\frac{\eta_w}{s_w-s_h}(t-t_{s_w})} & t_{s_w} \leq t < \infty, \end{cases} \quad (2.22)$$

where

$$t_{s_{fc}} = \frac{1}{\beta(m-\eta)}\left\{\beta(s_{fc}-s_0) + \ln\left[\frac{\eta-m+me^{\beta(s_0-s_{fc})}}{\eta}\right]\right\},$$

$$t_{s^*} = \frac{s_{fc}-s^*}{\eta} + t_{s_{fc}}, \quad (2.23)$$

$$t_{s_w} = \frac{s^*-s_w}{\eta-\eta_w}\ln\left(\frac{\eta}{\eta_w}\right) + t_{s^*}$$

represent the time to evolve, in the absence of rainfall, from s_0 to s_{fc}, s^*, and s_w respectively. Note that, since the moisture decays exponentially to s_h, the process will only be at $s = s_h$ if it starts at that value.

Figures 2.7a and 2.7b show examples of Eq. (2.22) starting from saturated conditions. One can clearly see the remarkable control exerted by both the soil texture and the active soil depth. Under the same climatic conditions and with the same vegetation type (i.e., same E_{max}, Ψ_{s,s_w}, and Ψ_{s,s^*}) the time to reach the wilting point in the absence of precipitation can vary from around 20 days for a shallow loamy sand, up to well beyond 60 days for a deep loamy soil. It is also interesting to note the different effects of soil texture on the water availability for vegetation. On one hand, because of its lower hydraulic conductivity, a finer soil increases soil water availability and thus the time to reach the wilting point; on the other hand, the higher values of s_h, s_w, and s^* lead to an increase in the amount of residual water that is not extractable for plant transpiration.

2.2 Probabilistic evolution of the soil moisture process

The stochastic rainfall forcing in Eq. (2.3) makes its solution meaningful only in probabilistic terms. The probability density function of soil moisture, $p(s, t)$, can be derived from the Chapman–Kolmogorov forward equation for the process under analysis (Rodríguez-Iturbe et al., 1999a; see also Cox and Miller, 1965).

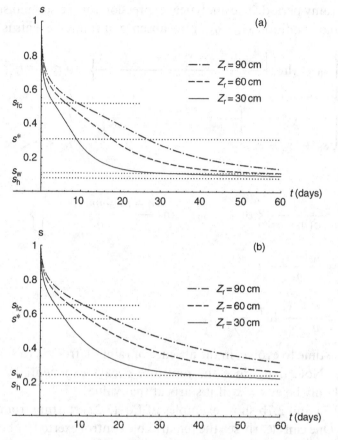

Figure 2.7 Plots of the solutions of the soil-drying process for a loamy sand (a) and loam (b) with different values of active soil depth (30, 60, and 90 cm). The parameters used are $E_{max} = 0.45$ cm day^{-1}, and $E_w = 0.01$ cm day^{-1}. The values of s_h, s_w, s^*, s_{fc}, n, β, and K_s are reported in Table 2.1. After Laio et al. (2001a).

Suppose that at time t the soil moisture level is $s(t)$ and consider a small time interval from t to $t + \Delta t$. The probability that no positive increment in soil moisture occurs is $1 - \lambda' \Delta t + o(\Delta t)$, where λ' is defined in Eq. (2.10). In this case soil moisture evolves deterministically as

$$s(t + \Delta t) = \begin{cases} s(t) - \Delta s & s(t) > \Delta s + s_h \\ s_h & s_h \leq s(t) \leq \Delta s + s_h, \end{cases} \qquad (2.24)$$

where $\Delta s = \int_t^{t+\Delta t} \rho[s(\tau)] d\tau = \rho[s(t)] \Delta t + o(\Delta t)$.

The probability that a positive increment in soil moisture takes place is $\lambda' \Delta t + o(\Delta t)$. In this case

$$s(t + \Delta t) = \begin{cases} s(t) + y - \Delta s & s(t) > \Delta s + s_h + y \\ s_h & s_h \leq s(t) \leq \Delta s + s_h + y, \end{cases} \qquad (2.25)$$

where y is the normalized soil moisture increment governed by the probability distribution $f_Y(y, s)$ described by Eq. (2.12).

In general, at time t the distribution of $s(t)$ consists of a discrete atom of probability $p_0(t)$ that the soil is at $s = s_h$ and a density $p(s, t)$ for $s > s_h$. The probability that the soil moisture takes a value in $(s, s + \Delta s)$ at the time $t + \Delta t$ can therefore be expressed as follows (e.g., Cox and Miller, 1965, page 241)

$$p(s, t + \Delta t)\Delta s = (1 - \lambda' \Delta t)p(s + \Delta s, t)\, d(s + \Delta s)$$
$$+ \lambda' \Delta t \int_{s_h}^{s} p(u + \Delta u, t)\, f_Y(s - u, u)\, d(u + \Delta u)\, ds \quad (2.26)$$
$$+ \lambda' \Delta t p_0(t)\, f_Y(s - s_h, s_h)\Delta s + o(\Delta t).$$

The first term on the r.h.s. of Eq. (2.26) describes the situation when the soil moisture reaches the value s at time $t + \Delta t$ given that no rainfall events have occurred in the interval Δt. The second term allows for the case when the soil moisture reaches s due to a positive increment resulting from the arrival of a rainfall event and the third term corresponds to the case when the process is at $s = s_h$ at time t and the arrival of a storm event makes the moisture content jump to s at time $t + \Delta t$.

Similarly, the atom of probability at s_h satisfies the equation

$$p_0(t + \Delta t) = (1 - \lambda' \Delta t)\, p_0(t) + (1 - \lambda' \Delta t) \int_{s_h}^{\rho(s_h)\Delta t} p(u, t)\, du + o(\Delta t)$$
$$= (1 - \lambda' \Delta t)\, [p_0(t) + p(s_h, t)\, \rho(s_h)\Delta t] + o(\Delta t),$$
$$(2.27)$$

where the second term on the r.h.s. accounts for the probability of moving to $s = s_h$ from a value infinitesimally above it, in the absence of rain.

Substituting now for Δs and Δu in the r.h.s. of Eq. (2.26), one obtains

$$p(s, t + \Delta t)\Delta s = (1 - \lambda' \Delta t)\, p[s + \rho(s)\Delta t, t]\, d[s + \rho(s)\Delta t]$$
$$+ \lambda' \Delta t \int_{s_h}^{s} p[u + \rho(u)\Delta t, t]\, f_Y(s - u, u)\, d[u + \rho(u)\Delta t]\, ds$$
$$+ \lambda' \Delta t p_0(t)\, f_Y(s - s_h, s_h)\, ds + o(\Delta t)$$
$$= (1 - \lambda' \Delta t)\left[p(s, t) + \rho(s)\Delta t \frac{\partial}{\partial s}p(s, t)\right]\left[1 + \frac{\partial}{\partial s}\rho(s)\Delta t\right] ds$$
$$+ \lambda' \Delta t \Delta s \int_{s_h}^{s} p(u, t)\, f_Y(s - u, u)\, du$$
$$+ \lambda' \Delta t \Delta s p_0(t)\, f_Y(s - s_h, s_h) + o(\Delta t).$$
$$(2.28)$$

Finally, dividing by Δs, subtracting $p(s, t)$ from both sides, dividing by Δt, and taking the limit as $\Delta t \to 0$, one gets

$$\frac{\partial}{\partial t} p(s, t) = \frac{\partial}{\partial s} [p(s, t)\rho(s)] - \lambda' p(s, t) + \lambda' \int_{s_h}^{s} p(u, t) \, f_Y(s - u, u) \, du$$
$$+ \lambda' p_0(t) \, f_Y(s - s_h, s_h). \tag{2.29}$$

The various terms on the r.h.s. of the integro-differential Eq. (2.29) represent the contributions to $p(s, t)$ of the different mechanisms of the soil moisture process. The first term is related to the gain of probability due to the drift of the pdf in the deterministic decay caused by $\rho(s)$, the second term represents the loss of probability due to possible jumps with frequency λ' which cause the process to leave the given trajectory moving to the level s at time t, and the last term is the positive contribution to the probability due to jumps to level s starting from lower soil moisture values.

Similarly, taking the limit as $\Delta t \to 0$, Eq. (2.27) yields

$$\frac{d}{dt} p_0(t) = -\lambda' p_0(t) + \rho(s_h) p(s_h, t). \tag{2.30}$$

Equations (2.29) and (2.30) are the Chapman–Kolmogorov forward equations for the evolution of the probability of s and are similar to those of Cox and Isham (1986). They are valid for general choices of the loss function $\rho(s)$ and jump distribution $f_Y(y, s)$ and thus describe both the classic Takacs (1955) problem, where the jumps can be of any size, and the present case, where the state of the system is bounded, as follows from the particular distribution of jump given by Eq. (2.12). Notice that the bound at $s = 1$ does not produce a probability mass for the soil moisture process at $s = 1$, since the process decays immediately from that state. The reason for this lies in the fact that the input is in the form of an instantaneous pulse rather than a pulse with a finite duration (as in the case considered by Gani, 1955). A further consequence of the saturation at $s = 1$ is that the right hand tail of the pdf of the soil moisture does not necessarily decay to zero, but instead has $p(1, t) > 0$. Since the soil moisture losses approach zero at s_h in a continuous manner (see Eq. (2.18) and Figure 2.5), the soil moisture approaches s_h asymptotically and thus the process will only be at $s = s_h$ if it starts at that value (Figure 2.7). As a consequence, the probability distribution of the soil moisture has no atom p_0 except when the process starts at s_h.

2.3 Steady-state probability distribution of soil moisture

The complete solution of Eqs. (2.29) and (2.30) presents serious mathematical difficulties. Only formal solutions in terms of Laplace transforms have been obtained for simple cases when the process is not bounded at $s = 1$ (Cox and Isham, 1986 and references therein). While some insights into the transient conditions for the soil moisture process will be given in Chapter 3 and in Sections 7.1 and 8.1, attention is focused here on the steady-state case.

Considering losses approaching zero at s_h in a continuous manner (Figure 2.5) the steady-state solution is continuous with no atom of probability at s_h. On taking the limit as $t \to \infty$ in Eqs. (2.29) and (2.30), one obtains

$$\frac{d}{ds}[\rho(s)p(s)] - \lambda'p(s) + \lambda' \int_{s_h}^{s} p(u)\, f_Y(s - u, u)\, du = 0, \qquad (2.31)$$

which, using Eq. (2.12) yields

$$\frac{d}{ds}[\rho(s)p(s)] - \lambda'p(s) + \lambda' \int_{s_h}^{s} p(u)\gamma e^{-\gamma(s-u)}\, du$$
$$+ \delta(s - 1)\lambda' \int_{s_h}^{s} p(u)\gamma e^{-\gamma(1-u)}\, du = 0. \qquad (2.32)$$

For $s < 1$, the delta functions disappear in Eq. (2.32) and, after multiplying by $e^{\gamma s}$ and differentiating with respect to s, it can be easily shown to be equivalent to the ordinary differential equation (Cox and Isham, 1986; Rodríguez-Iturbe et al., 1999a)

$$\frac{d}{ds}[\rho(s)p(s)] + \gamma\rho(s)p(s) - \lambda'p(s) = 0. \qquad (2.33)$$

Comparing Eqs. (2.32) and (2.33) one also obtains

$$\rho(s)p(s) = \lambda' \int_{s_h}^{s} p(u)\, e^{-\gamma(s-u)}\, du. \qquad (2.34)$$

For $s = 1$, Eq. (2.32) becomes

$$p(1)\frac{d}{ds}\rho(s)\Big|_{s=1} - \rho(1)p(1)\delta(0) - \lambda'p(1) + \lambda' \int_{s_h}^{1} p(u)\gamma e^{-\gamma(1-u)}\, du$$
$$+ \delta(0)\lambda' \int_{s_h}^{1} p(u)e^{-\gamma(1-u)}\, du = 0. \qquad (2.35)$$

Since $\rho(s)$ in 1 is continuous and therefore its derivative in 1 does not contain delta functions, by dividing Eq. (2.34) by $\delta(0)$ and neglecting the infinitesimal terms, one obtains

$$\gamma\rho(1)p(1) = \lambda' \int_{s_h}^{1} p(u)\, e^{-\gamma(1-u)}\, du, \qquad (2.36)$$

which turns out to be the same as Eq. (2.34) calculated in $s = 1$. Equations (2.33) and (2.34) are thus both valid for the bounded and unbounded case, whose solutions therefore only differ by an arbitrary constant of integration. Accordingly, the general form of the steady-state solution is easily obtained from Eq. (2.33) as

$$p(s) = \frac{C}{\rho(s)} e^{-\gamma s + \lambda' \int \frac{du}{\rho(u)}} \qquad \text{for } s_h < s \leq 1, \qquad (2.37)$$

where C is the normalization constant such that

$$\int_{s_h}^{1} p(s)\, ds = 1. \qquad (2.38)$$

Notice that all the complications brought by the presence of the bound at $s = 1$ are contained in the integration constant and the only effect of the saturation of the soil at $s = 1$ is the restricted range over which $p(s)$ is normalized in Eq. (2.38). The explanation for this lies in the Markovian nature of the soil moisture process (Rodríguez-Iturbe et al., 1999a). If excursions of the process above 1 are impossible, the process will spend more time in states $s \leq 1$ than would be the case otherwise, but the relative proportions of times in those states will be unchanged. Imagine two processes with, and without, the restriction that s is bounded to values less than 1. In the latter case, trajectories of the soil moisture process will jump above the level $s = 1$ and, eventually, drift down across this level once more. In the former case, these excursions are effectively excised, as the process jumps only to $s = 1$ and then immediately begins its downward decay. The trajectories below $s = 1$ in the two processes are indistinguishable.

The limits of the integral in the exponential term of Eq. (2.37) are chosen so as to assure the continuity of $p(s)$ at the end points of the four different components of the loss function described by Eq. (2.18) (Cox and Isham, 1986; Rodríguez-Iturbe et al., 1999a). In this manner the general solution is obtained as

$$p(s) = \begin{cases} \frac{C}{\eta_w} \left(\frac{s - s_h}{s_w - s_h} \right)^{\frac{\lambda'(s_w - s_h)}{\eta_w} - 1} e^{-\gamma s} & s_h < s \leq s_w \\[2ex] \frac{C}{\eta_w} \left[1 + \left(\frac{\eta}{\eta_w} - 1 \right) \left(\frac{s - s_w}{s^* - s_w} \right) \right]^{\frac{\lambda'(s^* - s_w)}{\eta - \eta_w} - 1} e^{-\gamma s} & s_w < s \leq s^* \\[2ex] \frac{C}{\eta} e^{-\gamma s + \frac{\lambda'}{\eta}(s - s^*)} \left(\frac{\eta}{\eta_w} \right)^{\frac{\lambda'(s^* - s_w)}{\eta - \eta_w}} & s^* < s \leq s_{fc} \\[2ex] \frac{C}{\eta} e^{-(\beta + \gamma)s + \beta s_{fc}} \left(\frac{\eta\, e^{\beta s}}{(\eta - m)e^{\beta s_{fc}} + m\, 2e^{\beta s}} \right)^{\frac{\lambda'}{\beta(\eta - m)} + 1} \left(\frac{\eta}{\eta_w} \right)^{\frac{\lambda'(s^* - s_w)}{\eta - \eta_w}} e^{\frac{\lambda'}{\eta}(s_{fc} - s^*)} & s_{fc} < s \leq 1. \end{cases}$$
$$(2.39)$$

The expression of the constant C can be obtained analytically, but it is quite involved due to both the piecewise form of the losses and the presence of the bound (see Appendix A at the end of this chapter). The expressions for the mean and variance of soil moisture are also quite cumbersome and are not reported here. Their behavior was discussed by Rodríguez-Iturbe et al. (1999a) for a simpler model (see Section 2.6.1), but the qualitative behavior is similar to that of the complete model.

Figure 2.8 shows some examples of the pdf's derived from Eq. (2.39). The two different types of soil, already used for Figures 2.6 and 2.7, are loamy sand

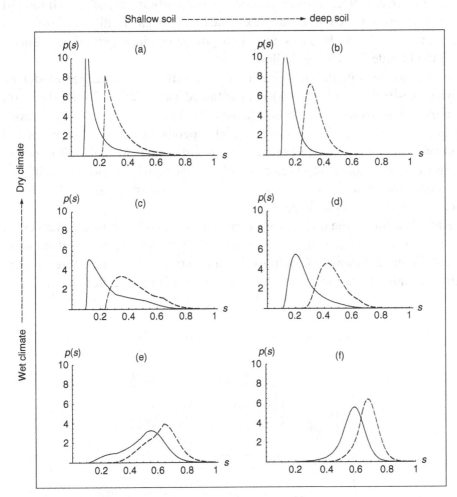

Figure 2.8 Examples of pdf's of relative soil moisture for different type of soil, soil depth, and mean rainfall rate. Continuous lines refer to loamy sand, dashed lines to loam (see Table 2.1 for the values of soil parameters). Left column corresponds to $Z_r = 30$ cm, right column to 90 cm. Top, center, and bottom graphs have a frequency of rainfall events λ of 0.1, 0.2, and 0.5 day^{-1} respectively. Common parameters to all graphs are $\alpha = 1.5$ cm, $\Delta = 0$ cm, $E_w = 0.01$ cm day^{-1}, and $E_{max} = 0.45$ cm day^{-1}. After Laio et al. (2001a).

and loam with two different values of active soil depth. These are chosen in order to emphasize the role of soil in the soil moisture dynamics. The parameters related to vegetation are kept fixed and correspond to those used in Table 2.1. A detailed discussion of the role of different functional vegetation types is postponed to the chapters that follow. Here that role of climate is studied only in relation to changes in the frequency of storm events λ, keeping fixed the mean rainfall depth α and the maximum evapotranspiration rate E_{max}. A coarser soil texture corresponds to a consistent shift of the pdf toward drier conditions, which in the most extreme case can reach a difference of 0.2 in the location of the mode. The shape of the pdf also undergoes marked changes, with the broadest pdf's for shallower soils.

Although the qualitative behavior of the pdf's of the earlier model by Rodríguez-Iturbe et al. (1999a) is maintained (see Section 2.6.1), there are important improvements especially in the tails of the distribution. At low levels of soil moisture the residual water content depends directly on the type of soil and vegetation present (see Figure 2.8a in particular). At high soil moisture levels the more realistic representation of leakage provides smoother pdf's for wet conditions and lower values of relative soil moisture as a result of an increased relevance of leakage.

An indication of the impacts of the probabilistic soil moisture dynamics on vegetation is provided by the analysis of particular realizations of soil moisture traces. Figure 2.9 shows a superposition of two traces of soil moisture with the same rainfall realization for the case of a loam with two different active soil

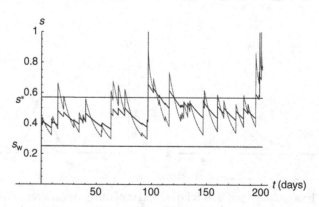

Figure 2.9 Example of traces of relative soil moisture for the same rainfall sequence in a loamy soil with different depths ($\alpha = 1.5\,\text{cm}$, $\lambda = 0.2\,\text{day}^{-1}$, $\Delta = 0\,\text{cm}$, $E_w = 0.01\,\text{cm day}^{-1}$, $E_{max} = 0.45\,\text{cm day}^{-1}$; see Table 2.1 for soil parameters). Continuous line refers to $Z_r = 90\,\text{cm}$, dashed line to $Z_r = 30\,\text{cm}$. The traces correspond to the pdf's reported with dashed lines in Figures 2.8c and 2.8d. After Laio et al. (2001a).

depths (the parameter values are those used for the dashed line pdf's in Figures 2.8c and 2.8d). In both cases the levels of soil moisture s^* and s_w are the same and so approximately is the mean soil moisture. The continuous line, corresponding to the deeper soil, is almost always between s^* and s_w, which is different from the shallower soil where the trace lively jumps and decays between low and high soil moisture levels. Without anticipating a discussion on vegetation response, the importance of such different soil moisture dynamics for different types of plants and the possible implications for the development of different strategies of adaptation to water stress are nevertheless clearly perceived. A discussion of the impact of climate, soil, and plant water uptake on vegetation response requires a more detailed analysis of the mechanisms of water stress and the definition of quantitative relationships to link such stress to soil moisture dynamics. This will be presented in Chapter 4.

The steady-state analysis is most appropriate for the study of water-controlled ecosystems where rainfall is mostly concentrated in a warm growing season, and winter is usually temperate and dry. As a consequence, in those cases the soil moisture condition at the beginning of the growing season is not very different from that during the rest of the season and transients due to seasonality are generally not significant. Different ecosystems that are examples of these conditions (see Chapter 5) are the savannas of South Africa (Scholes and Walker, 1993), the shrublands in Southern Texas, at least for most of the growing season (Archer et al., 1988; Scholes and Archer, 1997), and the shortgrass steppe in Colorado (Sala et al., 1992). Nevertheless, transient soil moisture dynamics and climatic seasonality can be important in other semi-arid environments, especially at the beginning of the growing season. As will be seen in Chapter 3 and Sections 7.1 and 8.1, this is related to the climatic conditions (e.g., seasonality), to the soil depth, and to the ratio between losses and rainfall input during the months preceding the growing season. Typical cases include Mediterranean climates (e.g., Major, 1963; Naveh, 1967; Ng and Miller, 1980) and the Patagonian steppe (Paruelo and Sala, 1995; Golluscio et al., 1998). Temperate forests in the northwest United States (e.g., Waring and Running, 1998) are also heavily controlled by transient soil moisture conditions. In these cases, rainfall and temperature are markedly out of phase, and the dormant season thus becomes a period of consistent soil-water storage to be used during the following growing season (see Chapter 8).

2.4 Water balance

It is interesting to discuss the soil water balance resulting from the model presented above. The different components of the balance are the long-term averages of the respective components of the soil moisture dynamics. As described in Eqs. (2.3) through (2.5), rainfall is first partitioned in interception, runoff, and infiltration. The latter is then divided into leakage and evapotranspiration. For the purposes of describing vegetation conditions, the amount of water transpired can be further divided into water transpired under stressed conditions and water transpired under unstressed conditions. This last distinction will be important in relation to the links among soil moisture dynamics, plant water stress, and the structure and productivity of ecosystems.

The long-term average of Eq. (2.3) is simply

$$\langle \varphi \rangle - \langle \chi \rangle = 0. \tag{2.40}$$

Using Eqs. (2.4) and (2.5), the mean infiltration rate is

$$\langle \varphi \rangle = \langle R \rangle - \langle I \rangle - \langle Q \rangle, \tag{2.41}$$

where $\langle R \rangle$, $\langle I \rangle$, and $\langle Q \rangle$ stand for the mean rates of rainfall, interception, and runoff respectively, while the average losses are

$$\langle \chi \rangle = nZ_{\mathrm{r}} \int_{s_{\mathrm{h}}}^{1} \rho(s)p(s)\, \mathrm{d}s. \tag{2.42}$$

The mean rainfall intensity is

$$\langle R \rangle = \alpha \cdot \lambda, \tag{2.43}$$

and the mean rate of interception is (see Section 2.1.2)

$$\langle I \rangle = \alpha \cdot (\lambda - \lambda') = \alpha\lambda\left(1 - \mathrm{e}^{-\frac{\Delta}{\alpha}}\right). \tag{2.44}$$

The water balance Eq. (2.40) can therefore be written as

$$\alpha \cdot \lambda' - \langle Q \rangle = \langle \chi \rangle. \tag{2.45}$$

Returning to the mean rate of water losses from the soil, $\langle \chi \rangle$, this can be also written as

$$\langle \chi \rangle = \langle E_{\mathrm{s}} \rangle + \langle E_{\mathrm{ns}} \rangle + \langle L \rangle = \int_{s_{\mathrm{h}}}^{s^*} E(s)\, p(s)\, \mathrm{d}s$$

$$+ \int_{s^*}^{1} E(s)\, p(s)\, \mathrm{d}s + \int_{s_{\mathrm{fc}}}^{1} L(s)\, p(s)\, \mathrm{d}s, \tag{2.46}$$

where $\langle E_{\mathrm{s}} \rangle$, $\langle E_{\mathrm{ns}} \rangle$, and $\langle L \rangle$ are the mean rate of evapotranspiration under stressed conditions, the mean rate of unstressed evapotranspiration, and the mean rate of leakage, respectively.

Analytical expressions for the terms of Eq. (2.46) can be obtained using the simplified forward Eq. (2.33). Integrating Eq. (2.33) from s' to s'', one obtains

$$\int_{s'}^{s''} \rho(s)p(s) \, ds = \frac{\lambda'}{\gamma}[P(s'') - P(s')] - \frac{1}{\gamma}[\rho(s'')p(s'') - \rho(s')p(s')], \quad (2.47)$$

where $P(s)$ is the cumulative probability distribution of soil moisture (see Appendix A at the end of this chapter). From Eq. (2.47) with $s' = s_h$ and $s'' = s^*$, the analytical expression of $\langle E_s \rangle$ can be calculated as

$$\langle E_s \rangle = \int_{s_h}^{s^*} E(s)p(s)ds = nZ_r \int_{s_h}^{s^*} \rho(s)p(s)ds = \alpha\lambda'P(s^*) - \alpha\eta p(s^*), \quad (2.48)$$

where the condition $\eta = \rho(s^*)$ (see Eq. (2.18)) has been used. The evapotranspiration under unstressed conditions can be obtained as

$$\langle E_{ns} \rangle = \int_{s^*}^{1} E(s)p(s)ds = nZ_r\eta \int_{s^*}^{1} p(s)ds = E_{max}[1 - P(s^*)], \quad (2.49)$$

and the average leakage losses are

$$\langle L \rangle = \int_{s_{fc}}^{1} L(s)p(s)ds = nZ_r \int_{s_{fc}}^{1} [\rho(s) - \eta]p(s)ds$$
$$= \alpha\left[\lambda' - \lambda'P(s_{fc}) - \left(\eta + \frac{K_s}{nZ_r}\right)p(1) + \eta p(s_{fc})\right] - E_{max}[1 - P(s_{fc})], \quad (2.50)$$

where Eq. (2.47) has been used with $s' = s_{fc}$ and $s'' = 1$; again from Eq. (2.18), $\rho(1) = \eta + \frac{K_s}{nZ_r}$ and $\rho(s_{fc}) = \eta$.

The total mean losses from the soil can now be calculated either by applying Eq. (2.47) with $s' = s_h$ and $s'' = 1$ (see Eq. (2.42)), or by summing Eqs. (2.46), (2.48), and (2.49) and simplifying the terms by using Eq. (2.47) with $s' = s^*$ and $s'' = s_{fc}$,

$$\langle \chi \rangle = \alpha\lambda' - \alpha\left(\eta + \frac{K_s}{nZ_r}\right)p(1). \quad (2.51)$$

Finally, Eq. (2.45) and Eq. (2.51) allow one to write the mean runoff as

$$\langle Q \rangle = \alpha\left(\eta + \frac{K_s}{nZ_r}\right)p(1). \quad (2.52)$$

Figure 2.10 presents examples of the behavior of the various components of the water balance normalized by the mean rainfall rate, for some specific

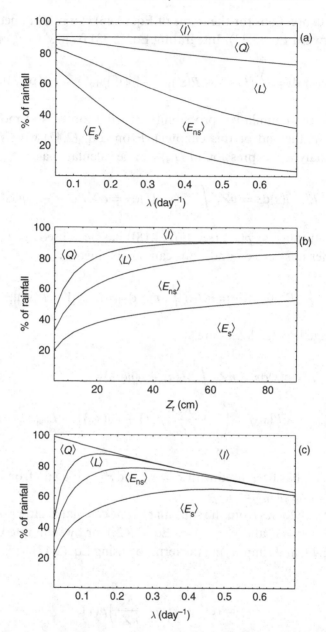

Figure 2.10 Components of the water balance normalized by the total rainfall. (a) Water balance as a function of the frequency of rainfall events λ, for a shallow loamy soil ($Z_r = 30$ cm, $\alpha = 2$ cm). (b) Water balance as a function of the soil depth Z_r, for a loamy sand ($\alpha = 2$ cm, $\lambda = 0.2$ day^{-1}). (c) Water balance for a loamy sand as a function of the frequency of rainfall events for a constant mean total rainfall during a growing season, $\Theta = 60$ cm. Other common parameters are $E_w = 0.01$ cm day^{-1}, $E_{max} = 0.45$ cm day^{-1}, and $\Delta = 0.2$ cm (see Table 2.1 for soil parameters). After Laio et al. (2001a).

rainfall, soil, and vegetation characteristics. The influence of the frequency of rainfall events, λ, is shown in Figure 2.10a for the case of a shallow loam. Since the amount of interception changes in proportion to the rainfall rate, it is not surprising that the fraction of water intercepted remains constant when normalized by the total rainfall, $\alpha\lambda$. The percentage of runoff increases almost linearly. More interesting is the interplay between leakage and the two components of evapotranspiration. The fraction of water transpired under stressed conditions rapidly decreases from $\lambda = 0.1$ to about $\lambda = 0.4$, while the evapotranspiration under unstressed conditions evolves in a much more gentle manner. As will be discussed in Chapter 4, this last aspect has interesting implications for vegetation productivity. It is clear that in semi-arid conditions most of the water that actually reaches the soil is lost by evapotranspiration (in particular transpiration), a result in agreement with many field observations (e.g., Sarmiento, 1984; Eagleson and Segarra, 1985; Sala et al., 1992; Scholes and Walker, 1993).

Figure 2.10b shows the role of the active soil depth in the water balance. For relatively shallow soils there is a strongly nonlinear dependence on soil depth of all components of the water balance (with the obvious exception of interception, which is constant because the rainfall is constant). For example, changing from $nZ_r = 5$ to $nZ_r = 20$ the amount of water transpired is practically doubled in this particular case.

Figure 2.10c shows the impact on water balance when the frequency and amount of rainfall are varied keeping constant the total amount of rainfall in a growing season. The result is interesting, because of the existence of two opposite mechanisms regulating the water balance. On one hand, runoff production, for a given mean rainfall input, strongly depends on the ratio between soil depth and the mean depth of rainfall events. The rapid decrease of runoff is thus somewhat analogous to that in the first part of Figure 2.10b, where a similar behavior was produced by an increase in soil depth. On the other hand interception increases almost linearly with λ. The interplay between these two mechanisms determines a maximum of both leakage and evapotranspiration at moderate values of λ (of course the position of the maxima changes according to the parameters used). This is particularly important from the vegetation point of view, since the mean transpiration rate is linked to productivity of ecosystems (e.g., Kramer and Boyer, 1995, page 383). The role not only of the amount, but also of the timing of rainfall in soil moisture dynamics (Noy-Meir, 1973), is made clear by the existence of an optimum for transpiration/productivity, which is directly related to the climate–soil–vegetation characteristics. The particular position of this maximum in the parameter space is governed by the interplay of all the mechanisms acting in the soil water balance; namely, the intensity and amount of rainfall, interception, the active soil depth, and the nonlinear losses due to evapotranspiration and leakage.

2.5 Comparison with field data

The analytical solution of the stochastic model of soil water balance is illuminating in many issues, because it allows a quantitative comprehensive view of the soil moisture dynamics and related phenomena. The results are in good agreement with physical expectations. Other investigations on soil moisture dynamics have used similar assumptions, concerning both the stochastic representation of rainfall (e.g., Eagleson, 1978a, 1978b; Cordova and Bras, 1981; Milly, 1993) and the behavior of losses due to evapotranspiration and leakage (e.g., Cordova and Bras, 1981; Paruelo and Sala, 1995). Here we discuss some recent field data that provide further support for the model assumptions. More detailed comparisons with natural ecosystems of different types will be given in Chapter 5.

The first set of information concerns the three-stage sequence of evapotranspiration losses from a shortgrass steppe, analyzed by Brutsaert and Chen (1995, 1996). Using data sampled at a daily time scale, which is also the time scale adopted here, they reported the existence of three phases of drying with distinct temporal behavior. Such phases could correspond to the three different types of soil moisture control of evapotranspiration of the present model (e.g., Eq. (2.14)). According to their analysis, during the first stage of drying after an abundant rainfall, soil water evaporates from both the soil surface and vegetation, at a rate that is governed by the available energy supply. This phase corresponds in our model to the range of soil moisture $\{s^*, 1\}$ or, according to Eq. (2.22), to the time interval $0 < t \leq t_{s^*}$. Then, after soil moisture content has decreased below a critical level, a transitional stage sets in, during which soil moisture is the controlling factor. This agrees with the assumptions of the model presented in this chapter, where soil moisture controls transpiration through stomatal closure between s_w and s^*, or in the interval $t_{s^*} < t \leq t_{s_w}$. Finally, when soil moisture decreases below a certain threshold corresponding to plant wilting, drying takes place mainly as evaporation from the soil surface ($s_h < s \leq s_w$, or $t_{s_w} < t \leq \infty$). Brutsaert and Chen (1995) refer to this phase as a desorption phase with a temporal decay proportional to $t^{-0.5}$, but they also maintain that different mathematical formulations could be possible for this phase. Moreover, the relevance of this last phase for modeling purposes is limited, except for extremely dry environments.

Some important field support for the theoretical results presented before comes from the recent analysis by Salvucci (2001). By noticing that the expected value of the change in soil moisture storage, conditioned on the storage *s*, is zero under statistically steady-state conditions, he deduced that the conditionally averaged net precipitation rate is equal to the moisture-dependent losses. Thus, using daily precipitation and average soil moisture data from a site in Illinois, and dividing the sample into two parts, one corresponding to the early growing season and the other to the late growing season, Salvucci (2001) estimated the conditionally averaged precipitation rate for various values of soil moisture. In this manner, from field data, he then obtained soil moisture daily losses whose functional form closely agrees with the one used here, namely Eq. (2.18). Figure 2.11a reproduces the loss functions obtained by Salvucci (2001). Employing these loss functions, jointly with the estimates of the frequency of occurrence and mean depth of daily rainfall, λ and α, Salvucci (2001) also computed the relative soil moisture pdf's with Eq. (2.37), obtaining an excellent match with the observed frequency distributions (Figure 2.11b).

2.6 Simpler models of soil moisture dynamics

The loss function previously used (see Eq. (2.18) and Figure 2.5) attempts a balance between a realistic representation of the processes involved and mathematical tractability. In some cases, however, when the main focus is not on the soil moisture–vegetation interaction, the loss function may be further simplified and still provide a useful description of the processes that are at the basis of the water balance. Three different types of loss functions with increasing level of simplification are considered in this section: the simplified model studied by Rodríguez-Iturbe et al. (1999a), the minimalistic model proposed by Milly (1993, 2001), and a simple alternative to the latter.

2.6.1 A simplified stochastic model of soil moisture dynamics

The model proposed by Rodríguez-Iturbe et al. (1999a) is an earlier version of the more complete model presented in this chapter. Although somewhat less suitable for ecohydrological studies, it still retains important features and is worth considering because of its simpler expressions for both the mean soil moisture and the water balance components.

The precipitation and infiltration models are the same as in Sections 2.1.1 through 2.1.4. The only part that differs from the previously described model is the loss function, namely the evapotranspiration at very low values of soil

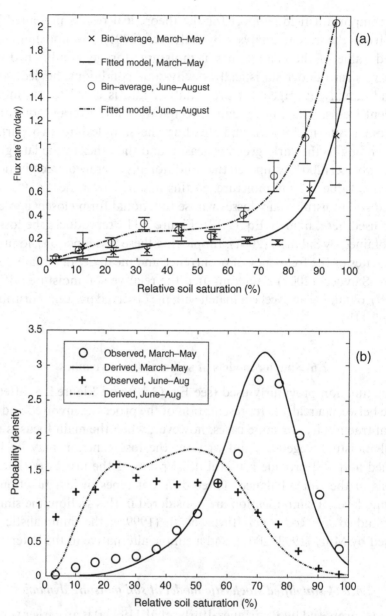

Figure 2.11 (a) Estimation of water loss from conditional mean precipitation for a site in Illinois. (b) Observed and derived probability distributions of relative soil saturation (after Salvucci, 2001).

moisture and the leakage. Evapotranspiration is assumed to increase linearly from 0 at $s = 0$ to E_{max} at s^* and then remain constant until saturation. Leakage is simply a linearly increasing function from 0 at a given threshold $s_1 > s_{fc}$ up to K_s at saturation.

Because of the form of the losses, the pdf is now

$$
p(s) = \begin{cases}
\frac{C'}{\eta} \left(\frac{s}{s^*}\right)^{\frac{\lambda s^*}{\eta}-1} e^{-\gamma s} & 0 < s \le s^* \\[2ex]
\frac{C'}{\eta} e^{-\frac{\lambda s^*}{\eta}} e^{-s\left(\gamma - \frac{\lambda}{\eta}\right)} & s^* < s \le s_1 \\[2ex]
\frac{C'}{\eta} \left[\frac{k(s-s_1)}{(1-s_1)\eta} + 1\right]^{\frac{\lambda(1-s_1)}{k}-1} e^{-\gamma s + \lambda \frac{s_1-s^*}{\eta}} & s_1 < s \le 1.
\end{cases}
\tag{2.53}
$$

The expression for C' resulting from the normalizing condition, Eq. (2.38), is

$$
C' = \frac{\eta\,\zeta\,k\,e^{\chi}\,R}{\eta\zeta k(e^{\chi - \gamma s^*} - e^{\lambda s_1/\eta + \gamma s^*}) + \eta\zeta\xi\,e^{\vartheta}(1-s_1)R(\Gamma_2 - \Gamma_1) + e^{\chi}ks^*R\Gamma_3}, \tag{2.54}
$$

where

$$
R = \gamma\eta - \lambda
$$

$$
\zeta = (\gamma s^*)^{\frac{\lambda s^*}{\eta}}
$$

$$
\vartheta = \frac{\gamma\eta(1-s_1)}{k} + \frac{\lambda s_1}{\eta} + \gamma s^*
$$

$$
\xi = \left[\frac{\gamma\eta(1-s_1)}{k}\right]^{\frac{-\lambda(1-s_1)}{k}}
$$

$$
\chi = \gamma(s_1 + s^*) + \frac{\lambda s^*}{\eta}
$$

$$
\Gamma_1 = \Gamma[\lambda(1-s_1)/k, \gamma\eta(1-s_1)/k]
$$

$$
\Gamma_2 = \Gamma[\lambda(1-s_1)/k, \gamma(\eta+k)(1-s_1)/k]
$$

$$
\Gamma_3 = \Gamma[\lambda s^*/\eta, \gamma s^*] \tag{2.55}
$$

and $\Gamma(\cdot, \cdot)$ stands for the generalized incomplete gamma function.

The expression for the mean, $\langle s \rangle = \int_0^1 s\,p(s)\,ds$, can be obtained as

$$
\langle s \rangle = c \left\{ \left[\varphi_1 e^{-\gamma s_1 + \frac{\lambda}{\eta}(s_1 - s^*) + \gamma s^*} - \varphi_2 e^{-\gamma s^*}\right] \right.
$$
$$
\left. + \xi e^{\vartheta - \chi} \frac{1-s_1}{k}\left[(\Gamma_1 - \Gamma_2)\frac{\eta(1-s_1) - ks_1}{k} - \frac{1}{\gamma}(\Gamma_4 - \Gamma_5)\right] + \frac{s^*}{\zeta\eta\gamma}\Gamma_6 \right\}, \tag{2.56}
$$

where the new symbols are defined as

$$\varphi_1 = \frac{-\eta - s_1(\gamma\eta - \lambda)}{(\gamma\eta - \lambda)^2}$$

$$\varphi_2 = \frac{-\eta - s^*(\gamma\eta - \lambda)}{(\gamma\eta - \lambda)^2}$$

$$\Gamma_4 = \Gamma[1 + \lambda(1 - s_1)/k, \gamma\eta(1 - s_1)/k]$$

$$\Gamma_5 = \Gamma[\,1 + \lambda(1 - s_1)/k, \gamma(\eta + k)(1 - s_1)/k]$$

$$\Gamma_6 = \Gamma[1 + \lambda s^*/\eta, \gamma s^*]. \tag{2.57}$$

Notice that Γ_4, Γ_5, and Γ_6 are related to Γ_1, Γ_2, and Γ_3 respectively (e.g., Abramowitz and Stegun, 1964), but this does not allow for further simplifications. The variance may also be obtained analytically but it is a long expression that adds little by itself.

Figure 2.12 shows the impact of evapotranspiration, rate of storm arrivals, and mean depth per event on the mean soil moisture. The soil characteristics

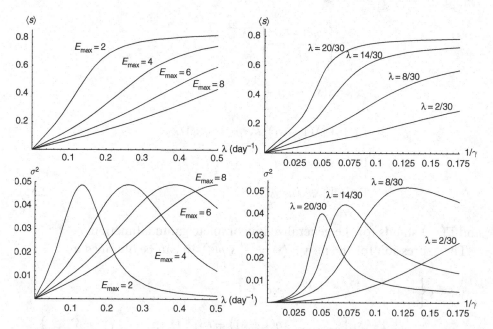

Figure 2.12 Mean and variance of soil moisture as a function of rainfall and evapotranspiration parameters ($s^* = 0.35$, $s_1 = 0.85$, $K_s = 100\,\text{cm day}^{-1}$, $nZ_r = 15\,\text{cm}$). Left column: $1/\gamma = 15/1.5$; right column: $E_{max} = 5\,\text{mm day}^{-1}$. Redrawn after Rodriguez-Iturbe et al. (1999a).

are taken in the middle range of their possible values. Notice how, when rain events occur very frequently, the mean value of the soil moisture becomes highly dependent on the mean storm depth up to a point after which it remains practically constant and close to saturation.

The variance of the soil moisture pdf is also shown in Figure 2.12 for a fixed set of soil conditions as a function of the characteristics of rainfall and evapotranspiration. One can observe the high sensitivity of the variance to rainfall and evapotranspiration rates. Except for very dry climates, the soil moisture variance has a well-defined maximum. Thus for a given mean frequency of rain events, the variance increases with increasing active soil depth up to a point after which any further increase in soil depth leads to a decrease in variance. The point at which the maximum occurs varies with the frequency of events and is also a function of the soil and vegetation parameters.

Rodríguez-Iturbe et al. (1999a) also give the expressions of the normalized components of the average water balance, which can be expressed in a simpler form than those of the full model (Section 2.4). As for the mean and variance, due to the averaging operation, the differences with the full model are even smaller.

2.6.2 Minimalistic models of soil moisture dynamics

In the minimalistic model proposed by Milly (1993, 2001) the losses are assumed to be constant and equal to E_{max} in the range $s_w < s < s_1$ and zero at s_w. Above s_1, leakage and runoff losses take place instantaneously (i.e., at an infinite rate, $K_s \to \infty$), so that s_1 is the effective upper bound of the process. Although Milly's model may be too simplified for a meaningful description of plant conditions, its parsimony in the number of parameters makes it interesting as a theoretical simplified description of the mean water balance. From a mathematical viewpoint Milly's model corresponds to the virtual waiting-time process with the addition of an upper bound at s_1. Such a process has been a well-studied one since Takacs pointed out its importance in the queuing and storage context, where the state variable is interpreted as the total time it would take to serve all customers in an office at time t, or the time-dependent amount of water in a reservoir depleted at a constant rate (e.g., Cox and Miller, 1965, page 241).

Since in this model the soil moisture fluctuates only between s_w and s_1, one can introduce a new variable

$$x = \frac{s - s_w}{s_1 - s_w},$$

(2.58)

which will be called effective relative soil moisture, and define the available water storage as $w_0 = (s_1 - s_w)nZ_r$. In the new variable x, the only relevant parameters are w_0, λ, α, and E_{max}. Using dimensional analysis, the problem can be expressed as a function of only two dimensionless variables. Possible choices (Milly, 2001) are either

$$\gamma = \frac{w_0}{\alpha} \quad \text{and} \quad \frac{\lambda}{\eta} = \frac{\lambda w_0}{E_{max}}, \tag{2.59}$$

which are similar to those used before, or

$$\gamma = \frac{w_0}{\alpha} \quad \text{and} \quad D_I = \frac{\gamma\eta}{\lambda} = \frac{E_{max}}{\langle R \rangle}, \tag{2.60}$$

where D_I is the dryness index of Budyko (1974), which represents the long-term ratio between the potential evapotranspiration and the rainfall rate.

Since in this case the losses approach s_w (i.e., $x = 0$) discontinuously, the soil moisture pdf is a mixed distribution with an atom of probability at $x = 0$. The normalized form of Eqs. (2.29) and (2.30) becomes

$$\frac{\partial}{\partial \tau}p(x,\tau) = \frac{\eta}{\lambda}\frac{\partial}{\partial x}p(x,\tau) - p(x,\tau) + \gamma\int_0^x p(u,\tau)\,e^{-\gamma(x-u)}\,du + \gamma p_0(\tau)\,e^{-\gamma x}, \tag{2.61}$$

$$\frac{d}{d\tau}p_0(\tau) = -p_0(\tau) + \frac{\eta}{\lambda}p(0,\tau), \tag{2.62}$$

where $\tau = \lambda t$. The steady-state solution can be obtained as

$$p(x) = \frac{\lambda}{\eta}\,p_0\,e^{\left(\frac{\lambda}{\eta}-\gamma\right)x}, \tag{2.63}$$

$$p_0 = \frac{\frac{\lambda}{\eta} - \gamma}{\frac{\lambda}{\eta}\,e^{\frac{\lambda}{\eta}-\gamma} - \gamma}. \tag{2.64}$$

In the special case when $\frac{\lambda}{\eta} = \gamma$ (e.g., $D_I = 1$),

$$p(x) = \frac{\gamma}{1+\gamma}, \quad p_0 = \frac{1}{1+\gamma}. \tag{2.65}$$

The continuous part of the steady-state pdf is thus an exponential function (see Figures 2.13 and 2.14a) which becomes uniform for $\frac{\lambda}{\eta} = \gamma$ (e.g., $D_I = 1$).

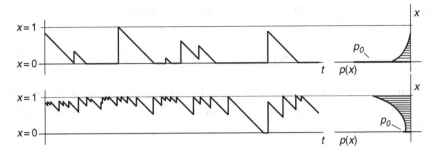

Figure 2.13 Examples of traces and steady-state pdf's for Milly's model.

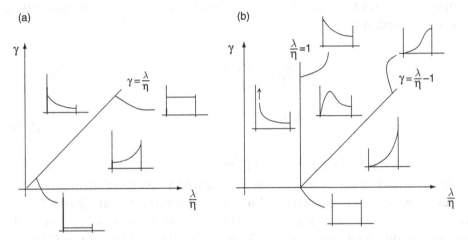

Figure 2.14 Different forms of the relative soil moisture steady-state pdf's for (a) the minimalistic model with constant losses and (b) with linearly increasing losses with soil moisture.

Although the resulting pdf's of these minimalistic models are too coarse a description of the soil moisture process, both the mean and variance as well as the mean components of the water balance show some resemblance to those of the full model described before (see Milly, 2001). The mean and variance are given by

$$\langle x \rangle = \frac{1}{\frac{\lambda}{\eta} - \gamma} + \frac{1 + \frac{\lambda}{\eta} \, e^{\frac{\lambda}{\eta} - \gamma}}{\frac{\lambda}{\eta} \, e^{\frac{\lambda}{\eta} - \gamma} - \gamma}, \tag{2.66}$$

$$\sigma_x^2 = \frac{1}{\left(\frac{\lambda}{\eta} - \gamma\right)^2} - \frac{1 + (\gamma + 2)\frac{\lambda}{\eta} \, e^{\frac{\lambda}{\eta} - \gamma}}{\left(\frac{\lambda}{\eta} \, e^{\frac{\lambda}{\eta} - \gamma} - \gamma\right)^2}. \tag{2.67}$$

In the special case $\frac{\lambda}{\eta} = \gamma$ (e.g., $D_I = 1$), the limit yields

$$\langle x \rangle = \frac{\gamma}{2(1+\gamma)}, \qquad \sigma_x^2 = \frac{\gamma^2 + 4\gamma}{12(\gamma+1)^2}. \tag{2.68}$$

Neglecting interception (e.g., $\lambda = \lambda'$, $\Delta = 0$), the water balance is simply given by the partition between long-term evapotranspiration $\langle E \rangle$ and leakage plus runoff $\langle LQ \rangle$, i.e.,

$$\langle R \rangle = \langle E \rangle + \langle LQ \rangle = E_{\max}(1 - p_0) + \langle LQ \rangle. \tag{2.69}$$

Normalizing with the mean total rainfall and introducing the dryness index, one readily obtains

$$\frac{\langle E \rangle}{\langle R \rangle} = D_I (1 - p_0), \tag{2.70}$$

$$\frac{\langle LQ \rangle}{\langle R \rangle} = 1 - D_I (1 - p_0), \tag{2.71}$$

where p_0 is given by Eqs. (2.64) and (2.65).

An alternative minimalistic model is obtained by assuming linear increasing evapotranspiration losses with soil moisture (Porporato et al., 2004). In this case soil moisture decays exponentially, so that x approaches zero only asymptotically. As a consequence, the corresponding steady-state pdf has no atom at zero and reads

$$p(x) = \frac{C''}{\eta} e^{-\gamma x} x^{\frac{\lambda}{\eta} - 1}, \tag{2.72}$$

where

$$C'' = \frac{\eta \gamma^{\frac{\lambda}{\eta}}}{\Gamma\left(\frac{\lambda}{\eta}\right) - \Gamma\left(\frac{\lambda}{\eta}, \gamma\right)}. \tag{2.73}$$

As shown in Figure 2.14, the pdf of this new minimalistic model shows a richer behavior than the corresponding model with constant losses.

The mean is given by

$$\langle x \rangle = \frac{\lambda}{\gamma \eta} - \frac{\gamma^{\frac{\lambda}{\eta} - 1} e^{-\gamma}}{\Gamma\left(\frac{\lambda}{\eta}\right) - \Gamma\left(\frac{\lambda}{\eta}, \gamma\right)}, \tag{2.74}$$

and the normalized water balance is easily obtained by noticing that $\langle E \rangle = E_{\max} \langle x \rangle$. Hence

$$1 = D_I \langle x \rangle + \frac{\langle LQ \rangle}{\langle R \rangle}. \tag{2.75}$$

Figure 2.15 shows a computation of the fraction of evapotranspiration losses for the two minimalistic models described above. Interestingly, the results reproduce qualitatively analogous empirical curves that have been used to describe observational data for the mean annual water balance over large regions (e.g., Budyko, 1974; Brutsaert, 1982) and, in particular, the well-known hydroclimatological relationship of Budyko (1974). The latter describes the average terrestrial water balance by means of a semi-empirical curve which represents the fraction of the rainfall that is evapotranspired as a nonlinear function of the dryness index,

$$\langle E/R \rangle = \{D_I[1 - \exp(-D_I)] \tanh(1/D_I)\}^{0.5}, \tag{2.76}$$

and synthesizes the average partitioning of the rainfall input into evapotranspiration and runoff plus deep infiltration, thus offering a parsimonious and universal picture of the terrestrial water cycle. As shown in detail in Figure 2.16, for $\gamma \sim 5.5$ the minimalistic model with linear losses reproduces very well Budyko's curve. Therefore, using typical values of the parameters (e.g., average rainfall depth per event $\alpha = 1.5$ cm, relative soil moisture at the wilting point $s_w = 0.2$, relative soil moisture threshold for deep infiltration and runoff

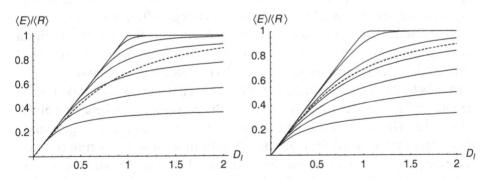

Figure 2.15 Comparison of the two minimalistic models. Fraction of precipitation that is evapotranspired as a function of the dryness index for various values of the dimensionless soil water-holding capacity, γ. Values of γ are from bottom to top: 0.5, 1, 2, 4, 8, 24, and 1000. Dashed line: Budyko's curve, Eq. (2.76). Left: minimalistic model with constant losses (modified after Milly, 1993); right: minimalistic model with losses linearly increasing with soil moisture.

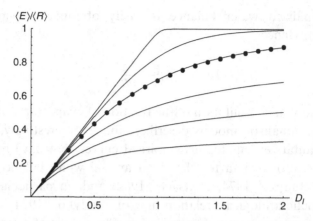

Figure 2.16 Fraction of total rainfall lost by evapotranspiration as a function of Budyko's dryness index for different values of the parameter γ for the minimalistic model with linear losses. The dots represent the semi-empirical curve of Budyko, while the continuous line underlying the dots corresponds to $\gamma = 5.5$. As explained in the text, this refers to an average effective rooting depth of approximately 35 cm. From the lowest to the highest one, the continuous curves refer to $\gamma = 0.5, 1, 2, 5.5, 20, 1000$, respectively. After Porporato et al. (2004).

$s_1 = 0.85$, and porosity $n = 0.4$), Budyko's curve corresponds to an average active soil depth of approximately 30–35 cm which, interestingly, is the average depth within which most of the roots are typically comprised (Jackson et al., 1996; Schenk and Jackson, 2002). The model interpretation also makes clear the effects of possible climate changes on Budyko's curve. As an example, depending on the degree to which evapotranspiration, rainfall regime and plant characteristics are affected by climate change, alterations in the mean depth of rainfall per event and in the rooting depth will imply a vertical shift in the diagram, while a shift along the x-axis will entail changes in potential transpiration and total amounts of rainfall (Porporato et al., 2004).

Thanks to the realistic representation of the soil water balance provided by the minimalistic model with linear losses, the corresponding behavior of the pdf of the effective relative soil moisture as a function of the governing parameters may be used for a general classification of soil moisture regimes. Accordingly, the boundaries between different shapes of the pdf may be used to define an "arid" regime (pdf's with zero mode), an "intermediate" regime (corresponding to soil moisture pdf's with a central maximum) and a "wet" regime (with the mode at saturation) as indicated in Figure 2.17. A further distinction within the intermediate regime can be made on the basis of plant response to soil moisture dynamics. Using the effective relative soil moisture value (x^*) as a threshold marking the onset of plant water stress (as will be seen

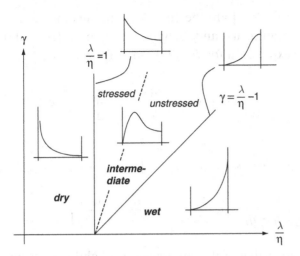

Figure 2.17 Classification of soil water balance, based on the shape of the effective relative soil moisture pdf, as a function of the two governing parameters, λ/η and γ, that synthesize the role of climate, soil, and vegetation. The dashed line, $\gamma = 1/x^*(\lambda/\eta - 1)$, is the locus of points where the mode of the soil moisture pdf is equal to the threshold x^*, which marks the onset of plant water stress. After Porporato et al. (2004).

in Chapter 4, x^* is typically in the order of 0.3–0.4), the dashed line of slope $1/x^*$ in Figure 2.17 becomes the place where the mode of the effective relative soil moisture pdf is equal to x^* and thus where plants are more likely to be at the boundary between stressed and unstressed conditions. Accordingly, it may be used to divide water-stressed (or semi-arid) types of water balance on the left side from unstressed ones on the right side (Figure 2.17).

2.7 Appendix A. Soil moisture cumulative probability distribution

The analytical expressions of the cumulative probability distribution $P(s)$ of soil moisture are reported here, using the parameters defined in Eqs. (2.19), (2.20), and (2.21). The expression is split into four parts resulting from the piecewise loss function Eq. (2.18). The expression for $P(s)$ is obtained by direct integration of the pdf of soil moisture given by Eq. (2.39).

For $s_h < s \leq s_w$ the cumulative distribution is

$$P(s) = \frac{C}{\eta_w} e^{-\gamma s_h}(s_w - s_h)[\gamma(s_w - s_h)]^{-\frac{\lambda'(s_w - s_h)}{\eta_w}} \left\{ \Gamma\left[\frac{\lambda'(s_w - s_h)}{\eta_w}\right] \right.$$
$$\left. + \Gamma\left[\frac{\lambda'(s_w - s_h)}{\eta_w}, \gamma(s - s_h)\right] \right\},$$

(2.77)

where $\Gamma[\,\cdot\,]$ and $\Gamma[\,\cdot\,,\,\cdot\,]$ are the complete and incomplete gamma function, respectively (Abramowitz and Stegun, 1964, n. 6.1.1 and n. 6.5.3). For $s_w < s \leq s^*$ the expression for $P(s)$ takes the form

$$P(s) = P(s_w) + \frac{C}{\eta - \eta_w}(s^* - s_w)\left[\frac{\eta_w \gamma(s^* - s_w)}{\eta - \eta_w}\right]^{-\lambda'\frac{s^* - s_w}{\eta - \eta_w}} e^{-\gamma\left[s_w - \frac{\eta_w(s^* - s_w)}{\eta - \eta_w}\right]}$$

$$\times \left\{\Gamma\left[\lambda'\frac{s^* - s_w}{\eta - \eta_w}, \frac{\eta_w \gamma(s^* - s_w)}{\eta - \eta_w}\right]\right.$$

$$\left. + \Gamma\left[\lambda'\frac{s^* - s_w}{\eta - \eta_w}, \gamma(s - s_w) + \frac{\eta_w \gamma(s^* - s_w)}{\eta - \eta_w}\right]\right\}, \tag{2.78}$$

where $P(s_w)$ is the value of the cumulative probability distribution calculated in $s = s_w$ from Eq. (2.77). When $s_w < s \leq s^*$ the cumulative probability distribution is

$$P(s) = P(s^*) + \frac{C}{\lambda' - \gamma\eta}\left(\frac{\eta}{\eta_w}\right)^{\lambda'\frac{s^* - s_w}{\eta - \eta_w}}\left[e^{-\gamma s + \frac{\lambda'}{\eta}(s - s^*)} - e^{-\gamma s^*}\right], \tag{2.79}$$

with the value of $P(s^*)$ calculated from Eq. (2.78). Finally, for $s_{fc} < s \leq 1$ the expression for $P(s)$ can be written as

$$P(s) = P(s_{fc}) + \frac{C}{\gamma(\eta - m) - \lambda'}\left(\frac{\eta}{\eta_w}\right)^{\lambda'\frac{s^* - s_w}{\eta - \eta_w}} e^{\frac{\lambda'}{\eta}(s_{fc} - s^*)}\left\{e^{-\gamma s_{fc}}\left(\frac{\eta}{\eta - m}\right)^{\frac{\lambda'}{\beta(\eta - m)}}\right.$$

$$\times {}_2F_1\left[\frac{\lambda'}{\beta(\eta - m)} + 1, \frac{\lambda'}{\beta(\eta - m)} - \frac{\gamma}{\beta}; \frac{\lambda'}{\beta(\eta - m)} + 1 - \frac{\gamma}{\beta}; \frac{m}{m - \eta}\right]$$

$$- e^{-\gamma s}\left[\frac{\eta}{\eta - m} \cdot \frac{e^{\beta(s - s_{fc})}m + \eta - m}{e^{-\beta(s - s_{fc})}(\eta - m) + m}\right]^{\frac{\lambda'}{\beta(\eta - m)}}$$

$$\left. \times {}_2F_1\left[\frac{\lambda'}{\beta(\eta - m)} + 1, \frac{\lambda'}{\beta(\eta - m)} - \frac{\gamma}{\beta}; \frac{\lambda'}{\beta(\eta - m)} + 1 - \frac{\gamma}{\beta}; \frac{e^{\beta(s - s_{fc})}m}{m - \eta}\right]\right\}, \tag{2.80}$$

where $P(s_{fc})$ is calculated from Eq. (2.79) and ${}_2F_1[\,\cdot\,,\,\cdot\,;\,\cdot\,;\,\cdot\,]$ is the Hypergeometric Function (Abramowitz and Stegun, 1964, n. 15.1.1).

The integration constant C in Eqs. (2.77)–(2.80) is the same of Eq. (2.39). The analytical expression of C can be derived by imposing the condition $P(1) = 1$ in Eq. (2.80), and noting that $P(s_{fc})$ in Eq. (2.80) is a function of C as well.

3

Crossing properties of soil moisture dynamics

The frequency and duration of excursions of the soil moisture process above and below some levels directly related to the physiological dynamics of plants are of crucial importance for ecohydrology. This chapter focuses on the crossing properties of specific levels of soil moisture that are important for vegetation. After a general mathematical derivation of the crossing characteristics typical of shot noise processes, attention is turned to the impact of climate, soil, and vegetation on the duration and frequency of soil moisture excursions below the wilting point, s_w, and below the point where transpiration is reduced, s^*. These properties will be basic for the study of plant water stress in Chapter 4.

The analysis of the crossing properties is also important for gaining insights into the transient dynamics of soil moisture, either at the beginning of the growing season, to evaluate the time to reach the stress levels after the soil winter storage is depleted, or after a drought, to estimate the time to recover from a situation of intense water stress. The expressions of the mean time to reach s^* from field capacity, s_{fc}, will be used in Chapter 7 to extend the definition of water stress to transient conditions and furthermore to analyze different strategies of water use when the winter water recharge is important. This chapter ends with the crossing analysis of the minimalistic models of soil moisture dynamics described in Section 2.6.2.

Although the results of this chapter are extensively used later, an understanding of the mathematical derivations presented in Section 3.1 is not necessary for their use.

3.1 Mean first passage times of processes driven by white shot noise

Several papers have recently dealt with the derivation of exact expressions for the mean first passage times (MFPT's) of specific stochastic processes (e.g., Hanggi and Talkner, 1985; Sancho, 1985; Masoliver et al., 1986; Rodríguez and Pesquera, 1986; Hernandez-Garcia et al., 1987; Masoliver, 1987; Porra and Masoliver, 1993). The particular case of systems driven by white shot noise has also received considerable attention (Hernandez-Garcia et al., 1987; Masoliver, 1987; Porra and Masoliver, 1993; Laio et al., 2001c), because of its analytical tractability and the large number of its possible applications. We follow here the work by Laio et al. (2001c), where simple interpretable results for the MFPT's of stochastic processes driven by white shot noise are derived for cases relevant to soil moisture dynamics.

Consider a general family of stochastic processes whose dynamical evolution is given by a stochastic differential equation of the type of Eq. (2.3). Referring to its normalized form and considering an upper bound, s_b, the process is described by a stochastic differential equation

$$\frac{ds}{dt} = Y[s(t), t] - \rho(s) \tag{3.1}$$

(see Eqs. (2.3) and (2.18)), where $\rho(s)$ is, in general, any function defining the deterministic losses of the process, and $Y(t)$ is the driving random process defined by a sequence of pulses at random times τ_i, each pulse having an independent random height y_i, i.e.,

$$Y[s(t), t] = \sum_i y_i \delta(t - t_i), \tag{3.2}$$

where $\delta(\cdot)$ is the Dirac delta function (see Sections 2.1.1 and 2.1.3). The random times $\{t_i\}$ form a Poisson sequence with rate λ, and the probability distribution of y_i is state dependent and reads (Eq. (2.12))

$$f_Y(y, s) = \gamma e^{-\gamma y} + \delta(y - s_b + s) \int_{s_b-s}^{\infty} \gamma e^{-\gamma u} du, \quad \text{for} \quad 0 \leq y \leq s_b - s, \tag{3.3}$$

where the bound at s_b has been explicitly indicated; for the case of relative soil moisture s_b is equal to 1, whereas the unbounded case is recovered in the limit $s_b \to \infty$.

Masoliver (1987), in the more general framework of non-Markovian processes, obtained closed exact expressions for the MFPT's of dynamic systems driven by white shot noise for the special cases of exponentially distributed and constant jump heights. The exact expressions for the general process described by Eqs. (3.1) to (3.3) were also derived by Laio et al. (2001c) in a more direct manner and written in a simpler form, thanks to the Markovian nature of the process. They obtained the linkage between the first passage times and the steady-state probability density function of the process.

As seen in Section 2.3, the pdf of the state variable s in the process described by Eq. (3.1) is, in general, a mixed distribution

$$p(s, t) = p(s, t|s_0, t_0) + \delta(s - s_h)p_0(t|s_0, t_0), \tag{3.4}$$

with $p_0(t|s_0, t_0)$ the atom of probability at the lower bound s_h, and where, for future convenience, the starting point s_0 at time t_0, i.e., $p(s, t = t_0) = \delta(s - s_0)$, has been explicitly indicated. As already noticed in Chapter 2, the atom of probability in $s = s_h$ is present only if the losses are discontinuous at the lower bound, $\rho(s_h) \neq \rho(s_h^-) = 0$. The resulting forward differential Chapman–Kolmogorov equations are given by Eqs. (2.29) and (2.30), i.e.,

$$\frac{\partial}{\partial t}p(s, t|s_0, t_0) = \frac{\partial}{\partial s}[p(s, t|s_0, t_0)\rho(s)] - \lambda p(s, t|s_0, t_0)$$
$$+ \lambda \int_{s_h}^{s} du\, p(s, t|s_0, t_0)\, f_Y(s - u; u) + \lambda p_0(t|s_0, t_0)\, f_Y(s - s_b, s_b),$$
$$\tag{3.5}$$

for the continuous part of the pdf and

$$\frac{d}{dt}p_0(t|s_0, t_0) = -\lambda p_0(t|s_0, t_0) + \rho(s_h)p(s_h, t|s_0, t_0), \tag{3.6}$$

for the atom of probability in $s = s_h$.

In order to analyze the crossing properties of the Markovian process just described, from the forward Eqs. (3.5) and (3.6) one obtains the corresponding backward or adjoint equations (Gardiner, 1990). Using the backward equation it is then possible to write the differential equation that describes the evolution of the probability density, $g_T(s_0, t)$, that a particle starting from s_0 inside an interval $\{\xi', \xi\}$ leaves for the first time the interval at a time t (Figure 3.1). This is the usual procedure for obtaining an equation for the MFPT statistics when the process is Markovian (e.g., Rodríguez and Pesquera, 1986; Gardiner, 1990; Van Kampen, 1992). For the process under consideration (see the Appendix to this chapter), the resulting equation for $g_T(s_0, t)$ is

$$\frac{\partial g_T(s_0, t)}{\partial t} = -\rho(s_0)\frac{\partial g_T(s_0, t)}{\partial s_0} - \lambda g_T(s_0, t) + \lambda \int_{s_0}^{\xi} f_Y(z - s_0, s_0)g_T(z, t)dz. \tag{3.7}$$

One does not need to solve the partial integro-differential Eq. (3.7) to obtain the distribution of the first passage times of the process. An expression involving the mean time for exiting the interval $\{\xi', \xi\}$, $\overline{T}_{\xi'\xi}(s_0)$, is obtained from

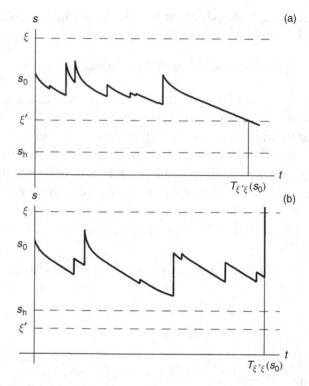

Figure 3.1 Trajectories and first passage times when the thresholds ξ' and ξ are both greater than the fixed point s_h (a) and when $\xi' < s_h$ (b). After Laio et al. (2001c).

Eq. (3.7) multiplying by t and then integrating between 0 and ∞. Using the definition of the mean time of crossing, $\overline{T}_{\xi'\xi}(s_0) = \int_0^\infty t g_T(s_0, t) dt$ and integrating by parts the term with the time derivative, one gets

$$-1 = -\rho(s_0)\frac{d\overline{T}_{\xi'\xi}(s_0)}{ds_0} - \lambda\overline{T}_{\xi'\xi}(s_0) + \lambda\int_{s_0}^{\xi}\gamma e^{-\gamma(z-s_0)}\overline{T}_{\xi'\xi}(z)dz, \qquad (3.8)$$

where the exponential part of the jumps' distribution, $f_Y(z - s_0, s_0)$, has been used in the integral on the r.h.s. because, in the hypothesis that $\xi < s_b$, the presence of the bound at s_b becomes irrelevant. The integro-differential Eq. (3.8) was also obtained by Masoliver (1987, Eq. (A4) therein) in a different and more general way. A similar procedure may be followed to derive the equations for the higher-order moments of $g_T(s_0, t)$.

Differentiating Eq. (3.8) with respect to s_0 and reorganizing the terms, the following second-order differential equation is obtained

$$\rho(s_0)\frac{d^2\overline{T}_{\xi'\xi}(s_0)}{ds_0^2} + \left[\lambda + \frac{d\rho(s_0)}{ds_0} - \gamma\rho(s_0)\right]\frac{d\overline{T}_{\xi'\xi}(s_0)}{ds_0} + \gamma = 0. \qquad (3.9)$$

Equation (3.9) needs two boundary conditions: the first is obtained from Eq. (3.8) evaluated at $s_0 = \xi$,

$$\rho(\xi)\frac{d\overline{T}_{\xi'\xi}(s_0)}{ds_0}\bigg|_{s_0=\xi} = 1 - \lambda\overline{T}_{\xi'\xi}(\xi). \qquad (3.10)$$

For the second boundary condition one has to consider whether the lower limit ξ' is above or below s_h. In the first case, ξ' is a real absorbing barrier, so that the boundary condition is $\overline{T}_{\xi'\xi}(\xi') = 0$ (see Figure 3.1a). In contrast, when $\xi' < s_h$, ξ' cannot be reached by the trajectory (see Figure 3.1b), and the average exiting time from the interval becomes the mean first passage time of the threshold ξ. In this case the second boundary condition is obtained by setting $s_0 = s_h$ in Eq. (3.8), i.e.,

$$\overline{T}_{\xi'\xi}(s_h) = \int_{s_h}^{\xi} \gamma e^{-\gamma(z-s_h)}\overline{T}_{\xi'\xi}(z)dz + \frac{1}{\lambda}. \qquad (3.11)$$

3.1.1 MFPT's of a threshold ξ above the initial point s_0

Consider first the case when $\xi' < s_h$ (Figure 3.1b). In this case the MFPT is independent of ξ' and will be indicated as $\overline{T}_{\xi}(s_0)$. The solution of Eq. (3.9) with boundary conditions given by Eqs. (3.10) and (3.11) is (Masoliver, 1987, Eq. (5.34) therein)

$$\begin{aligned}
\overline{T}_{\xi}(s_0) = &-\gamma\int_{s_0}\frac{1}{\rho(u)}e^{M(u)}\int_u e^{-M(z)}dzdu \\
&+ C_1(\xi)\int_{s_0}\frac{1}{\rho(u)}e^{M(u)}du + C_2(\xi),
\end{aligned} \qquad (3.12)$$

where $C_1(\xi)$ and $C_2(\xi)$ are integration constants and

$$M(u) = \gamma u - \lambda\int_u\frac{1}{\rho(z)}dz. \qquad (3.13)$$

Equations (3.12) and (3.13) present some difficulties of application due to the involved form of the boundary conditions (3.10) and (3.11) which define $C_1(\xi)$ and $C_2(\xi)$. However, $C_1(\xi)$ and $C_2(\xi)$ can be directly calculated when the jump heights are exponentially distributed. From Eqs. (3.12) and (3.13) one may easily determine the value of the integration constant $C_2(\xi)$, which allows us to rewrite Eq. (3.12) as

$$
\bar{T}_\xi(s_0) = \frac{1}{\lambda} - \frac{C_1(\xi)}{\lambda} e^{M(\xi)} + \frac{\gamma}{\lambda} e^{M(\xi)} \int_\xi e^{-M(x)} dx - C_1(\xi) \int_{s_0}^\xi \frac{e^{M(x)}}{\rho(x)} dx
$$
$$
+ \gamma \int_{s_0}^\xi \frac{e^{M(x)}}{\rho(x)} \int_x e^{-M(x')} dx dx'.
$$
(3.14)

Equation (3.14) can now be inserted into Eq. (3.11). Integration by parts and reorganization of the terms lead to

$$
C_1(\xi) \left[\int_{s_h}^\xi e^{-\gamma(u-s_h)} \frac{e^{M(u)}}{\rho(u)} du + e^{-\gamma(\xi-s_h)} \frac{e^{M(\xi)}}{\lambda} \right]
$$
$$
= -\frac{1}{\lambda} + \frac{e^{-\gamma(\xi-s_h)}}{\lambda}
$$
$$
+ \gamma \int_{s_h}^\xi e^{-\gamma(u-s_h)} \frac{e^{M(u)}}{\rho(u)} \int_u e^{-M(z)} dz du
$$
$$
+ \frac{\gamma}{\lambda} e^{-\gamma(\xi-s_h)} e^{M(\xi)} \int_\xi e^{-M(u)} du.
$$
(3.15)

By noting from Eq. (3.13) that $\frac{e^{-\gamma u} e^{M(u)}}{\rho(u)} = -\frac{1}{\lambda} \frac{d[e^{-\gamma u} e^{M(u)}]}{du}$ one can proceed with the direct integration of the first term on the l.h.s. of Eq. (3.15) and with the integration by parts of the third term on the r.h.s. of the same equation. A further reorganization of terms leads to

$$
C_1(\xi) = \gamma \int_{s_h} e^{-M(u)} du.
$$
(3.16)

The value of C_1, which is independent of ξ, can be substituted in Eq. (3.14) yielding

$$
\bar{T}_\xi(s_0) = \frac{1}{\lambda} + \frac{\gamma}{\lambda} e^{M(\xi)} \int_{s_h}^\xi e^{-M(u)} du + \gamma \int_{s_0}^\xi \frac{e^{M(u)}}{\rho(u)} \int_{s_h}^u e^{-M(z)} dz du.
$$
(3.17)

Equation (3.17) represents a first simplification of the result that was given in Masoliver (1987) as a combination of Eqs. (3.10)–(3.12). The linkage between the MFPT's and the steady-state pdf of the process allows further simplification of Eq. (3.17). Consider Eqs. (2.37) and (3.13): one can write $p(s) = \frac{C}{\rho(s)} e^{-M(s)}$, so that Eq. (3.17) becomes

$$\overline{T}_\xi(s_0) = \frac{1}{\lambda} + \frac{\gamma}{\lambda p(\xi)\rho(\xi)} \int_{S_h}^\xi p(u)\rho(u)\mathrm{d}u + \gamma \int_{s_0}^\xi \frac{1}{p(u)\rho^2(u)} \int_{S_h}^u p(z)\rho(z)\mathrm{d}z\mathrm{d}u.$$

(3.18)

Equation (2.33) can now be used to simplify the above expression. Integration of Eq. (2.33) and substitution in Eq. (3.18), after using Eq. (2.30) under steady-state conditions, yields

$$\overline{T}_\xi(s_0) = \frac{P(\xi)}{p(\xi)\rho(\xi)} + \int_{s_0}^\xi \left[\frac{\lambda P(u)}{p(u)\rho^2(u)} - \frac{1}{\rho(u)} \right] \mathrm{d}u.$$

(3.19)

Note that, when the starting point s_0 coincides with the threshold ξ, the integral on the r.h.s. cancels out and the mean crossing time reads

$$\overline{T}_\xi(\xi) = \frac{P(\xi)}{p(\xi)\rho(\xi)}.$$

(3.20)

As a consequence, under steady-state conditions the mean rate of occurrence or frequency of the upcrossing (or downcrossing) events of the threshold ξ can be obtained from Eq. (3.20) as (see also Vanmarke, 1983)

$$\nu(\xi) = p(\xi)\rho(\xi).$$

(3.21)

Returning to the MFPT's, some manipulation of Eq. (3.19) leads to the synthetic expression

$$\overline{T}_\xi(s_0) = \overline{T}_{s_0}(s_0) + \gamma \int_{s_0}^\xi \overline{T}_u(u)\mathrm{d}u,$$

(3.22)

which, along with Eq. (3.20), completely defines the MFPT from s_0 to ξ for $s_0 < \xi$. Similar relationships between MFPT's and steady-state pdf's were obtained by Balakrishnan et al. (1988) for processes driven by Gaussian white noise.

Some important properties of the MFPT's become clear from this formulation: both $\overline{T}_\xi(\xi)$ and $\overline{T}_\xi(s_0)$ in Eqs. (3.20) and (3.22) can be expressed as functions of the ratio $\frac{P(u)}{p(u)\rho(u)}$, where $P(u)$ is the steady-state cumulative probability distribution calculated at a certain level u, $p(u)$ is the steady-state probability density function at the same level, and $\rho(u)$ is the loss function, always at the level u. As pointed out in Section 2.3, all the changes induced in $p(s)$ from the presence of the bound at $s = s_b$ are embedded in the constant of normalization C. However, the constant C is present in both $P(u)$ and $p(u)$, so that it cancels out from the expressions of the MFPT. The validity of the latter is thus independent of the presence of the bound in $s = s_b$, and, for similar reasons, of the shape of the loss function above the threshold ξ. The opposite is true for the frequency of crossings, $\nu(\xi)$, which contains the constant C through $p(\xi)$ and therefore also depends on the part of the dynamics above ξ.

3.1.2 MFPT's of a threshold ξ' below the initial point s_0 and $\xi \to \infty$

Consider now the MFPT's when $\xi' > s_h$ (see Figure 3.1a) in the special case when $\xi \to \infty$ ($\xi > s_b$ in the bounded case). The latter condition implies that ξ is never reached so that the MFPT only depends on ξ' and will thus be called $\overline{T}_{\xi'}(s_0)$. It represents the average time that a particle starting from $s = s_0 > \xi'$ takes to arrive at ξ'. Equation (3.9) now needs to be integrated with the boundary conditions given by Eq. (3.10) and $\overline{T}_{\xi'}(\xi'+) = 0$, where the plus sign is to evidence the discontinuity $\overline{T}_{\xi'}(\xi') = \frac{P(\xi')}{\rho(\xi')p(\xi')} \neq \overline{T}_{\xi'}(\xi'^+) = 0$. The procedure for calculating the resulting integration constants is analogous to that used before. The final result is (see also Porra and Masoliver, 1993)

$$\overline{T}_{\xi'}(s_0) = \gamma \int_{\xi'}^{s_0} \frac{e^{M(u)}}{\rho(u)} \int_u^\infty e^{-M(z)} dz du. \tag{3.23}$$

Splitting the integral on the r.h.s. into two parts corresponding to the region below and that above the bound s_b and considering again the relationship between $M(u)$ and the steady-state pdf, one obtains

$$
\begin{aligned}
\overline{T}_{\xi'}(s_0) &= \int_{\xi'}^{s_0} \frac{1}{\rho^2(u)p(u)} [\lambda - \lambda P(u) + p(u)\rho(u)] du \\
&= \overline{T}_{s_0}(s_0) - \overline{T}_{\xi'}(\xi') + \frac{1}{\nu(\xi')} - \frac{1}{\nu(s_0)} + \gamma \int_{\xi'}^{s_0} \left[\frac{1}{\nu(u)} - \overline{T}_u(u) \right] du,
\end{aligned}
\tag{3.24}
$$

where $\overline{T}_{s_0}(s_0)$, $\overline{T}_{\xi'}(\xi')$, and $\overline{T}_u(u)$ are calculated from Eq. (3.20) and $\nu(\xi')$, $\nu(s_0)$, and $\nu(u)$ from Eq. (3.21).

Differently from $\overline{T}_\xi(s_0)$, $\overline{T}_{\xi'}(s_0)$ depends on the presence of the bound at $s = s_b$ and on the shape of $\rho(s)$ for $s > s_0$. This is clear from the presence in Eq. (3.24) of $\nu(\xi')$, $\nu(s_0)$, and $\nu(u)$, which in turn contain the normalization constant C. In fact, the trajectory from s_0 to ξ' can take any value above ξ' with the presence of the upper bound decreasing the first passage time of ξ' for all those trajectories that would have taken values above s_b in unbounded conditions.

3.2 Crossing properties of the soil moisture process

Two random variables originating from the analysis of soil moisture crossing properties will be particularly important for dealing with the temporal dimension of water stress (see Chapter 4): the length T_ξ of the time intervals in which soil moisture is below a threshold ξ, and the number n_ξ of such intervals during a growing season (Figure 3.2). The analytical expressions for the mean of such variables were obtained in the previous section. For simplicity of notation, the mean time of an excursion below ξ will be indicated as $\overline{T}_\xi$, rather than the more precise notation $\overline{T}_\xi(\xi)$ used throughout the mathematical derivation of the previous section.

The influence of different kinds of climate, soil, and vegetation on $\overline{T}_\xi$ and $\overline{n}_\xi$ was studied by Ridolfi et al. (2000a) using the simplified model of Section 2.6.1, while Porporato et al. (2001) employed the more realistic loss function described in Chapter 2 (see Eq. (2.18) and Figure 2.5), which allows for a better assessment of the role of plant and soil characteristics. In particular, the explicit inclusion of the wilting and hygroscopic points in the loss function leads to changes sometimes relevant to $\overline{T}_\xi$ and $\overline{n}_\xi$, especially for the mean duration of an excursion below the wilting point.

If one is interested only in the crossing properties of a single given level of soil moisture, a much simpler and more intuitive derivation than the one given in the previous section may be followed. Consider first the mean rate of occurrence, or frequency, of upcrossing events of a threshold ξ of soil moisture, ν_ξ. Given the definition of the soil moisture process, $s(t)$, as instantaneous

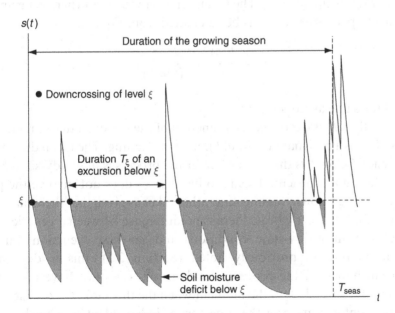

Figure 3.2 Temporal evolution of the soil moisture process and definition of some of its crossing properties with respect to the threshold ξ. After Ridolfi et al. (2000a).

upward jumps followed by continuous decays with rate $\rho(s)$, the time spent by the soil moisture process between the levels ξ and $\xi + d\xi$ is zero during an upcrossing and equal to $d\xi/\rho(\xi)$ during each downcrossing. Thus, since the total fraction of time spent by the process between ξ and $\xi + d\xi$ is by definition $p(\xi)\,d\xi$, the frequency of downcrossings (or upcrossings) can simply be obtained as the ratio between this fraction of time and the time spent during a single downcrossing, i.e.,

$$\nu(\xi) = \frac{p(\xi)d\xi}{\frac{d\xi}{\rho(\xi)}}, \tag{3.25}$$

which immediately gives Eq. (3.21).

The mean number of upcrossings (or downcrossings) during a growing season of length T_{seas} is then readily obtained as

$$\bar{n}_\xi = \nu_\xi T_{\text{seas}} = \rho(\xi)p(\xi)T_{\text{seas}}, \tag{3.26}$$

while the mean duration of an excursion below the soil moisture level ξ can be found by combining the steady-state pdf and the frequency of crossing. The product of ν_ξ and $\overline{T}_\xi$ gives the percentage of time the trajectory spends below ξ, which in steady-state conditions is equal to the cumulative probability distribution $P(\xi)$ evaluated in ξ. The mean time between a downcrossing and the subsequent upcrossing can thus be calculated from Eq. (3.21) as

$$\overline{T}_\xi = \frac{P(\xi)}{\nu_\xi} = \frac{P(\xi)}{\rho(\xi)p(\xi)}, \tag{3.27}$$

which is the same as Eq. (3.20).

The fact that $\overline{T}_\xi$ is expressed as a function of the steady-state characteristics of the soil moisture dynamics should not be misleading. The mean duration of an excursion below ξ is the same whether the process is in steady-state conditions or during a transient. This is so because $\overline{T}_\xi$ does not involve the probability of having a downcrossing, but rather takes as given that there has been a downcrossing. In other words, there is no difference between a generic excursion below ξ in steady-state conditions and the first excursion during a transient period. As previously noted (Section 2.3), this is due to the Markovian nature of the process: the dynamics does not have memory of the past in the sense that, in the moment when the threshold ξ is downcrossed, the subsequent evolution of the trajectory is independent of what happened before. For similar reasons, the value of $\overline{T}_\xi$ is independent of the shape of the loss function above the level ξ.

The results of the crossing analysis can now be applied to two important thresholds for vegetation, the soil moisture at wilting point (s_w), and the point at which transpiration starts being reduced (s^*). If Eq. (3.20) is evaluated at $\xi = s_w$, the analytical expression for the mean duration of an excursion below the wilting point can be obtained using the results of Chapter 2 (e.g., Eqs. (2.18), (2.39), and (2.78) for $\rho(s_w)$, $p(s_w)$, and $P(s_w)$, respectively) as

$$
\overline{T}_{s_w} = \frac{1}{\lambda'} + \frac{1}{\lambda'} \, e^{\gamma(s_w - s_h)} [\gamma(s_w - s_h)]^{-\frac{\lambda'(s_w - s_h)}{\eta_w}}
$$
$$
\times \left\{ \Gamma\left[1 + \frac{\lambda'(s_w - s_h)}{\eta_w}\right] - \Gamma\left[1 + \frac{\lambda'(s_w - s_h)}{\eta_w}, \gamma(s_w - s_h)\right] \right\},
$$
(3.28)

where $\Gamma[\cdot]$ and $\Gamma[\cdot, \cdot]$ are the complete and incomplete gamma function respectively (Abramowitz and Stegun, 1964). Analogously, from Eq. (3.21) and from Eqs. (2.18) and (2.39), one can easily get the mean number of upcrossings of s_w,

$$
\overline{n}_{s_w} = C \, e^{-\gamma s_w} T_{\text{seas}},
$$
(3.29)

where C is the integration constant of the soil moisture pdf defined in the Appendix of Chapter 2. Note that, as said before for the general case, C is present in the expression of $\overline{n}_{s_w}$ but not in that of $\overline{T}_{s_w}$.

The values of $\overline{T}_{s_w}$ and $\overline{n}_{s_w}$ are plotted in Figure 3.3 as a function of the storm frequency λ for typical values of the other parameters. A clear nonlinearity is present in the behavior of both variables. Notice that there are values of λ below which the mean duration of an excursion is very long, while the mean number of crossings is still relatively high. This combination of conditions may be disastrous for vegetation ecosystems. As will be explained in Chapter 4, when soil moisture goes below s_w, the cells lose their turgor and the plant is no longer able to extract water from the soil. The damage has become permanent and the consequences worsen with time. Some authors have suggested that the "survival time" of a plant could be related to the ratio between the internal water storage and the rate of cuticular transpiration (e.g., Levitt, 1980), but, due to the complexity of the mechanisms involved, no clear relationship seems to be plausible (Larcher, personal communication, 1999). The prolonged permanence below the wilting point is nevertheless a rare event except for very dry climates (see the values of $\overline{n}_{s_w}$ well below 1 in Figure 3.3). For wet climates, $\overline{T}_{s_w}$ approaches the limit value $1/\lambda'$, since the trace of soil moisture never goes below the hygroscopic point. In such conditions, almost every rainfall event is sufficient to produce an upcrossing of the threshold s_w. Due to the relevance of this bound at $1/\lambda'$, the most significant trends regarding $\overline{T}_{s_w}$ are those with respect to λ, which is functionally related to λ' by Eq. (2.10), and

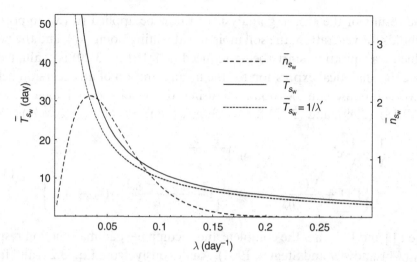

Figure 3.3 Average duration of an excursion below the wilting point, $\overline{T}_{s_w}$, and the average number of downcrossings of s_w during the growing season, $\overline{n}_{s_w}$, versus the frequency of storm arrivals λ (the latter can also be interpreted as λ' after interception is taken into account). $\alpha = 1.5\,\text{cm}$, $T_{\text{seas}} = 200$ days, $Z_r = 40\,\text{cm}$, $E_w = 0.01\,\text{cm day}^{-1}$, $E_{\text{max}} = 0.45\,\text{cm day}^{-1}$; the soil is a loam (see Table 2.1). After Porporato et al. (2001).

the effect of the other parameters is restricted to relatively small fluctuations close to the limit value.

Setting $\xi = s^*$ in Eq. (3.20), one obtains the mean duration of an excursion below s^* as

$$\overline{T}_{s^*} = \frac{1}{\lambda'} + \left(\overline{T}_{s_w} - \frac{1}{\lambda'}\right) e^{\gamma(s^*-s_w)} \left(\frac{\eta}{\eta_\omega}\right)^{-\frac{\lambda'(s^*-s_w)}{\eta-\eta_w}}$$

$$+ \frac{1}{\lambda'} e^{\frac{\gamma\eta(s^*-s_w)}{\eta-\eta_w}} \left[\frac{\gamma\eta(s^*-s_w)}{\eta-\eta_w}\right]^{\frac{\lambda'(s^*-s_w)}{\eta-\eta_w}}$$

$$\times \left\{\Gamma\left[1 + \frac{\lambda'(s^*-s_w)}{\eta-\eta_w}, \frac{\gamma\eta_w(s^*-s_w)}{\eta-\eta_w}\right]\right.$$

$$\left. - \Gamma\left[1 + \frac{\lambda'(s^*-s_w)}{\eta-\eta_w}, \frac{\gamma\eta(s^*-s_w)}{\eta-\eta_w}\right]\right\},$$

(3.30)

while, from Eq. (3.21) evaluated in $\xi = s^*$, the mean number of downcrossings becomes

$$\bar{n}_{s^*} = C\left(\frac{\eta}{\eta_{\rm w}}\right)^{\frac{\lambda'(s^*-s_{\rm w})}{\eta-\eta_{\rm w}}} e^{-\gamma s^*} \, T_{\rm seas}. \tag{3.31}$$

Both $\bar{T}_{s^*}$ and $\bar{n}_{s^*}$ will be used in Chapter 4 in the study of the temporal structure of plant water stress.

Figure 3.4 shows the behavior of $\bar{T}_{s^*}$ and $\bar{n}_{s^*}$ for different climatic conditions and soil depths. An increase of the frequency of the rain events results in a strong nonlinear decrease of $\bar{T}_{s^*}$, even more marked than that of $\bar{T}_{s_{\rm w}}$ shown in

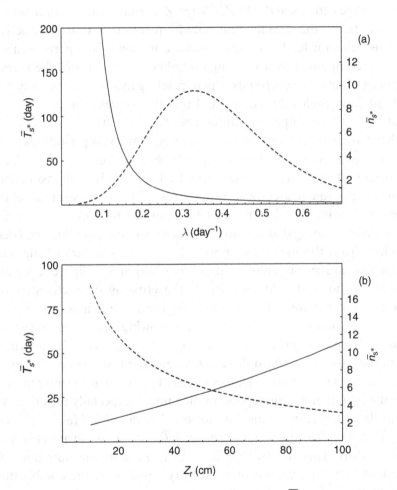

Figure 3.4 Average duration of an excursion below s^*, $\bar{T}_{s^*}$ (continuous lines), and average number of downcrossings of s^* during the growing season, $\bar{n}_{s^*}$ (dashed lines) versus: (a) rainfall rate λ (or λ' if interception is taken into account), and (b) active soil depth Z_r. $\alpha = 1.5\,\mathrm{cm}$, $T_{\rm seas} = 200$ days, $E_{\rm w} = 0.01\,\mathrm{cm\,day^{-1}}$, $E_{\rm max} = 0.45\,\mathrm{cm\,day^{-1}}$; the soil is a loam (see Table 2.1); in (a) $Z_r = 60\,\mathrm{cm}$, while in (b) $\lambda = 0.2\,\mathrm{day^{-1}}$. After Porporato et al. (2001).

Figure 3.3. As far as $\bar{n}_{s^*}$ is concerned, the fact that s^* is much higher than s_w causes a rightward shift of the maximum mean number of upcrossings (or downcrossings) when compared to Figure 3.3. For low values of λ the mean soil moisture is lower than s^*, and the trajectory tends to remain almost always below s^*. Increasing λ, the mean soil moisture becomes closer to s^*, with frequent crossings of the threshold, until it moves above s^* and $\bar{n}_{s^*}$ begins to decrease.

In contrast to when $\xi = s_w$, in the case of $\xi = s^*$ parameters other than λ have an important influence on $\overline{T}_{s^*}$ and $\bar{n}_{s^*}$. Figure 3.4 shows an example of the relevance of the active soil depth Z_r. When Z_r is small, the soil moisture tends to react very fast to the rainfall input, with frequent saturation of the active soil layer. This behavior leads to a small mean duration of the permanence time below s^*, accompanied by a very high number of crossings of the threshold. The opposite is true for deep soils which, reacting much slower to any rainfall or drought, have high values of $\overline{T}_{s^*}$ and very low values of $\bar{n}_{s^*}$.

Figure 3.4b shows important differences between shallow and deep active soils: shallow-rooted plants face short and frequent stress periods, as opposed to deep-rooted species, which on average experience longer and less frequent stress periods. One could argue that this difference may be responsible for very different adaptation strategies to water stress, favoring resistance when the rooting depth is large and resilience (the ability to recover after a period of stress) when the rooting depth is small. Notice also that this role of soil depth is not evident from the mean soil moisture value, which varies little with Z_r although there are serious changes in the distribution (see Figures 2.8 and 2.9).

Figure 3.5 shows the dependence of the crossing characteristics of the threshold s^* on the interplay between the timing and amount of rainfall, when the total amount of rainfall during a growing season, Θ, is kept fixed. $\overline{T}_{s^*}$ and $\bar{n}_{s^*}$ have a minimum and a maximum respectively for intermediate values of the α to λ ratio; both these points tend to move toward the right part of the diagram as the climate becomes wetter. The fact that the minimum of $\overline{T}_{s^*}$ and the maximum of $\bar{n}_{s^*}$ correspond to the same λ, expecially for dry climates, can be explained by considering that, for small values of λ, $P(\xi = s^*)$ is close to 1 in Eq. (3.20), so that the product between $\overline{T}_{s^*}$ and $\bar{n}_{s^*}$ is always close to T_{seas} (see Eq. (3.26)). This means that $\bar{n}_{s^*} \simeq T_{\text{seas}}/\overline{T}_{s^*}$ and the minimum of $\overline{T}_{s^*}$ corresponds to the maximum of $\bar{n}_{s^*}$. In dry conditions, the combination of smaller values of $\overline{T}_{s^*}$ and higher frequencies of crossing seems more suitable for plant survival than the opposite situation, because even a brief interruption of the stress period can have positive effects, possibly inhibiting the apparency of permanent damage. Thus, for a fixed amount of total rainfall, better vegetation conditions seem to be associated with intermediate values of

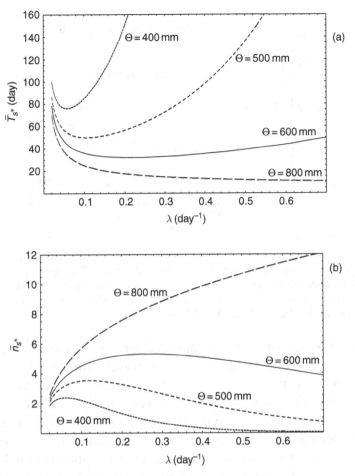

Figure 3.5 Influence of timing and amount of rainfall on (a) the average duration of an excursion below s^*, $\overline{T}_{s^*}$, and (b) the average number of down-crossings of s^* during the growing season, $\overline{n}_{s^*}$, keeping fixed the total rainfall during the growing season, Θ. λ and α change along the abscissa, but their product remains constant ($\Theta = T_{\text{seas}} \alpha \lambda$). $T_{\text{seas}} = 200$ days and $Z_{\text{r}} = 60\,\text{cm}$; see caption of Figure 3.3 for soil and vegetation parameters. After Porporato et al. (2001).

depth and frequency of rainfall. In wetter climates, the optimal condition moves toward larger rates of rainfall arrival. The mean value of soil moisture moves closer to s^* and very intense events (left part of the diagram) are no longer fundamental for crossing the threshold.

In Figure 3.6, $\overline{T}_{s^*}$ is studied as a function of the frequency of the rainfall events λ and of the mean rainfall depth α, in such a way that the product $\alpha\lambda$ remains constant for different values of the maximum evapotranspiration rate. This is to compare environments with the same total rainfall during a growing

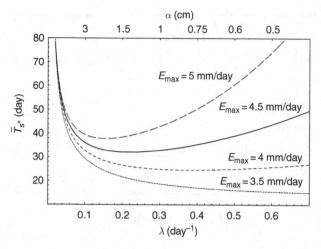

Figure 3.6 Mean duration of excursions below s^*, $\overline{T}_{s^*}$, as a function of the frequency of the rainfall events λ when the total rainfall during the growing season is kept fixed at 650 mm for different values of the maximum evapotranspiration rate. The root depth is $Z_r = 60$ cm, the soil is a loam (see Table 2.1). After Laio et al. (2001c).

season, but with differences in the timing and average amount of the precipitation events. In the case of high maximum transpiration rates, plants may experience longer periods of stress either where the rainfall events are very rare but intense or where the events are very frequent and light. From a physical viewpoint, this is due to the relevant water losses by leakage, runoff or canopy interception, indicating possible optimal conditions for vegetation. Both Figures 3.5 and 3.6 point out the existence of an optimum ratio between λ and α. The same behavior was encountered in Section 2.4 in connection with the water balance and will be further discussed in Chapter 4 in relation to the plant water stress.

3.3 Duration of excursions between two different levels of soil moisture

The analysis of excursions between two different levels of soil moisture is useful in two important cases: the first case concerns the average time of recovery after a period of drought, while the second, and perhaps the most important, is the average time for soil moisture to reach stress levels starting from high initial soil moisture values.

Because soil moisture conditions below s^* entail the occurrence of vegetation water stress (see Chapter 4), the crossing properties related to s^* are especially important for characterizing plant conditions. Figure 3.7 analyzes the impact of rooting depth on the MFPT of s^* starting from different initial

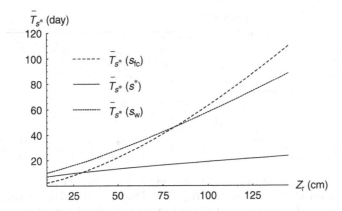

Figure 3.7 Impact of plant rooting depth on the mean duration of a water stress period, $\overline{T}_{s^*}$, on the mean duration of a period without water stress at the beginning of the growing season, $\overline{T}_{s^*}(s_{fc})$, and on the mean time plants need to recover after a period of intense stress, $\overline{T}_{s^*}(s_w)$. The mean rainfall frequency is $\lambda = 0.2\,\mathrm{day}^{-1}$, the mean rainfall depth is $\alpha = 2\,\mathrm{cm}$. $E_{max} = 4.5\,\mathrm{mm\,day}^{-1}$ and the soil is a loam (see Table 2.1). After Laio et al. (2001c).

conditions at s_w, s^*, and s_{fc}. $\overline{T}_{s^*}(s_w)$ may be obtained from Eq. (3.22) and is important for the analysis of plant recovery from a period of intense stress. The MFPT of s^* from s_{fc}, $\overline{T}_{s^*}(s_{fc})$, which is calculated from Eq. (3.24), is useful for analyzing the mean duration of periods without water stress at the beginning of the growing season, especially in regions where a relatively dry growing season is preceded by a wet winter season. The depth of active soil, nZ_r, damps the fluctuations in soil moisture induced by rainfall events and the trajectories of soil moisture are smoother for deeper soils. Thus $\overline{T}_{s^*}(s_w)$ and $\overline{T}_{s^*}(s_{fc})$ rapidly increase with Z_r. A higher value of $\overline{T}_{s^*}(s_{fc})$ is favorable for plants because, in the presence of moisture, it implies a longer unstressed period at the beginning of the growing season. On the other hand, a higher value of $\overline{T}_{s^*}(s_w)$ is problematic for plants, which then need a long time to recover after a period of intense stress. These features may lead to important differences in the patterns of water use of deep and shallow rooted plants, with advantages and drawbacks in different situations that affect how favorable a given environment is to different species.

The focus now shifts to the crossing times of s^* starting from high soil moisture values. Figure 3.8a shows how the MFPT of s^* does not change noticeably when the initial soil moisture, s_0, is increased above field capacity. Because of this, $T_{s^*}(s_{fc})$ provides an objective and general indication of the time taken to reach stress conditions during the growing season after a wet winter season and will be used later to define plant stress under transient

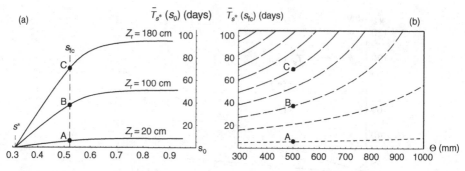

Figure 3.8 (a) Mean time to cross s^* starting from different initial conditions, $s_0 > s^*$, for different rooting depths ($\alpha = 1.25\,\mathrm{cm}$, $\lambda = 0.2\,\mathrm{day}^{-1}$). (b) Mean time of crossing between s_{fc} and s^* as a function of rate of rainfall events (λ is varied keeping $\alpha = 1.25\,\mathrm{cm}$) for different rooting depths (starting from the bottom curve with $Z_r = 20\,\mathrm{cm}$, Z_r is progressively increased by 40 cm). The three dots in each figure are shown to locate the same conditions in the two diagrams. For both figures the soil is a loamy sand (see Table 2.1) and $E_{\mathrm{max}} = 4.5\,\mathrm{mm}\,\mathrm{day}^{-1}$.

conditions as well as to analyze different plant strategies of water use (Section 7.1). In particular, the crucial role of Z_r will be studied in detail in relation to two different strategies of water use: an extensive one (i.e., with deep roots and lower transpiration rates) for the best exploitation of the winter soil water storage as opposed to an intensive one (i.e., with shallow roots and high transpiration) that takes advantage of the highly fluctuating water resource during the growing season.

3.4 MFPT's for minimalistic models of soil moisture

The analysis of the crossing properties for the minimalistic models discussed in Section 2.6.2 provides specific examples in which the general expressions for the MFPT's attain interesting and relatively simple form.

3.4.1 The minimalistic model with constant losses

The particular expressions for Milly's minimalistic model (Milly, 1993, 2001) are immediately obtainable using the results of Section 3.1 with the bound at one (i.e., $s_b = 1$) together with the expressions for the steady-state pdf of Section 2.6.2, Eqs. (2.63), (2.64), and (2.65).

From Eq. (3.20), the mean duration of an excursion below ξ (using the more precise notation of Section 3.1) is

$$\overline{T}_\xi(\xi) = \frac{\gamma\eta}{\lambda(\gamma\eta - \lambda)} \, e^{\xi\left(\gamma - \frac{\lambda}{\eta}\right)} - \frac{1}{\gamma\eta - \lambda}. \tag{3.32}$$

Equation (3.21) yields the frequency of upcrossings of ξ as

$$\nu(\xi) = \lambda p_0 \, e^{-\xi\left(\gamma - \frac{\lambda}{\eta}\right)}, \tag{3.33}$$

and the MFPT of ξ when $x_0 < \xi$ is, from Eq. (3.22),

$$\overline{T}_\xi(x_0) = \frac{\gamma\eta}{\gamma\eta - \lambda} \left[\frac{\gamma\eta}{\lambda(\gamma\eta - \lambda)} \, e^{\xi\left(\gamma - \frac{\lambda}{\eta}\right)} - \frac{1}{\gamma\eta - \lambda} \, e^{x_0\left(\gamma - \frac{\lambda}{\eta}\right)} - \frac{1}{\eta}(\xi - x_0) - \frac{1}{\gamma\eta} \right]. \tag{3.34}$$

Finally, the MFPT of ξ', with $x_0 > \xi'$, reads from Eq. (3.24)

$$\overline{T}_{\xi'}(x_0) = \frac{\lambda p_0 \gamma\eta - \lambda(\gamma\eta - \lambda)}{\lambda p_0 (\gamma\eta - \lambda)^2} \left[e^{\xi'\left(\gamma - \frac{\lambda}{\eta}\right)} - e^{x_0\left(\gamma - \frac{\lambda}{\eta}\right)} \right] + \frac{\gamma}{\gamma\eta - \lambda}(x_0 - \xi'), \tag{3.35}$$

which, in the unbounded case, assumes the simple form

$$\overline{T}_{\xi'}(x_0) = \frac{\gamma}{\gamma\eta - \lambda}(x_0 - \xi'), \quad \gamma > \frac{\lambda}{\eta}. \tag{3.36}$$

Equations (3.34) and (3.35) are plotted in Figure 3.9, taking the initial condition x_0 as variable. Four curves are shown for different values of the threshold ξ (or ξ') and a black circle is placed on each curve where $x_0 = \xi$ (the position of the circles is therefore also described by Eq. (3.32)). On the left of these circles we have $x_0 < \xi$ and Eq. (3.34) is valid, while on their right $x_0 > \xi'$ and Eq. (3.35) is used. Equations (3.34) and (3.35) result from the imposition of different boundary conditions for the differential Eq. (3.9); this leads to the already mentioned inequality $\overline{T}_\xi(\xi) \neq \overline{T}_{\xi'}(\xi'^+) = 0$ and to the discontinuity of each curve at x_0. Also note that, due to the presence of the upper bound at 1, the r.h.s. parts of the curves are bent downward with respect to the linear expression given by Eq. (3.35). In fact, as pointed out before, the values of $\overline{T}_{\xi'}(x_0)$ decrease as a consequence of the restriction imposed on the trajectories by the presence of the bound. This effect is more evident when the starting point x_0 is closer to the bound.

From a physical standpoint, Figure 3.9 refers to a case with a shallow soil and very wet climate in which the difference between the crossing times of ξ from below (i.e., curves on the left side of the dots) is not much longer than the time to cross the same threshold from above (curves on the right side of the

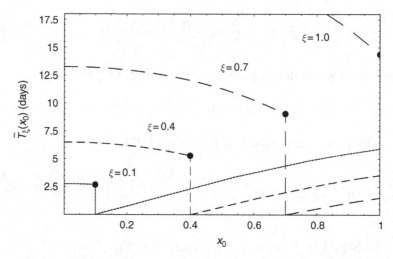

Figure 3.9 MFPT's for the threshold ξ or ξ' as a function of the initial condition x_0 for the minimalistic model with constant losses. Four curves for different thresholds ξ are shown. Black circles are placed where $x_0 = \xi$ (Eq. (3.32)); on the left of these circles $x_0 < \xi$ and Eq. (3.34) is valid, on their right $x_0 > \xi$ and Eq. (3.35) is used. In all cases the parameter values are $\lambda = 0.5\,\mathrm{day}^{-1}, \gamma = 3.2, \eta = 0.25\,\mathrm{day}^{-1}$. Modified after Laio et al. (2001c).

dots). Considering the threshold at 0.4, it is seen that for any initial condition below 0.4 the time of crossing is always between 5.5 and 6.5 days. This is due to the joint effect of the abundant rainfall and the shallow soil; for the same reasons, the descent from $x = 1$ to 0.4 takes place in just three and a half days.

If one analyzes instead a typical condition of water-controlled ecosystems, the times of crossing the same threshold change considerably (Figure 3.10).

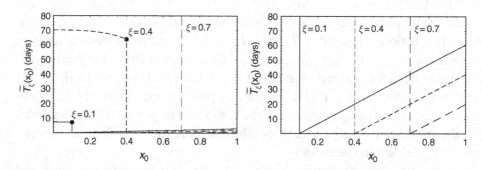

Figure 3.10 MFPT's for the threshold ξ or ξ' as a function of the initial condition x_0 for the minimalistic model with constant losses. The parameter values are $\lambda = 0.3\,\mathrm{day}^{-1}, \alpha = 1.5\,\mathrm{cm}, E_{\max} = 0.4\,\mathrm{cm}\ \mathrm{day}^{-1}$. Left figure: $Z_r = 30\,\mathrm{cm}$; right figure: $Z_r = 60\,\mathrm{cm}$. Modified after Laio et al. (2001c).

Still focusing on the threshold at 0.4, which can be seen as a typical level of the beginning of water stress (see Chapter 4), one notices the importance of the rooting depth in the soil moisture dynamics: when the roots are shallow, the time of crossing from below, although considerable, is always shorter than 70 days, while for very deep roots the times become much longer. At the beginning of the growing season, shallow-rooted plants lose the winter water supply in a few days, while very deep-rooted plants can use it for more than 40 days before the stress begins and for much longer before reaching the wilting point. This is a very important fact, which is at the basis of the different strategies of water use that will be studied in depth in Section 7.1

In the special case when $\gamma = \frac{\lambda}{\eta}$ the steady-state probability distribution becomes uniform (Eq. (2.65)), and the MFPT's are obtained either by taking the limit in Eqs. (3.32)–(3.33) or by directly applying Eqs. (3.19), (3.22), and (3.23) with the probability distribution of Eq. (2.65). One thus obtains

$$\overline{T}_\xi(\xi) = \frac{1}{\lambda} + \frac{\xi}{\eta}, \tag{3.37}$$

$$\nu(\xi) = \lambda p_0 = \frac{\lambda\eta}{\eta + \lambda}, \tag{3.38}$$

$$\overline{T}_\xi(x_0) = \frac{1}{\lambda} + \frac{\gamma\xi}{\lambda} + \frac{\gamma}{2\eta}(\xi^2 - x_0^2), \tag{3.39}$$

$$\overline{T}_{\xi'}(x_0) = \frac{1+\gamma}{\eta}(x_0 - \xi') - \frac{\gamma}{2\eta}(x_0^2 - \xi'^2). \tag{3.40}$$

In unbounded conditions, $\overline{T}_\xi(\xi)$ and $\overline{T}_\xi(s_0)$ remain unchanged, the frequency $\nu(\xi)$ tends to zero, and $\overline{T}_{\xi'}(x_0)$ tends to infinity, due to the nonstationarity of the process in this case.

3.4.2 The minimalistic model with linear losses

In the case of the shot noise process with linear losses bounded at 1, the MFPT of the threshold ξ or ξ' with starting point x_0 can be calculated from Eqs. (3.20), (3.22), and (3.24), using the expression for the steady-state pdf, Eqs. (2.72) and (2.73). For $x_0 = \xi$ one obtains

$$\overline{T}_\xi(\xi) = \frac{1}{\lambda} + \frac{1}{\lambda} e^{\gamma\xi}(\gamma\xi)^{-\frac{\lambda\xi}{\eta}}\left\{\Gamma\left[1 + \frac{\lambda\xi}{\eta}\right] - \Gamma\left[1 + \frac{\lambda\xi}{\eta}, \gamma\xi\right]\right\}$$
$$= \frac{1}{\lambda}{}_1F_1\left[1, 1 + \frac{\lambda}{\eta}, \gamma\xi\right],$$

(3.41)

where ${}_1F_1[\cdot, \cdot, \cdot]$ is the confluent hypergeometric function or Kummer function (Abramowitz and Stegun, 1964). Notice that Eq. (3.41) is the same as Eq. (3.28). When $x_0 < \xi$ the MFPT of ξ reads

$$\overline{T}_\xi(x_0) = \frac{1}{\lambda}{}_1F_1\left[1, 1 + \frac{\lambda}{\eta}, \gamma s_0\right] + \frac{\gamma\xi}{\lambda}{}_2F_2\left[1, 1; 2, 1 + \frac{\lambda}{\eta}; \gamma\xi\right]$$
$$- \frac{\gamma s_0}{\lambda}{}_2F_2\left[1, 1; 2, 1 + \frac{\lambda}{\eta}; \gamma x_0\right],$$

(3.42)

where ${}_2F_2[\cdot, \cdot; \cdot, \cdot; \cdot]$ is the generalized hypergeometric function (Prudnikov et al., 1986). Finally, the MFPT from x_0 to ξ' is, when $x_0 > \xi'$,

$$\overline{T}_{\xi'}(x_0) = \frac{1}{\eta}(\gamma s_0)^{-\frac{\lambda}{\eta}} e^{\gamma s_0}\left\{\Gamma\left[\frac{\lambda}{\eta}, \gamma x_0\right] - \Gamma\left[\frac{\lambda}{\eta}, \gamma\right]\right\}$$
$$- \frac{1}{\eta}(\gamma\xi')^{-\frac{\lambda}{\eta}} e^{\gamma\xi'}\left\{\Gamma\left[\frac{\lambda}{\eta}, \gamma\xi'\right] - \Gamma\left[\frac{\lambda}{\eta}, \gamma\right]\right\}$$
$$+ \frac{\gamma\xi'}{\lambda}{}_2F_2\left[1, 1; 2, 1 + \frac{\lambda}{\eta}; \gamma\xi'\right] - \frac{\gamma s_0}{\lambda}{}_2F_2\left[1, 1; 2, 1 + \frac{\lambda}{\eta}; \gamma x_0\right]$$
$$+ \frac{1}{\eta}(-1)^{\frac{\lambda}{\eta}}\left\{\Gamma\left[1 - \frac{\lambda}{\eta}, -\gamma x_0\right] - \Gamma\left[1 - \frac{\lambda}{\eta}, -\gamma\xi'\right]\right\}\left\{\Gamma\left[\frac{\lambda}{\eta}\right] - \Gamma\left[\frac{\lambda}{\eta}, \gamma\right]\right\}.$$

(3.43)

$\overline{T}_\xi(\xi)$ and $\nu(\xi)$ are plotted as a function of ξ in Figure 3.11. Common features for all the curves are the increase of $\overline{T}_\xi(\xi)$ with ξ and the presence of a maximum in the crossing frequency, $\nu(\xi)$. In particular, as the maximum evapotranspiration increases, so does the duration of excursions below any given threshold ξ, while the value ξ_{max} for which $\nu(\xi)$ has a maximum is usually very close to the mean value $\langle x \rangle$ of the steady-state distribution. This is because they both represent levels of x around which the trajectory preferably evolves. However, only in the unbounded case do the two values coincide: in fact, one can set $\rho(x)p(x) = \nu(x)$ and, from Eq. (2.33), obtain the equation for the threshold that experiences the maximum crossing frequency. When the loss function is linear, one obtains $\xi_{max} = \frac{\lambda}{\gamma\eta}$. Only in the unbounded case is such a value equal to the mean steady-state value given by Eq. (2.74).

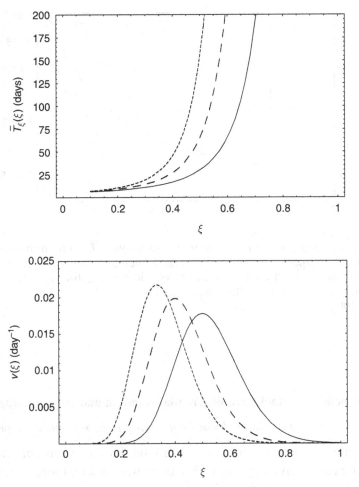

Figure 3.11 Mean duration of an excursion below ξ, $\overline{T}_\xi(\xi)$ (top), and frequency of upcrossing of ξ, $\nu(\xi)$ (bottom), as a function of the threshold value ξ for the minimalistic soil moisture process with linear losses. $\lambda = 0.2\,\text{day}^{-1}$ and $\alpha = 1\,\text{cm}$, $Z_r = 100\,\text{cm}$. The three curves refer to different E_{max}: $0.4\,\text{cm}\,\text{day}^{-1}$ (continuous line), $0.5\,\text{cm}\,\text{day}^{-1}$ (dashed line), and $0.6\,\text{cm}\,\text{day}^{-1}$ (dotted line).

It is also interesting to note that the mean time of crossing, $\overline{T}_\xi(\xi)$, does not decrease monotonically as ξ goes to zero, since these are very low soil moisture values below which the process very seldom goes. Although such levels (not shown in Figure 3.11) are usually very low and thus of little practical interest in water-controlled ecosystems, they could be higher and thus more relevant when the conditions are humid, especially in the case of very low transpiration rates. An example of such a case is shown in Figure 3.12.

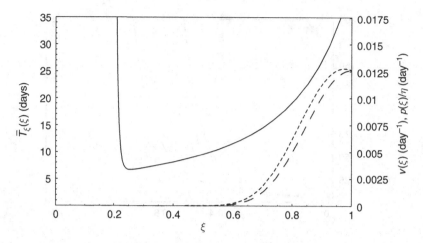

Figure 3.12 Mean duration of an excursion below ξ, $\overline{T}_\xi(\xi)$ (continuous line), frequency of upcrossing of ξ, $\nu(\xi)$ (dashed line), and soil moisture pdf, $p(\xi)$ (dotted line), as a function of the threshold value ξ for the minimalistic soil moisture process with linear losses. $\lambda = 0.2\,\mathrm{day}^{-1}$ and $\alpha = 1\,\mathrm{cm}$, $Z_r = 100\,\mathrm{cm}$, $E_{\max} = 0.2\,\mathrm{cm}\,\mathrm{day}^{-1}$.

3.5 Appendix B. Backward differential equation and crossing properties

3.5.1 Derivation of the backward equation for the soil moisture process

The Chapman–Kolmogorov forward differential equations for the general shot noise process given by Eq. (3.5) can be written in a more compact form in terms of the total mixed distribution

$$p^*(s, t|s_0, t_0) = p(s, t|s_0, t_0) + \delta(s - s_h)p_0(t|s_0, t_0). \tag{3.44}$$

It is easy to show that Eqs. (3.5) and (3.6) are equivalent to the following equation

$$\frac{\partial}{\partial t}p^*(s, t|s_0, t_0) = \frac{\partial}{\partial s}[p^*(s, t|s_0, t_0)\rho(s)] - \lambda p^*(s, t|s_0, t_0)$$
$$+ \lambda \int_0^s p^*(u, t|s_0, t_0)f_Y(s - u, u)\, du. \tag{3.45}$$

Gardiner (1990, page 51) shows that, for the most general univariate, time-homogeneous Markov process, the forward differential equation for the transition probability $p^*(s, t|s_0, t_0)$ can be written as

$$\frac{\partial}{\partial t}p^*(s, t|s_0, t_0) = -\frac{\partial}{\partial s}[A(s)p^*(s, t|s_0, t_0)] + \frac{1}{2}\frac{\partial^2}{\partial s^2}[B(s)p^*(s, t|s_0, t_0)]$$

$$+ \int_{-\infty}^{\infty} W(s|u)p^*(u, t|s_0, t_0)\, du - \qquad (3.46)$$

$$p^*(s, t|s_0, t_0) \int_{-\infty}^{\infty} W(u|s)\, du,$$

where $A(s)$ is a drift factor, $B(s)$ is a diffusion factor, and $W(s|u)$ is the probability of having a jump (i.e., a discontinuity in the trajectory of s) from u to s. In Eq. (3.46), $t_0 < t$, and the initial condition for the forward equations is usually

$$p^*(s, t = t_0|s_0, t_0) = \delta(s - s_0). \qquad (3.47)$$

The so-called backward differential equation for the evolution of the same probability $p^*(s, t|s_0, t_0)$ can be obtained analogously to the forward equation, considering s_0 and t_0 as the new variables while holding s and t fixed. The general form for the backward equation is given by (Gardiner, 1990, page 56)

$$\frac{\partial}{\partial t_0}p^*(s, t|s_0, t_0) = -A(s_0)\frac{\partial}{\partial s_0}p^*(s, t|s_0, t_0) - \frac{1}{2}B(s_0)\frac{\partial^2}{\partial s_0^2}p^*(s, t|s_0, t_0)$$

$$- \int_{-\infty}^{\infty} W(u|s_0)p^*(s, t|u, t_0)\, du + \qquad (3.48)$$

$$p^*(s, t|s_0, t_0) \int_{-\infty}^{\infty} W(u|s_0)\, du,$$

with $t_0 < t$ and where the initial condition at $t_0 = t$ is usually

$$p^*(s, t|s_0, t_0 = t) = \delta(s - s_0). \qquad (3.49)$$

Since Eq. (3.49) is solved backward in time (hence its name), the condition (3.49) is actually the final condition for the backward equation. Notice that, when the stochastic process has specific boundary conditions in the forward formulation, these need to be modified appropriately when the problem is set using the backward formulation (e.g., Gardiner, 1990, page 128).

It is important to note that the forward and the backward differential equations are equivalent, in that they both describe the evolution in time of the same stochastic process (see e.g., Gardiner, 1990; Van Kampen, 1992). Nevertheless, their difference lies in the way they describe the evolution of the transition probability: the forward equation assumes the initial condition to be fixed $\{s_0, t_0\}$ allowing $\{s, t\}$ to vary, while the backward equation, fixing the final condition $\{s, t\}$, uses $\{s_0, t_0\}$ as variables. Which one is more convenient to use depends on the particular application under examination: due to its more intuitive meaning, the forward equation is usually employed to analyze the evolution in time of the stochastic process, while the backward equation turns out to be more useful in other problems, such as the analysis of crossing times.

By comparing Eqs. (3.45) and (3.46), it is not difficult to show that for the shot noise process described by Eq. (3.1) one has

$$
\begin{aligned}
A(s) &= -\rho(s), \\
B(s) &= 0, \\
W(s|u) &= \lambda f_Y(s - u, u) H(s - u),
\end{aligned}
\tag{3.50}
$$

as can be easily checked by substitution, where $H(1)$ is the Heaviside function.

The backward differential equation for the soil moisture process described by Eq. (3.1) follows directly by using the specific expression Eq. (3.50) in the general Eq. (3.48), i.e.,

$$
\begin{aligned}
\frac{\partial p^*(s, t|s_0, t_0)}{\partial t_0} = {} & \rho(s_0) \frac{\partial p^*(s, t|s_0, t_0)}{\partial s_0} + \lambda p^*(s, t|s_0, t_0) \\
& - \lambda \int_{s_0}^{\infty} f_Y(z - s_0, s_0) p^*(s, t|z, t_0) \, dz.
\end{aligned}
\tag{3.51}
$$

3.5.2 *From the backward equation to the crossing properties*

The exceedance probability for the time T in which the trajectory leaves the interval $\{\xi', \xi\}$ starting from s_0 is given by (e.g., Gardiner, 1990, page 137)

$$
\text{Prob}(T \geq t) = \int_{\xi'}^{\xi} p^*(u, t|s_0, t_0) \, du \equiv 1 - G_T(s_0, t),
\tag{3.52}
$$

where $G_T(s_0, t)$ is the cumulative distribution of the times of crossing of one of the two thresholds, ξ' or ξ, starting from s_0, and where the pdf $p^*(s, t|s_0, t_0)$ is found by erecting absorbing barriers at ξ' and ξ, so that the "particle" is removed from the system when it reaches ξ' or ξ. Equation (3.52) expresses the fact that the exceedance probability of T is equal to the probability that at time t the trajectory described by s is still in the interval $\{\xi', \xi\}$.

The probability density $g_T(s_0, t)$ of the time for a particle, starting from s_0 inside the interval $\{\xi', \xi\}$, to leave for the first time the interval $\{\xi', \xi\}$ is thus given by

$$g_T(s_0, t) = \frac{d}{dt}[1 - G_T(s_0, t)] = -\frac{d}{dt}\left[\int_{\xi'}^{\xi} p^*(u, t|s_0, t_0)\, du\right]. \qquad (3.53)$$

Since the evolution equation for $g_T(s_0, t)$ is related to that of $p^*(s, t|s_0, t_0)$, both the backward and the forward equations can in principle be used to obtain it. It is easy to see, however, that the backward equation is much more useful, because of the presence of the s_0 derivatives, which allows the integration with respect to s without changing the form of the equation. In particular, using Eq. (3.52) and provided the limits in the integrals are changed appropriately, it is straightforward to show that both $G_T(s_0, t)$ and $g_T(s_0, t)$ obey an equation similar to the backward differential equation but with opposite signs in the r.h.s., i.e.,

$$\frac{\partial}{\partial t_0} g_T(s_0, t) = + A(s_0)\frac{\partial}{\partial s_0} g_T(s_0, t) + \frac{1}{2}B(s_0)\frac{\partial^2}{\partial s_0^2}g_T(s_0, t)$$

$$+ \int_{s_0}^{\infty} W(u|s_0)g_T(u, t)\, du - g_T(s_0, t)\int_{s_0}^{\infty} W(u|s_0)\, du. \qquad (3.54)$$

For the case of the soil moisture process, the upper limit of integration becomes ξ in Eq. (3.51) so that one obtains

$$\frac{\partial g_T(s_0, t)}{\partial t} = -\rho(s_0)\frac{\partial g_T(s_0, t)}{\partial s_0} - \lambda g_T(s_0, t) + \lambda \int_{s_0}^{\xi} f_Y(z - s_0, s_0)g_T(z, t)\, dz,$$

$$\qquad (3.55)$$

which is exactly Eq. (3.7).

4

Plant water stress

The reduction of soil moisture content during droughts lowers the plant water potential and leads to a decrease in transpiration. This in turn causes a reduction of cell turgor and relative water content in plants, which brings about a sequence of damages of increasing seriousness. After a discussion of the links between soil moisture and plant conditions, the mean crossing properties of soil moisture, analytically derived in Chapter 3, are used here to define an index of plant water stress that combines the intensity, duration, and frequency of periods of soil-water deficit. Plant water stress is then studied under different climatic conditions, to analyze how the interplay between plant, soil, and environment impacts on vegetation conditions. The analysis of plant water stress presented in this chapter closely follows the paper by Porporato et al. (2001).

4.1 Soil-water deficit and plant water stress

From the level of the single plant to that of the entire ecosystem, the action of climate, soil, and vegetation is linked to plant response by two fundamental processes: the first one is centered around the soil moisture dynamics and controls the intensity and duration of the periods of soil-water deficit, while the second one regulates the impacts of water deficit on plant physiology (Figure 4.1).

Since plants get their water from the soil, many of the impacts of climate and soil are felt by plants through the filter of soil moisture dynamics. Effective drought conditions for plants are determined by soil moisture availability and not necessarily by precipitation scarcity (Stephenson, 1990, 1998). Thus plants can experience drought even under favorable climatic conditions, due to poor soil characteristics (Newman, 1967), and different vegetation types can be found within a short distance simply because of changes in soil moisture levels

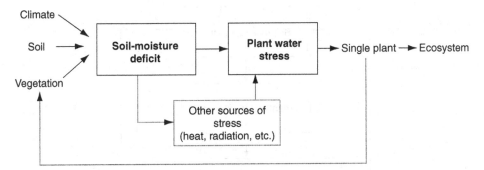

Figure 4.1 Schematic representation of the linkage between plant response and climate, soil, and vegetation conditions. After Porporato et al. (2001).

(Daubenmire, 1968; Branson et al., 1970, 1976; Harrington, 1991; Lauenroth et al., 1994).

An ecological stress is a condition produced in an organism by potentially harmful levels of environmental factors (Lauenroth et al., 1978). In many ecosystems, and especially in arid and semi-arid climates, soil-moisture deficit is often the most important stress factor for vegetation. Grazing and fire are also present, but their impact is frequently modulated by the effect of soil-water deficit on the existing vegetation (Western and van Praet, 1973; Scholes and Archer, 1997). Likewise, heat or radiation stress mostly take over after the cooling effect of transpiration has been reduced by the presence of water deficit (Nilsen and Orcutt, 1998). Other important sources of stress, like those related to nutrient limitation, are frequently initiated by the occurrence of water stress (e.g., Larcher, 1995; Nilsen and Orcutt, 1998) and are related to soil moisture dynamics (see Chapter 10).

Plants need to maintain an adequate level of water in their tissues to assure both growth and survival; they also require a continuous flux of water to perform vital processes such as photosynthesis and nutrient uptake. Figure 4.2 presents a simplified scheme of the steps linking soil-moisture deficit to plant water stress. As will be seen, the main control of soil-moisture deficit on plant condition is exerted through the plant water potential, which in turn affects cell turgor and the relative water content of the plant's living cells.

The plant response originates at the molecular level and, through the control of growth and reproduction, determines the condition of the entire ecosystem. Although the interconnection between the effects of different types of resources in plant physiology is quite complex (e.g., Bloom et al., 1985), plants often show similar responses to different sources of stress, especially when the triggering cause is water deficit (Chapin et al., 1987; Chapin, 1991; Nilsen and Orcutt, 1998). Thus, despite the fact that the precise mechanisms through

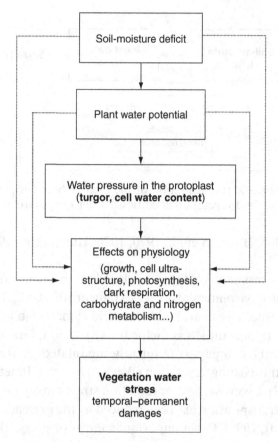

Figure 4.2 Schematic representation of the processes linking soil-moisture deficit to plant water stress. After Porporato et al. (2001).

which this control is exerted are extremely complicated and plant specific, the existence of important features of water stress that are common to many species gives hope for a meaningful general modeling scheme from a hydrologic perspective.

A large number of studies on physiological plant ecology have dealt with the problem of plant water stress (e.g., Hsiao, 1973; Lange et al., 1976; Levitt, 1980; Bradford and Hsiao, 1982; Smith and Griffith, 1993; Larcher, 1995; Ingram and Bartels, 1996; Nilsen and Orcutt, 1998). In this section we attempt a brief synthesis of its most important aspects, emphasizing the links between vegetation response and soil-moisture deficit. Many aspects of plant water stress are not yet completely understood, and much progress is to be expected in the near future from improvements in the experimental capabilities (e.g., Smith and Griffith, 1993) as well as from advances in the understanding of the molecular basis of water stress (Ingram and Bartels, 1996).

4.1.1 The components of plant water potential

Water movement from soil to the atmosphere is driven in plants by differences in water free energy. The total water potential, Ψ, is generally used to characterize and compare energy levels of water in the soil–plant–atmosphere system. Ψ is defined as the difference in free energy per unit volume calculated with respect to that of pure water at a zero reference level and under standard temperature and pressure conditions. So defined, Ψ has units of pressure (e.g., MPa), but sometimes water potential is also referred to unitary weight, in which case it takes the units of length.

In general, water potential is determined by the sum of the contributions of different kinds of energy. For liquid water these are

$$\Psi = M + \Pi + \Omega + G, \qquad (4.1)$$

where M, Π, Ω, and G are the matrix, pressure, osmotic, and gravitational potential, respectively. The matrix potential is linked to capillary and adhesion forces, which decrease the free energy of the moisture at the water–solid interfaces. The pressure potential is simply the positive pressure of the liquid water and the value of the osmotic potential is a function of solute concentration, being zero for pure water and decreasing with solute concentration. The gravitational potential increases with height, proportionally to the density of water, i.e., $G = \rho_w g z$, with ρ_w water density, g gravitational acceleration, and z elevation above the reference level.

Plant physiologists usually distinguish two kinds of water in plants: the apoplastic water, located outside the plasma membranes and relatively free to move from roots to leaves through the xylem conduits, and the symplastic water, which is contained in the protoplast of living cells. The relationship between these two types of water is fundamental for linking plant conditions to soil moisture levels.

With particular reference to the different kinds of water involved in the soil–plant–atmosphere system, Ψ_s, Ψ_x, Ψ_p, and Ψ_a indicate the water potential of soil, xylem, protoplast, and atmosphere, respectively.

As far as the soil water potential Ψ_s is concerned, in unsaturated conditions and with low solute concentrations, its most important component is the matrix potential. As discussed in Chapter 2, soil–water matrix potential is linked to relative soil moisture by the so-called soil moisture retention curves (Eq. (2.15)). Typical values of soil water potential range from around -0.01 MPa for well-watered soils, up to about -10 MPa at the hygroscopic point.

The most important component of apoplastic water potential Ψ_x is the pressure potential. Xylem sapwood is, in fact, quite diluted along its whole path, so that osmotic potential is modest (~ -0.1 MPa) and practically constant everywhere (e.g., Nobel, 1999). Matrix potential is usually negligible, whereas gravitation potential is relevant only when tall plants are considered. Xylem water potential can thus be written as

$$\Psi_x(z,t) = \Pi_x(z,t) + \Omega_x + G_x(z), \tag{4.2}$$

where we have put in evidence the dependence on time t (through soil moisture evolution) and elevation z.

The main components of the symplastic water potential Ψ_p are the osmotic and pressure potential. Gravitational potential is relevant only for tall trees. The relationship between the symplastic water potential components is summarized as

$$\Psi_p(z,t) = \Pi_p(z,t) + \Omega_p(z,t) + G_p(z) \simeq \Psi_x, \tag{4.3}$$

where the approximate equality results from the fact that the total symplastic water potential, Ψ_p, is very close to that of the surrounding apoplastic water (in general slightly lower in order to allow water flow into the cells). Under normal transpiration conditions in well-watered soils, the typical values of Ψ_p are only slightly negative, ranging from -0.1 MPa to -1 MPa. Living cells contain high concentrations of solute within them and consequently the osmotic potential, Ω_p, may be quite low, with typical values from -0.5 MPa to -3 MPa (e.g., Salisbury and Ross, 1992, page 103).[1] Low values of osmotic potential are very important for plant cells: when the xylem water potential, Ψ_x, is largely negative, only a low value of Ω_p can ensure that the pressure potential, Π_p, is maintained positive (see Eq. (4.3)), thus providing the essential turgor for plant growth. We will return to the relationship between the symplastic water potential components in the next section.

The water potential of the atmosphere is the potential of its water vapor. It is given by the sum of the gravitational potential G_a and the relative atmospheric pressure Π_a (i.e., pressure above the standard atmospheric pressure), plus a term related to temperature and relative humidity of water vapor, i.e.,

$$\Psi_a = \Pi_a + \frac{RT}{V_w}\ln\frac{e}{e_s} + G_a, \tag{4.4}$$

where R is the universal gas constant, T is the absolute temperature in K, e is the absolute pressure of water vapor, V_w is the partial molal volume of pure water at temperature T, and e_s is the absolute pressure of water vapor at

---

[1] As a reference value, the osmotic potential of sea water is of the order of -2.4 MPa.

saturation. Π_a is usually very close to zero (standard atmospheric pressure), and G_a is the same as the one of liquid water, $G_a = \rho_w g z$ (see Nobel, 1999, pages 68–71).

The second term on the r.h.s. of Eq. (4.4) is related to the change in internal energy, U, because of the work involved in the volume expansion of air against an imposed pressure

$$dU = dQ - p_j \, dV, \tag{4.5}$$

where dQ represents the heat exchange, which is equal to zero if the system is adiabatic. An expression for dV is obtained from the law of the perfect gases

$$p_j V = n_j RT, \tag{4.6}$$

where p_j is the partial pressure of gas j, n_j is the number of moles of gas j, and R is the universal gas constant, $R = 8314 \, \mathrm{J \, kmol^{-1} \, K^{-1}}$. Calculating dV from Eq. (4.6) and substituting into Eq. (4.5), one obtains the change in energy when going from a reference state with p_0 to a generic state with p_j

$$U = \int_{p_0}^{p} dU = n_j RT \ln\left(\frac{p_j}{p_0}\right). \tag{4.7}$$

The reference state is that at which air is in equilibrium with pure water and p_0 is then the saturation vapor pressure at air temperature, i.e., $p_0 = e_s$, while p_j is the ambient vapor pressure, i.e., $p_j = e$. Since Ψ is energy per unit volume, the second term of the r.h.s. of Eq. (4.4) is obtained directly from Eq. (4.7).

As the ratio between vapor pressure and saturation vapor pressure is nearly equal to the relative humidity, Eq. (4.4) can be written as

$$\Psi_a \sim \frac{RT}{V_w} \ln \frac{\% \text{relative humidity}}{100} + G_a, \tag{4.8}$$

where Π_a has been set to zero. For normal air conditions the value of Ψ_a is usually very low, with strong sensitivity to relative humidity. Figure 4.3a shows the behavior of Eq. (4.8) as a function of relative humidity for different temperatures ($T = 0, 20, 40°$ C). The water vapor potential drops very quickly to low values as soon as the relative humidity is reduced from saturation conditions. Notice that the direct effect of temperature is minimum when compared to that of relative humidity. Temperature is quite important indirectly, however, since it controls the value of relative humidity. The role of

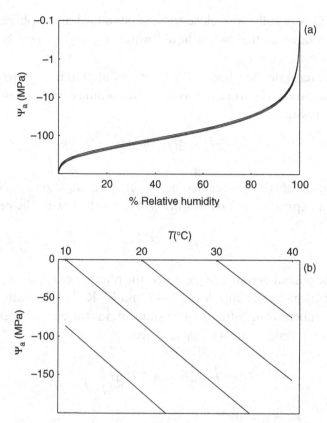

Figure 4.3 Atmosphere water potential Ψ_a versus (a) relative humidity for different values of air temperature (from top to bottom $T = 0, 20, 40°$ C) and (b) air temperature for the four values of absolute humidity, ρ_v, that give saturation at 0, 10, 20, and 30° C ($\rho_v = 0.048, 0.094, 0.17, 0.30$ g m^{-3}, respectively). After Porporato et al. (2001).

temperature in atmospheric water-vapor potential can be studied assuming constant absolute humidity in Eq. (4.4). Water vapor pressure is related to temperature and absolute humidity by the state equation of water vapor

$$e = \rho_v R_v T, \tag{4.9}$$

where R_v is the gas constant for water vapor and the absolute humidity, or vapor concentration, ρ_v, is the density of water vapor in moist air (kg m^{-3}). Substituting Eq. (4.9) in Eq. (4.4) and making explicit the dependence of $e_s = e_s(T)$ on temperature through an approximate solution of the Clausius–Clapeyron equation, Ψ_a can be studied as a function of temperature

for different values of absolute humidity. We use here the Richards expression
(Brutsaert, 1982, page 42)

$$e_s = 1013.25 \exp\left(13.3185\, t_r - 1.9760\, t_r^2 - 0.6445\, t_r^3 - 0.1299\, t_r^4\right), \quad (4.10)$$

with $t_r = 1 - 373.15/T$. Assuming $z = 0$ (reference level) one has

$$\Psi_a = \frac{RT}{V_w} \ln \frac{\rho_v R_v T}{e_s(T)} \quad (4.11)$$

with $e_s(T)$ given by Eq. (4.10). The above relationship is plotted in Figure 4.3b
for values of absolute humidity corresponding to saturation conditions under
atmospheric pressure at 0, 10, 20, and 30° C.

4.1.2 Plant water potential and soil moisture dynamics

A brief description of the mechanisms of water ascent in plants is useful for
understanding the relationship between soil moisture and plant water potential.
For more details on this fascinating phenomenon we refer to Zimmerman (1983)
and Nobel (1999). As previously said, water moves in the soil–plant–atmosphere
system by gradients of water potential. Figure 4.4 presents a schematic representation of the process using the scheme of the soil–plant–atmosphere continuum. The
particular case shown in this figure refers to a ten-meter-tall plant with the specific
values of the water potential components taken from Nobel (1999, page 384).

As water moves from the soil to the atmosphere, its potential follows a
decreasing path along the energy line. Soil water enters the roots because of the
lower plant water potential, and moves along the xylem conduits up to the leaves,
where it is evaporated and released through the stomata to the atmosphere. It is
commonly believed that in the xylem water flux takes place as a laminar, creeping
flow, driven by negative pressure gradients (Nobel, 1999). Along the upward
path, energy losses progressively reduce the level of plant water potential, so that
apoplastic water attains large negative pressure values, especially in the highest
parts of the plant. In these parts, under normal conditions, cavitation is only
hampered by the combined effect of the tensile strength of water and the small
cross-section of the conduits. Water can therefore ascend the tree like a continuous column, a concept that is the base of the so-called cohesion theory for
explanation of plant water ascent (e.g., Salisbury and Ross, 1992).[2]

---

[2] Recent high-precision measurements have suggested the possibility of other driving mechanisms for
water ascent acting in addition to the pressure gradient, such as the Marangoni convection caused by the
movement of small gaseous bubbles in the xylem, or the osmotic pumping due to the presence of poles of
solute concentration along the xylem (e.g., Zimmerman et al., 1993, 1995).

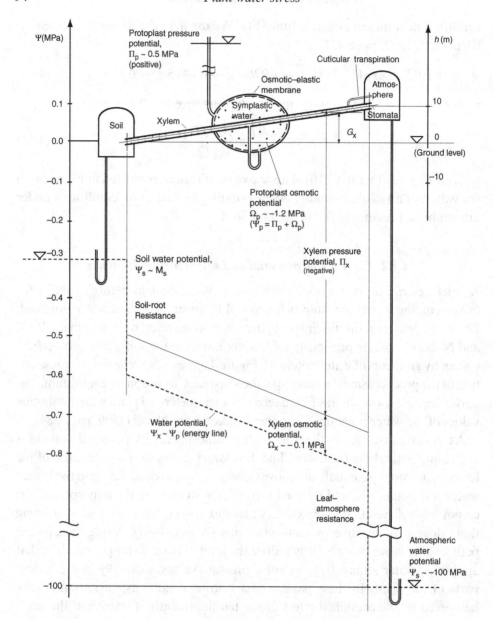

Figure 4.4 Schematic representation of various components of the water potential under the framework of the soil–plant–atmosphere continuum (see text for details). After Porporato et al. (2001).

Typically a soil-water-potential value of -0.3 MPa corresponds to relative soil moisture values of 0.2 and 0.4 for a loamy sand and a loam respectively (e.g., Figure 2.6). Such values correspond to conditions of moderate water stress, where the stomata are partially closed during the central part of the day (for plants in semi-arid ecosystems typical values of soil water potential for the start of stomata closure are of the order of -0.03 MPa (Scholes and Walker, 1993; see also Chapter 6)). Except during rainfall events, the atmospheric water potential is usually much lower than that of soil water. For average daytime conditions Ψ_a has a typical value of -100 MPa, which corresponds to a relative humidity of about 40% (Figure 4.3).

The water flow along the soil–plant–atmosphere system has to overcome a number of different resistances, as will be seen in detail in Chapter 6. A strong localized resistance is found at the soil–root passage: its specific value is controlled in a complicated way by soil moisture and root characteristics (Nobel, 1999, page 390). When water reaches the xylem, the energy losses are mostly caused by the laminar flow through the small pits that connect the vessels and tracheids of the xylem (see Figure 4.4). Finally, a major drop in water potential takes place at the leaf–atmosphere interface, when water evaporates in the intercellular pores and diffuses out through the stomata. This resistance is in part controlled by the plant, varying the stomatal opening. The decrease in xylem water potential, Ψ_x, and the increase in gravitation potential, G_x, make the xylem pressure, Π_x, increasingly more negative towards the leaves, while the level of osmotic potential, Ω_x, remains practically constant along the xylem. Figure 4.4 also shows the symplastic water, connected to the apoplastic water through an osmotic membrane. Its total potential, Ψ_p, follows the same behavior of that of the apoplastic water, but the existence of a more concentrated solution (i.e., more negative osmotic potentials, Ω_p) ensures positive pressure potentials, Π_p, that are essential to maintain the necessary cell turgor.

Assuming that the level of the atmospheric water potential does not change, with the aid of Figure 4.4 one can analyze the effects of soil moisture fluctuations on plant water potential. Consider first a condition with higher soil moisture than the one shown in Figure 4.4. In such conditions, the soil water potential Ψ_s is markedly increased, whereas the resistance at the soil–root interface is reduced (Nobel, 1999, page 390). As a consequence, plant water potential is relatively high, so that the negative pressure in the xylem is reduced, and the plant tissues are in full-turgor conditions. At the leaf–atmosphere interface, the resistance is reduced by an increase in the stomatal opening. Because of both the decreased resistances to flow and the increased soil–atmosphere water-potential gradient, the water flux is considerably increased. This produces higher energy losses along the xylem thereby steepening the energy line.

When the soil moisture level is lower than the example in Figure 4.4, the situation is opposite to the one discussed before. The soil water potential quickly drops to very negative values and the soil–root resistance increases considerably. Stomata are now almost completely closed and the resistance to water vapor efflux is very high. The water flux is scanty and the energy line, which is now shifted down, becomes almost flat. In the xylem the resulting low pressure values make increasingly more likely the appearance of cavitation phenomena, starting from the highest parts of the plant, while in the cells the low plant water potential induces a decrease in turgor and relative water content. If soil moisture is further depleted, the stomata completely close: at this point, soil water potential is so low that the plant is no longer able to take up water. However, since small water losses keep going on through cuticular transpiration, plant cells progressively lose their turgor and drought damages set in.

It is interesting to describe in more detail the consequences that variations in the plant water potential have on cell pressure and osmotic potential. Π_p and Ω_p are related to the osmotic and elastic properties of the cell membranes in a combined way that is well synthesized by the Höfler diagram shown in Figure 4.5 (see also Jones, 1992; Nobel, 1999). At full-turgor conditions, the plant water potential is quite high (slightly below zero) and the pressure and relative

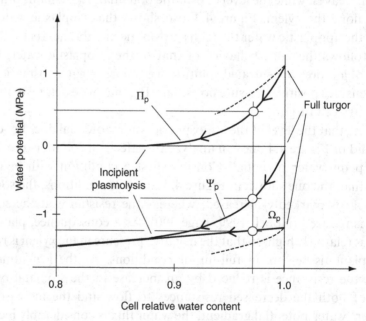

Figure 4.5 Höfler diagram relating the cell relative water content to the protoplast water potential Ψ_p, the protoplast turgor potential Π_p, and the protoplast osmotic potential Ω_p. The dashed lines show the variations in Ψ_p and Ω_p when osmotic adjustment takes place. The open circles correspond to the situation depicted in Figure 4.4. After Porporato et al. (2001).

water content of the cells are at their maximum values. As the soil moisture is depleted and plant water potential lowers, the osmotic potential decreases only slightly because of the solute concentration increase caused by water losses. According to Eq. (4.3), a reduction in water potential, not accompanied by an equivalent decrease in osmotic potential, produces a decrease of pressure potential and turgor loss. The three circles on the curves in Figure 4.5 correspond to the situation depicted in Figure 4.4. As the drought continues, water potential and turgor keep decreasing until the plasma membranes of the cells start pulling away from the cell walls (i.e., incipient plasmolysis). At this point, plant conditions have become very critical and permanent damage is produced.

By controlling the steepness of the curves and the relationship between the volume and pressure of the cells, different elastic properties of the cell walls can delay or anticipate the sequence leading to plasmolysis. Some plants in semi-arid ecosystems also have the capability of adjusting the levels of cell osmotic potential by increasing cell solute concentration. This allows them to maintain higher turgor (from Eq. (4.3), a decrease of Ω_p causes an increase of Π_p, when Ψ_x and G_p are kept fixed) and reduce the level of water stress (see dashed lines in Figure 4.5). This phenomenon is called osmotic adjustment, and is considered an important form of adaptation to water stress (Nilsen and Orcutt, 1998; see also Section 7.3).

4.1.3 Physiological effects of a reduction in plant water potential

Plant physiological response to water stress is quite complex and the effects of water stress differ from plant to plant and also depend on the timing of drought during the growing season. However, there are common features that can justify an attempt to a general modeling of plant water stress. Usually, the most important variables controlling the level of water stress of a plant are the turgor pressure, Π_p, and the relative water content of the living tissues (e.g., Hsiao, 1973; Bradford and Hsiao, 1982; Nilsen and Orcutt, 1998). As seen before, these variables are interdependent and are also related to the tissue water potential, Ψ_p, by the Höfler diagram (Figure 4.5).

Figure 4.6 reproduces a typical sequence of effects caused by a decrease in cell water potential. As can be seen, during drought conditions this decrease triggers a series of harmful events on plant physiology whose number and seriousness grow with the intensity and duration of water deficit. As Ψ_p is reduced, the first effect is a reduction of cell growth and wall-cell synthesis; then nitrogen uptake from soil is diminished due to the sharp decrease of nitrate reductase, an enzyme which catalyzes the reduction of nitrate to nitrite

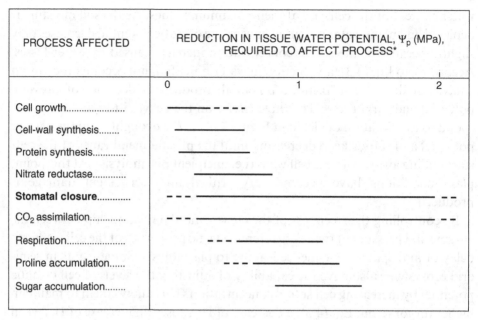

PROCESS AFFECTED	REDUCTION IN TISSUE WATER POTENTIAL, Ψ_p (MPa), REQUIRED TO AFFECT PROCESS*
	0 1 2
Cell growth......................	
Cell-wall synthesis........	
Protein synthesis............	
Nitrate reductase............	
Stomatal closure............	
CO_2 assimilation..............	
Respiration......................	
Proline accumulation......	
Sugar accumulation........	

*With Ψ_p of well-watered plants under mild evaporative demand as the reference point

Figure 4.6 Typical sequence of effects on plant physiology caused by a decrease in cell water potential. Redrawn after Hsiao (1973).

(which is the first internal step for nitrogen assimilation after nitrate has been uptaken by the plant). The stomatal closure and the related reduction in CO_2 assimilation also begin very soon: at first, stomatal closure is mostly a way to optimize water losses and carbon assimilation and does not produce permanent damage to the plant. Soon after, however, respiration is affected. Changes in patterns of resource allocation are then induced, including the accumulation of the amino acid proline, whose drastic increase is a characteristic disturbance in protein metabolism. Sugar accumulation, along with possible flowering reduction, inhibition of seed production, and fruit abortion are possible further effects of higher levels of water stress (Nilsen and Orcutt, 1998). At the same time, the reduction of the cooling effect of transpiration makes the insurgence of heat and radiation stress more likely. Finally, at very low water potentials, complete stomatal closure takes place, and this usually also marks the point of complete turgor loss and wilting (Bradford and Hsiao, 1982, page 309).

Stomatal closure is an important process for the description of plant response to water deficit. It is controlled by the turgor level of the guard cells surrounding the stomata (Figure 4.7): high guard-cell turgor produces stomatal opening, while low turgor induces stomatal closure (e.g, Salisbury and Ross, 1992; Larcher, 1995). When water stress is not too strong, the

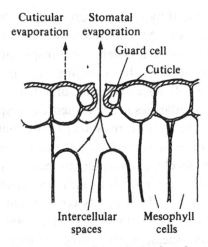

Figure 4.7 Pathways for water loss from the leaf surface. After Jones (1992).

control of guard-cell turgor is mainly an osmoregulatory process similar to the osmotic adjustment described above. The decrease in guard-cell turgor is brought about by an increase of abscisic acid (ABA) which, changing the permeability characteristics of the cell membranes, induces a lower solute concentration in the guard cells (Larcher, 1995; Nilsen and Orcutt, 1998). Light presence controls the diurnal cycle of stomatal opening and closure in the morning and at dusk.[3] The degree of stomatal opening during daytime is mainly controlled by soil moisture and relative humidity through different mechanisms, many of which are not yet completely clear (e.g., Bradford and Hsiao, 1982; Schulze and Hall, 1982; Schulze, 1986, 1993; Nilsen and Orcutt, 1998; Whitehead, 1998). One type of regulation is believed to act as a feedback control induced by low levels of plant water potential (which are in turn produced by soil-moisture deficit and low atmospheric humidity; see Figure 4.4). This control takes place when the plant status has already been affected by water stress and it does not involve the production of ABA. Stomatal closure can also be induced as a preventive measure to reduce internal water losses and risk of cavitation, before plant water potential is seriously lowered. In fact, roots have been shown to be able to induce a feed-forward control on stomata by producing ABA in conditions of incipient water stress (e.g., Schulze, 1993). Similarly, a reduction in relative atmospheric humidity can induce stomatal closure to prevent water stress. In general, however, stomatal regulation by atmospheric humidity acts at short time scales and is less

---

[3] Only desert plants with crassulacean acid metabolism (CAM) open their stomata at night when transpiration demand is lower (e.g., Larcher, 1995; Nobel, 1999). During the day, when the stomata are closed, these acids are hydrolyzed and the resulting CO_2 is used by photosynthesis.

important than that induced by soil moisture in the reduction of the average daily values of transpiration. At a daily time scale soil moisture control of stomatal opening determines a dependence of transpiration on soil moisture of the form shown in Figure 2.4. The modeling of stomatal response to the environmental variables will described in detail in Chapter 6.

Notwithstanding the similarities in the sequence of response to water deficit among different plants, the levels of resistance to drought are species-specific. Many species in semi-arid ecosystems have adapted to frequent and prolonged periods of water deficit, developing various strategies to delay the appearance of water stress (e.g., Ehleringer and Monson, 1993; Larcher, 1995; Nilsen and Orcutt, 1998). The ways by which plants compensate for water limitation range from mechanisms for drought escape (such as rapid phenological development, drought deciduousness, or extended dormancy) to drought tolerance by osmotic adjustment, desiccation tolerance, and reduction in water loss with enhancement of water accumulation.

The sequence of effects shown in Figure 4.6 is mostly related to the intensity of the soil-moisture deficit, without specifically considering its duration. It is clear that this "static" description of the physiological effects is not sufficient to fully describe the development of plant sufferance: water stress is a gradually intensifying process, in which the time dimension is particularly important. This is especially true for semi-arid ecosystems, where drought is often a prolonged and frequent phenomenon, more than an isolated event. The duration and frequency of periods of water stress are therefore essential for understanding vegetation response. For example, it has been shown that even mild stress, when it lasts long enough, is sufficient to produce substantial effects on canopy size (Bradford and Hsiao, 1982, page 305).

4.1.4 Modeling plant water stress: from water deficit to vegetation response

From the previous discussion on the processes connecting plant water stress with soil-moisture deficit, some characteristics that should be included in a simple but realistic model of water stress have emerged. In this section, a first quantification of plant water stress, related to the soil moisture conditions at the time under consideration and thus referred to as "static" water stress, will be presented. Later on, using the statistical analysis of the duration and frequency of water deficit periods, the temporal dimension will be included in the definition of water stress, introducing a "dynamic" water stress, θ.

As can be seen from Figure 4.6, the process of stomatal closure covers the entire scale of water stress. On one hand, the incipient stomatal closure is among the first effects of water deficit on plant physiology, while, on the other hand, complete stomatal closure only takes place at the very end of the sequence of the effects on physiology when the plant starts wilting. These properties make stomatal closure the ideal candidate to delimit both the starting and the maximum point of water stress. Moreover, the use of incipient and complete stomatal closure as indicators of water stress is also very convenient because these points are frequently measured in field experiments. Following Porporato et al. (2001), we will assume that (at the daily time scale) the static stress ζ is zero when soil moisture is above the level of incipient reduction of transpiration, s^*, and reaches a maximum value equal to one when soil moisture is at the level of complete stomatal closure (wilting), i.e.,

$$\zeta = 0 \quad \text{for} \quad s > s^*,$$
$$\zeta = 1 \quad \text{for} \quad s < s_w. \tag{4.12}$$

We refer to Section 2.1.5 and Chapter 6 for a discussion of typical values for s^* and s_w and their relation to the soil and plant characteristics.

As for the behavior of static water stress for soil moisture values between s_w and s^*, the effects described in Figure 4.6 clearly suggest a nonlinear increase of plant water stress with soil-moisture deficit. Figure 4.8a, adapted from Bradford and Hsiao (1982), may be used as a reference sketch for the modeling of ζ, whose form can thus be taken as

$$\zeta(t) = \left[\frac{s^* - s(t)}{s^* - s_w} \right]^q, \quad \text{for} \quad s_w \leq s(t) \leq s^*, \tag{4.13}$$

where q is a measure of the nonlinearity of the effects of soil-moisture deficit on plant conditions (Figure 4.8b). The nonlinearity is evident if one considers the sequence of effects and the increase of intensity of each effect with soil-moisture deficit. As in the case of s_w and s^*, the value of q can also vary with plant species and, to a smaller extent, with the soil type.

4.2 Probabilistic description of static water stress

The simple relationship between the static water stress $\zeta(t)$ and the soil moisture content $s(t)$ allows one to invert Eq. (4.13) and obtain the probability density function $f_Z(\zeta)$ of ζ as a derived distribution of the steady-state pdf of soil moisture $p(s)$ (Chapter 2, Eq. (2.39)). Due to the particular form of the stress function ζ, Eqs. (4.12) and (4.13), $f_Z(\zeta)$ has an atom of probability both at $\zeta = 0$ and $\zeta = 1$. The probability of having no stress corresponds to the probability of soil moisture above s^*, that is

$$F_Z(0) = 1 - P(s^*), \tag{4.14}$$

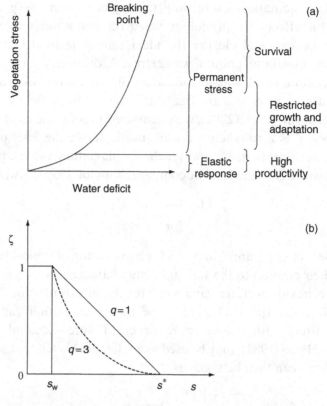

Figure 4.8 Relationship between soil water content and plant water stress: (a) qualitative representation of the effect of water deficit on vegetation; redrawn from Bradford and Hsiao (1982); (b) static water stress ζ versus soil moisture s, for two values of the nonlinearity parameter q; see Eq. (4.13). After Porporato et al. (2001).

where $P(s^*)$ is the value of the cumulative distribution of soil moisture calculated in $s = s^*$ (see the Appendix to Chapter 2 for the analytical expression of $P(s)$). Likewise, the atom of probability at $\zeta = 1$ is equal to the probability of having soil moisture below the wilting point, or

$$F_Z(1) = P(s_w). \tag{4.15}$$

The continuous part of the pdf is obtained from the soil moisture pdf for $s_w < s \leq s^*$, Eq. (2.39). After some calculations, it can be written as

$$f_Z(\zeta) = \frac{C_\zeta}{\eta_w}\left[\left(1 - \frac{\eta}{\eta_w}\right)\zeta^{\frac{1}{q}} + \frac{\eta}{\eta_w}\right]^{\frac{\lambda'(s^*-s_w)}{\eta-\eta_w}-1} e^{\gamma[(s^*-s_w)\zeta^{\frac{1}{q}}-s^*]}, \tag{4.16}$$

using the same parameters introduced in Chapter 2, that is

$$\eta = \frac{E_{\max}}{nZ_r}, \qquad \eta_w = \frac{E_w}{nZ_r},$$

$$\lambda' = \lambda e^{-\frac{\Delta}{\alpha}}, \qquad \gamma = \frac{nZ_r}{\alpha}, \tag{4.17}$$

where, as usual, $E_{\max}$ is the maximum daily evapotranspiration rate, Z_r the active soil depth, n the porosity, E_w the daily evaporation rate at wilting point, λ the mean frequency of rain events, Δ the interception depth, and α the mean depth of rainfall events.

The constant of integration C_ζ in Eq. (4.16) can be deduced by imposing the condition

$$\int_0^1 f_Z(\zeta) \, d\zeta = P(s^*) - P(s_w). \tag{4.18}$$

It can be calculated analytically only when q takes integer values, but the expression is cumbersome and thus is not reported here.

Figure 4.9 shows some examples of pdf's of static water stress for different types of climate and active soil depth (the atoms of probability at $\zeta = 0$ and $\zeta = 1$ are also shown using a different scale). Water stress decreases with increasing precipitation: this is evident from the significant increase of $F_Z(0)$, the decrease of $F_Z(1)$, and the leftward shift of the mode of the continuous part, both for $q = 1$ and for $q = 3$ (continuous and dashed lines in Figure 4.9, respectively). An increase in the probability of occurrence of low values of water stress is found in the continuous part when q increases. Mathematically, this is due to the normalization of Eq. (4.13), that is to the condition that all the curves linking water stress to soil moisture have to pass through the points $(s_w, 1)$ and $(s^*, 0)$ in the plane (s, ζ), as can be seen in Figure 4.8b. Apart from this, the shapes of the pdf's have the same type of dependence on climatic and soil conditions for the two values of q. Notice also that a change in q does not affect the two atoms of probability at $\zeta = 0$ and $\zeta = 1$.

The probabilistic dependence of water stress on active soil depth can also be studied from Figure 4.9. Except for the case of wet climates, shallow-rooted plants tend to experience higher stress conditions than deep-rooted species (greater values of $F_Z(1)$ and modes of the continuous part at higher ζ), but, importantly, these conditions are considerably less frequent as implied by the larger atom of probability at $\zeta = 0$. This behavior results from the temporal dynamics of soil moisture, an example of which is shown in Figure 4.10. In

Plant water stress

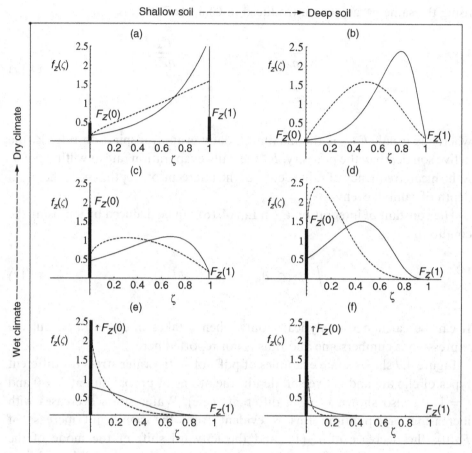

Figure 4.9 Examples of pdf's of static water stress ζ for different active soil depths and frequency of rainfall events λ (the atoms of probability at $\zeta = 0$ and $\zeta = 1$ have a different scale from $f_Z(\zeta)$). Continuous lines refer to $q = 1$, dashed lines to $q = 3$. Left panels correspond to $Z_r = 30$ cm, right panels to $Z_r = 90$ cm. Top, center, and bottom graphs have a frequency of rainfall events λ of 0.1, 0.2, and 0.5 day^{-1} respectively. Common parameters to all graphs are $\alpha = 1.5$ cm, $E_w = 0.01$ cm day^{-1}, $E_{max} = 0.45$ cm day^{-1}; the soil is a loam (see Table 2.1 for soil parameters). After Porporato et al. (2001).

deeper soils (continuous line in Figure 4.10a) the trace of soil moisture tends to remain almost always between s_w and s^*, so that in this example vegetation does not experience very high stress conditions (Figure 4.10b). In contrast, for shallow-rooted plants, the trace of stress frequently reaches high values, but also remains close to zero during longer periods. This situation is accurate while the mean value of soil moisture remains below s^*. In contrast, for wet climates the mean is above s^* and the trajectory of soil moisture for deep soils

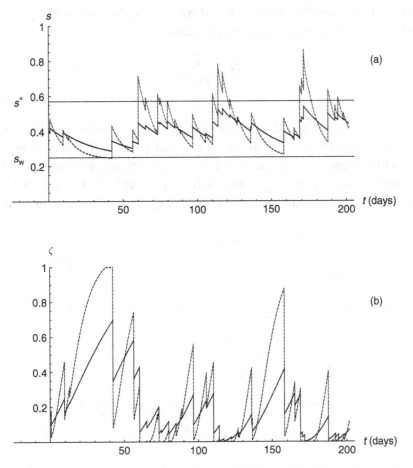

Figure 4.10 Example of traces of soil moisture (a) and static water stress (b) for the same rainfall sequence ($\alpha = 1.5\,\mathrm{cm}$, $\lambda = 0.2\,\mathrm{day}^{-1}$) in a loamy soil with a typical vegetation of semi-arid environments (see the caption of Figure 4.9 for other parameters). Continuous line refers to $Z_r = 90\,\mathrm{cm}$, dashed line to $Z_r = 30\,\mathrm{cm}$. After Porporato et al. (2001).

hardly ever crosses s^*, thus reducing the level of water stress with respect to that of shallow-rooted plants (Figures 4.9e and 4.9f).

The mean water stress can be calculated from Eqs. (4.15) and (4.16) as

$$\bar{\zeta} = \int_0^1 \zeta f_Z(\zeta)\,\mathrm{d}\zeta + F_Z(1). \tag{4.19}$$

The value of $\bar{\zeta}$ obviously takes into account the periods when ζ is 0 and hence it is not very indicative of the actual vegetation conditions. The mean value of water stress given that the plant is under stress is more meaningful for our purposes. To obtain the latter, denoted by $\overline{\zeta'}$, only the part of the pdf corresponding to ζ above zero should be considered, i.e.,

$$\overline{\zeta'} = \frac{\overline{\zeta}}{P(s^*)}.$$ (4.20)

The sensitivity of $\overline{\zeta'}$ to different rainfall conditions is shown in Figure 4.11a: the relationship between the mean conditional stress and λ is strongly non-linear, with a rapid increase for dry climates. The difference between $q = 1$ and $q = 3$ remains approximately constant for all the values of λ, except for the lowest ones when s is almost always below s_w and differences in q have a

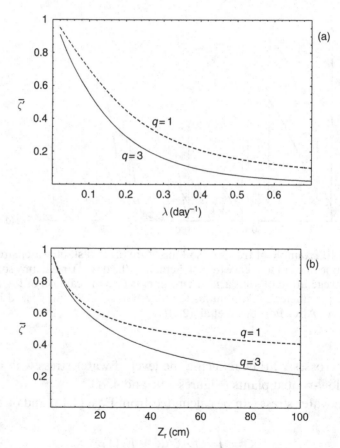

Figure 4.11 Sensitivity of the mean static stress $\overline{\zeta'}$ with respect to: (a) rainfall frequency λ, and (b) active soil depth Z_r ($\alpha = 1.5\,\text{cm}$, $T_{seas} = 200\,\text{days}$; the active soil depth in (a) is $Z_r = 60\,\text{cm}$, while the rainfall frequency in (b) is $\lambda = 0.2\,\text{day}^{-1}$; see caption of Figure 4.9 for other parameters). After Porporato et al. (2001).

smaller effect (see Figure 4.8b). The value of $\overline{\zeta'}$ remains recognizably different from 0 also for very wet climates, since $\overline{\zeta'}$ by definition does not take into account the probability of occurrence of water stress, which for wet climates is very low. The influence of the soil depth, Z_r, is analyzed in Figure 4.11b for moderate values of the average intensity and frequency of rainfall. As already noticed in Figure 4.9, for this climate water stress decreases with soil depth, both for $q = 1$ and, more rapidly, for $q = 3$.

Since $\overline{\zeta'}$ gives the mean value of water stress provided that the plant is under stress, its definition is not restricted to steady-state conditions but is also valid for transient conditions, such as those taking place when the water content of the soil at the beginning of the growing season is relevantly different from the mean steady-state soil moisture. If a plant experiences stress during this transient, the intensity of the stress is the same on average as the one that would occur under steady-state conditions. Mathematically, this is due to the memory-less (i.e. Markovian) nature of the dynamics provided by the Poisson process of rainfall. Thus, after the trajectory has reached a certain level (for example s^*), statistically it will always behave in the same manner, without any dependence on past conditions (see Chapter 3). The only quantity changing because of the transient condition is the probability of occurrence of a stress event. This change affects the value of $\overline{\zeta}$ and $P(s^*)$, but not that of $\overline{\zeta'}$.

4.3 Dynamic water stress

The linkage of plant water stress and soil moisture dynamics is a much more complex problem than what is simply accounted for by the static water stress considered before. A measure of water stress that combines the previously defined "static" stress, $\overline{\zeta'}$, representing the mean plant water stress during an excursion below s^*, with the mean duration and frequency of water stress through the variables $\overline{T}_{s^*}$ and $\overline{n}_{s^*}$ was proposed by Porporato et al. (2001). This, referred to as "dynamic" water stress or mean total dynamic stress during the growing season $\overline{\theta}$, was defined as

$$\overline{\theta} = \begin{cases} \left(\frac{\overline{\zeta'}\,\overline{T}_{s^*}}{kT_{\text{seas}}}\right)^{\overline{n}_{s^*}^{-r}} & \text{if } \overline{\zeta'}\,\overline{T}_{s^*} < kT_{\text{seas}} \\ 1 & \text{otherwise.} \end{cases} \tag{4.21}$$

The role of the parameters k and r is now discussed along with the rationale behind the definition of $\overline{\theta}$. Plant water stress occurs over time as a chain of events triggered by a decrease in soil moisture. As previously discussed, the starting point of this sequence is signaled by the stomatal closure induced by soil-moisture deficit (see Figure 4.6), the last stage involves the wilting of the plant and the occurrence of permanent damage. The value of soil moisture s^* when stomatal closure begins can thus be employed as a threshold for the occurrence of water stress. The intensity of plant water stress depends on the mean water deficit during the period of stress and on how such a deficit affects the plant.

The definition of the average static water stress, $\overline{\zeta'}$, Eq. (4.20), takes into account the mean intensity of the water deficit, but contains no information on its duration and frequency. Thus the same value of $\overline{\zeta'}$ may have a very different impact on a given type of vegetation, depending on the mean duration of the periods of stress; this indicates that $\overline{T}_{s^*}$ should be included in the definition of total water stress. Assuming a linear dependence between time under stress and intensity of dynamic stress, the mean amount of plant water stress during a period of stress becomes a function of the product $\overline{\zeta'}\,\overline{T}_{s^*}$. One could also include some degree of nonlinearity in the relationship between $\overline{\zeta'}$ and $\overline{T}_{s^*}$, but unfortunately no data are available to justify possibly more realistic (but also more complicated) expressions.

The actual plant water stress, however, cannot increase indefinitely with $\overline{\zeta'}\,\overline{T}_{s^*}$, since there must be a point, corresponding to the onset of permanent damage, from where on the stress is at its maximum level. Plant physiology literature frequently makes reference to this upper value of water stress (e.g., Levitt, 1980; Bradford and Hsiao, 1982). The parameter k in Eq. (4.21) allows one to fix the value of such a threshold: permanent damage appears when $\overline{\zeta'}\,\overline{T}_{s^*} > k\,T_{\text{seas}}$, with k representing an index of plant resistance to water stress. When no specific information on the resistance of different plants is available, the same value for all the species will be assumed (e.g., $k = 0.5$). The value of k may also be interpreted as the average static stress $\overline{\zeta'}$ a plant can experience without suffering permanent damage, when the duration of the period of stress is the whole growing season. According to the previous considerations, the normalized dynamic water stress due to a single excursion below s^* can be written as $\frac{\overline{\zeta'}\,\overline{T}_{s^*}}{k\,T_{\text{seas}}}$, which is 0 when there is no water stress and 1 when plants experience permanent damage.

The quantity $\frac{\overline{\zeta'}\,\overline{T}_{s^*}}{k\,T_{\text{seas}}}$ does not depend on the number of periods of stress during a growing season but only refers to an average single period of stress. However, for a reasonable definition of water stress it is important to take into account the effect that multiple periods of stress may have on the plant status. It is plausible that the role of the number of periods of water stress falls between the following two extreme conditions: (i) the plant completely recovers immediately after any period of water stress; (ii) the plant does not experience any recovery during the periods of high soil moisture, and each new occurrence of a period of water deficit contributes to increase the existing stress of the plant. In the former case, the frequency of periods of water stress has no importance, and a good measure of the total dynamic stress during the growing season is simply $\frac{\overline{\zeta'}\,\overline{T}_{s^*}}{k\,T_{\text{seas}}}$. In the latter case, assuming a simply additive accumulation of stress throughout time or, equivalently, a linear relationship between number of crossings and dynamic stress, the total water stress during the growing season is $\overline{n}_{s^*} \cdot \frac{\overline{\zeta'}\,\overline{T}_{s^*}}{k\,T_{\text{seas}}}$. However, with such formulation the dynamic water stress would be

$$\overline{n}_{s^*} \cdot \frac{\overline{\zeta'}\,\overline{T}_{s^*}}{kT_{\text{seas}}} = \frac{\overline{\zeta}}{k} \qquad (4.22)$$

(see Eqs. (4.20) and (3.27)), so that the information on the duration and frequency of water stress cancels out, resulting again in a "static" formulation of water stress.

None of the above two extreme formulations can be considered to be adequate. There is experimental evidence that "for at least several days after the end of the drying cycle, the photosynthetic rate does not recover to that prior to the commencement of the drying cycle" (Turner et al., 1985; see also Larcher, 1995, page 117), while on the other hand a long enough period of high soil moisture causes a bettering of plant conditions and, if the period is long enough and the damage is not permanent, even complete recovery will take place (elastic response to water stress, e.g., Levitt, 1980). An intermediate, more realistic description of water stress is to be sought, therefore, that properly blends the information on the frequency of water stress. One possible way to mathematically express this role is to use a decreasing function of $\overline{n}_{s^*}$ as the exponent of $\frac{\overline{\zeta'}\,\overline{T}_{s^*}}{k\,T_{\text{seas}}}$. Porporato et al. (2001) found that the particular choice $r = 1/2$ well tempers the importance of very high values of $\overline{n}_{s^*}$ (see Figure 3.4), thus avoiding erroneously high values of total dynamic stress in the case of very short, but frequent, stress periods. Different r values or other analytical functions with a similar behavior lead to an analogous behavior of the total dynamic stress $\overline{\theta}$. The simpler choice of $\overline{n}_{s^*}^{-1}$ turned out not to be feasible, because of the over-sensitivity to high $\overline{n}_{s^*}$ values mentioned before.

The behavior of the dynamic stress with respect to $\frac{\overline{\zeta'}\,\overline{T}_{s^*}}{k\,T_{\text{seas}}}$ and $\overline{n}_{s^*}$ is shown in Figure 4.12. When $\overline{\zeta'}\,\overline{T}_{s^*}$ exceeds the threshold value $k\,T_{\text{seas}}$, the resulting total dynamic stress is equal to 1, independently of $\overline{n}_{s^*}$. When $\frac{\overline{\zeta'}\,\overline{T}_{s^*}}{k\,T_{\text{seas}}} < 1$, the total dynamic stress increases with the intensity of the dynamic stress of the single average excursion below s^*, in a way that depends on the value of $\overline{n}_{s^*}$. If on average there is one downcrossing (or upcrossing) per growing season, the dynamic stress $\overline{\theta}$ equals $\frac{\overline{\zeta'}\,\overline{T}_{s^*}}{k\,T_{\text{seas}}}$; for $\overline{n}_{s^*}$ greater than one, $\overline{\theta}$ is greater than the

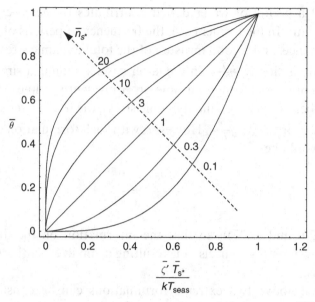

Figure 4.12 Dependence of the mean dynamic water stress $\bar{\theta}$ on $\frac{\bar{\zeta}'\,\bar{T}_{s*}}{k\,T_{seas}}$ and $\bar{n}_{s*}$, as described by Eq. (4.21). After Porporato et al. (2001).

dynamic stress of a single excursion below s^* (the curves are above the main diagonal obtained for $\bar{n}_{s*} = 1$), but lower than $\bar{n}_{s*}\frac{\bar{\zeta}'\,\bar{T}_{s*}}{k\,T_{seas}}$ (except for extremely low values of $\frac{\bar{\zeta}'\,\bar{T}_{s*}}{k\,T_{seas}}$). The value of $\bar{\theta}$ is thus between the two limit cases previously defined. Notice that $\bar{n}_{s*}$ has very little importance when $\frac{\bar{\zeta}'\,\bar{T}_{s*}}{k\,T_{seas}}$ tends to 1. In fact, when the dynamic stress of a single excursion below s^* is very high, the resulting total dynamic stress is already so high that having more than one period of stress becomes irrelevant.

Similar considerations apply to the case when $\bar{n}_{s*} < 1$, a condition which (for opposite reasons) is typical of both very wet and very dry climates (see Figure 3.4a). For very wet climates, $\frac{\bar{\zeta}'\,\bar{T}_{s*}}{k\,T_{seas}}$ is low and the resulting $\bar{\theta}$ is much lower than the one that would be obtained with $\bar{n}_{s*} = 1$. For very dry climates, $\frac{\bar{\zeta}'\,\bar{T}_{s*}}{k\,T_{seas}}$ approaches, and even surpasses, the threshold at 1: $\bar{T}_{s*}$ tends to be greater than T_{seas} and lower values of $\bar{n}_{s*}$ are just a consequence of a greater $\bar{T}_{s*}$ (see Eqs. (3.26) and (3.27)), without implying an improvement of plant conditions.

4.4 Impact of environmental conditions on dynamic water stress

The above formulation of the total dynamic water stress permits a quantitative study of its dependence on climate, soil, and vegetation. Figure 4.13 shows how the value of $\bar{\theta}$ decreases with an increase in the frequency of storms, λ (the mean depth of rainfall per storm is kept fixed). As rainfall becomes more

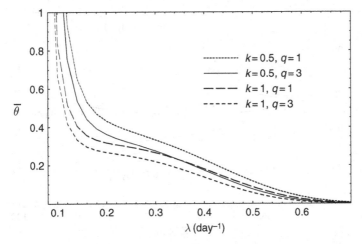

Figure 4.13 Mean dynamic water stress $\bar{\theta}$ versus frequency of rainfall events λ for four different choices of the parameters k and q ($T_{\text{seas}} = 200$ days, $\alpha = 1.5$ cm, and $Z_r = 60$ cm; see caption of Figure 4.9 for soil and vegetation parameters). After Porporato et al. (2001).

frequent, soil water becomes more abundant and the water stress tends to disappear. The behavior of the curves is interesting: a rapid decay of the dynamic stress for low values of λ is first followed by a plateau for intermediate frequencies of rainfall events and then by a slow decay to zero stress for very wet climates. The steep increase of stress for very arid climates is a sort of threshold effect of λ: below a certain frequency of rainfall events the duration of an excursion below s^* increases dramatically (see Figure 3.4), and this controls the value of $\bar{\theta}$. The shift to better conditions ($\bar{\theta} \sim 0.5$) is thus relatively fast in terms of λ.

The presence of a plateau for intermediate frequencies of rainfall events may be related to the fact that, in this range of λ, $\overline{T}_{s^*}$ and $\overline{\zeta'}$ decrease but the number of crossings tends to increase (see Figures 4.11a and 3.4). The plant status is thus in a transition from one condition of stress to another and the resulting dynamic stress decreases only slowly. Finally, for very wet climates, all the components of the dynamic stress, namely $\overline{\zeta'}$, $\overline{T}_{s^*}$, and $\bar{n}_{s^*}$, decrease with λ (see again Figures 4.11a and 3.4), so that the rate of decrease of $\bar{\theta}$ again becomes more marked.

Notice also from Figure 4.13 that the behavior of $\bar{\theta}$ is relatively robust to changes of the parameters k and q. Nevertheless, there are some changes from one case to another. Considering for example the part on the left-hand side of the graph, the region of maximum stress moves toward drier conditions if either k increases or q increases. It is in this region, representing drier

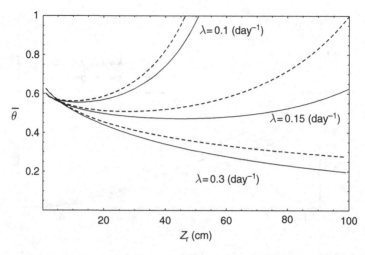

Figure 4.14 Mean dynamic water stress $\bar{\theta}$ versus active soil depth Z_r, for three different values of λ and two values of the parameter q ($q = 1$ dashed lines, $q = 3$ continuous lines). $T_{\text{seas}} = 200$ days, $k = 0.5$, and $\alpha = 1.5$ cm; see caption of Figure 4.9 for soil and vegetation parameters. After Porporato et al. (2001).

conditions, where small changes in the frequency of rainfall events, λ, lead to major changes in the total dynamic water stress. This is an important characteristic because λ undergoes random fluctuations throughout the growing seasons of different years and these fluctuations are especially strong in semi-arid climates (this will be studied in Chapter 8).

Figure 4.14 shows the influence on $\bar{\theta}$ of the effective soil depth (or rooting depth) Z_r. The response of plants to Z_r is very different for dry and wet climates. When the climate is wet, deep soils are generally more favorable for vegetation (or, from another viewpoint, deep-rooted species seem more fitted than shallow-rooted ones). The opposite is true for very dry climates, when deep-rooted plants experience permanent damage much earlier. The reason for this is that in dry climates the soil layers near the surface are the ones generally wetted by weak storm events, so that it becomes fundamental for a plant to have mostly superficial roots to be able to compete with the rapidly occurring evaporation losses. When the climate is wet, the surface soil layers are frequently not much wetter than for drier climates, due to the storage limit imposed by the field capacity and the soil evaporation occurring mainly in those layers (which explains why the curves are so close to one another for shallow soil depths). Nevertheless a sizeable amount of water infiltrates to the deeper layers, so that the presence of deep roots becomes very important. Notice that this contrasting behavior causes the curve for an intermediate climate to develop a broad minimum for a value of Z_r close to 30 cm. Following the suggestive implications of Figures 4.13 and 4.14, the dependence of vegetation type on the soil and climate characteristics will be explored in the examples of Chapter 5.

It is important to emphasize that all the above mechanisms, including the mathematical framework previously developed, do not consider any type of interaction between the active soil layer and the regional water table. Leakage from the active soil layer goes into deep percolation and the water table is assumed to be deep enough not to intersect the vegetation root zone. Obviously in some cases such an assumption may be unrealistic, especially in wet climates, where the analytical framework should be extended to incorporate the possibility of an interacting water table.

Finally, the interplay between the timing and amount of rainfall can be studied considering as variable the storm frequency λ while keeping fixed the total precipitation per growing season, $\Theta = \lambda \alpha T_{seas}$. Figure 4.15 shows that, except for very wet climates, there is an optimal partition between the timing and amount of rainfall (minimum of the dynamic water stress) which provides the best condition for vegetation. The shape of the curves is similar to that found for $\overline{T}_{s^*}$ in Figure 4.14, with the minimum decreasing and moving toward the right-hand side of the diagram for greater values of Θ.

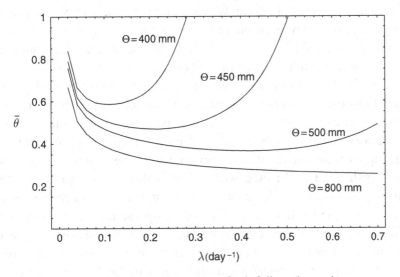

Figure 4.15 Impact of timing and amount of rainfall on dynamic water stress. Each curve corresponds to a constant total rainfall during the growing season, $\Theta = T_{seas}\alpha\lambda$. $T_{seas} = 200$ days, $q = 2$, $k = 0.5$, and $Z_r = 60$ cm; see caption of Figure 4.9 for other soil and vegetation parameters. After Porporato et al. (2001).

4.5 Optimal plant conditions

It was shown in Figure 2.10c that for a given total rainfall during the growing season there is an intermediate value of the storm frequency λ that produces a maximum of evapotranspiration and possibly optimal conditions for plant growth, and analogous conditions were found to give rise to a minimum in the duration of the excursions under stressed conditions (see Figures 3.5 and 3.6). Along the same lines, Figure 4.15 shows that, for a constant seasonal rainfall, the interplay between λ and α produces a minimum of dynamic water stress, again for intermediate values of λ. Such evidence provides a richer connotation to the concept of effective rainfall (i.e., an event that is intense enough to stimulate biological processes, particularly growth and reproduction), whose importance is well known in the ecology of arid and semi-arid ecosystems (e.g., Noy-Meir, 1973). The number of such effective events during the growing season is strongly linked to the interplay between λ and α in controlling the total rainfall. Moreover, the concept of vegetation-effective rainfall (and thus also the definition of optimal environmental conditions for vegetation) is not just related to climatic conditions, but is intimately coupled with soil and plant properties, such as rooting depth, soil texture, and physiological plant characteristics. The same rainfall event, in fact, can be effective for shallow-rooted plants and completely ineffective for deep-rooted ones. An interesting piece of field evidence for the importance of Z_r is given by Sala and Lauenroth (1982), who found that an event of 0.5 cm, which is often considered not very important for some deep-rooted vegetation (e.g., Coupland, 1950), is fundamental for *Bouteloua gracilis*, a short C_4 grass of the Colorado steppe which has most of its roots in the first centimeters of the soil (see also Section 5.3).

An example of the connection between rainfall regime and rooting depth is shown in Figure 4.16a, where the vegetation dynamic water stress is calculated for different values of Z_r and λ, keeping fixed the total rainfall during the growing season at $\Theta = 50$ cm. The dynamic stress is equal to 1 in a large region near the upper right corner of the diagram, where the values of α are low and the soil depth is large. In fact, with a low α the percentage of vegetation-effective rainfall decreases because interception becomes more important. In such conditions plants that are unable to develop a shallow rooting system end up suffering permanent damage, because in deeper soils the average level of s is consistently too low to sustain effective transpiration. The dynamic stress $\bar{\theta}$ is also high for very low values of λ, mainly because in this case the rainfall events become too rare and plant survival becomes very difficult. For the specific plant parameters and the particular total rainfall considered in this example, the area of best fitness (minimum dynamic stress) covers a wide range of values

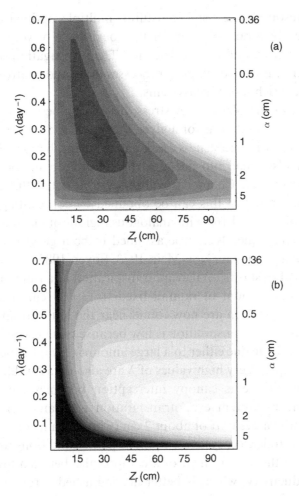

Figure 4.16 Optimal vegetation conditions in terms of plant water stress and total evapotranspiration. (a) Mean dynamic water stress and (b) total evapo-transpiration during the growing season versus frequency of rainfall events, mean depth of events, and active soil depth, keeping fixed the total amount of rainfall per growing season. In (a), $\bar{\theta} = 1$ in the white part of the diagram, $\bar{\theta} < 0.5$ in the darker area of the diagram, and the fixed interval between two adjacent contour lines is 0.05. In (b), evapotranspiration is more than 92% of incoming rainfall in the white part of the diagram, less than 50% in the black area, with a distance of 3% between adjacent contour lines. T_{seas} is 200 days, $q = 2$, and $k = 0.5$. Canopy interception is included here with a value of $\Delta = 0.1$ cm. See caption of Figure 4.9 for other parameters. After Porporato et al. (2001).

of λ but tends to be limited to medium–small rooting depths (dark gray area in Figure 4.16a). Of course, the region of optimal condition for a given amount of rainfall may be different when the transpiration characteristics of the plants or the soil properties are changed. Therefore, even when dealing with only a single plant resource (e.g., soil moisture), multiple optimal conditions for vegetation may be a possible way for the coexistence of very diverse species and the maximization of species diversity. The investigation of such aspects could help clarify how hydrological processes drive and control many aspects of the biological richness of ecosystems.

A minimum of the plant water stress, however, does not alone suffice to define the optimal conditions for a given type of vegetation. A more comprehensive measure of favorableness of an environment, besides the plant water stress, should also take into account the effective plant productivity and reproduction capacity. Chapter 6 will analyze in detail this connection. A preliminary indication of this can be obtained by confronting the optimum fitness region of Figure 4.16a with analogous regions of maximum evapotranspiration, which frequently can be assumed to be a good surrogate for the productivity of a plant (e.g., Noy-Meir, 1973; Boyer, 1982; Kramer and Boyer, 1995). Figure 4.16b shows total evapotranspiration as a function of Z_r and λ, keeping the total amount of rainfall fixed to $\Theta = 50$ cm. The unfavorable conditions for vegetation are now found near the left bottom corner of the diagram, where evapotranspiration is low because of an excessive production of runoff and leakage due either to a large amount of rainfall per event or to a small rooting depth. Very high values of λ are also unfavorable, because with very light rainfall events canopy interception becomes increasingly more important. The maximum evapotranspiration is attained in this case for deeper soils with values of α of about 2 cm (white area in Figure 4.16b).

In water-controlled ecosystems, optimal plant conditions are likely to be subordinated to the achievement of a compromise between low water stress and high productivity, which is better accomplished through some specific combinations of climate, soil, and vegetation parameters. For the particular example of Figure 4.16, in the case of very frequent but light rainfall events (say, $\lambda > 0.3$) the controlling factor is the plant water stress. Thus shallow-rooted species are preferred both because of a better exploitation of the incoming water (the amount of water transpired is approximately the same except for very low values of Z_r) and because of less severe conditions of water deficit. In the case of more intense and infrequent rainfall events (e.g., $\lambda \leq 0.3$) the preferable range of Z_r shifts toward deeper values where one finds both high transpiration (e.g., productivity) and low plant water stress.

This conceptual approach linking hydrological processes and plant responses will be pursued in the next chapters to investigate the optimal environmental condition for different functional types of vegetation.

5

Applications to natural ecosystems

Different water-controlled ecosystems for which valuable and extensive experimental data are available are studied in this chapter using the stochastic description of soil moisture dynamics and vegetation water stress developed before.

Previous chapters have pointed out how some physiological features of plants interacting with soil and climate characteristics may produce very different features in soil moisture dynamics. These are frequently observed in field studies where the intricate interconnection between causes and effects has been found to give rise to the development of strategies for optimum soil moisture exploitation and drought adaptation (e.g., Kemp and Williams, 1980; Sala et al. 1982a,b; Joffre and Rambal, 1993; Larcher, 1995; Linhart and Grant, 1996). Perhaps one of the most typical examples of the links between physiological characteristics of vegetation and soil moisture is the interplay among plant rooting depth, soil moisture fluctuations, and water use efficiency. For example, it was seen in Figure 4.10 that shallow-rooted plants produce high-variance soil moisture dynamics requiring high water use efficiency to exploit the ephemeral soil moisture availability. These high transpiration rates, in turn, cause an even faster decay of soil moisture that further enhances the effect of soil shallowness. The more slowly varying soil moisture dynamics of soils with deep-rooted vegetation seems instead to privilege strategies of careful soil moisture use, which contribute to make soil moisture dynamics even slower (see Section 7.1).

This chapter focuses on some specific water-controlled ecosystems in order to investigate the environmental and physiological origin of their distinctive features. In particular, the definition of a measure of the mean vegetation water stress, proposed in Chapter 4, is employed here to quantitatively assess the favorableness of environmental conditions to some plant species and thus to illuminate the conditions for their survival and/or coexistence.

117

Despite the fact that all of these ecosystems are driven by soil moisture scarcity, water-controlled ecosystems may be very different from one another, each presenting characteristic controlling features. In some cases the external climatic conditions may be so controlling that similar vegetation patterns are found regardless of the specific soil characteristics or initial historic conditions of vegetation distribution. Mediterranean types of climates are a typical example of this case, wherein dry growing seasons following relatively cold and wet winters give rise to ecosystems with strikingly similar characteristics, whether in North Africa, Middle East, California, or Chile. In other cases, the actual ecosystem structure is strongly dependent on very specific properties or initial conditions. Thus the existing functional vegetation type may self-sustain its presence, while different (and in great part unpredictable) scenarios would set in if vegetation conditions were altered. Many North American grasslands are believed to be in a similar fragile meta-stable equilibrium, threatened by either forest encroachment in wetter regions (Seastesdt and Knapp, 1993) or desert and shrub invasion in the dryer ones (Schlesinger et al., 1990; see also Figure 1.11). Tree–grass coexistence in savannas is another example of delicate equilibrium. Understanding the conditions of stability, or lack of it, of an ecosystem with respect to variations of soil moisture availability is important for predicting how impacting human interference or climate change can be.

Following the theoretical analysis of the previous chapters, the description is limited here to spatially lumped conditions and does not consider temporal variability other than the rainfall stochasticity during the growing season. Chapters 7–9 and in particular Chapter 11 will extend the applications to include other forms of spatial and temporal variability. The cases studied in this chapter are taken from Laio et al. (2001b), Porporato et al. (2003b), Caylor et al. (2004), Kiang (2002), and Baldocchi et al. (2004).

1. The first case study refers to the warm savanna of Nylsvley in South Africa (Scholes and Walker, 1993), where the relatively homogeneous soil texture and similar rooting depths of trees and grasses, as well as the presence of different woody and herbaceous plants with well-documented physiological characteristics, make it an ideal ecosystem for assessing the role of different vegetation types on the soil water balance. The analyses show that, despite the different plant physiological characteristics, the water stress for all the species turns out to be very similar, suggesting that indeed coexistence might be attained through differentiation of water use. The different patterns of soil moisture exploitation are thus at the origin of temporal and spatial niche selection, as well as of overall water stress minimization to sustain high species diversity and productivity.

2. The second ecosystem considered is the savanna of La Copita (Texas) which provides an example of a semi-arid ecosystem, where the issue of tree–grass

coexistence is further complicated by a marked sensitivity to interannual climatic variability. The two dominant species, having different phenological characteristics and rooting depth, produce very different soil moisture dynamics and water balances, which may be both the cause and the effect of the development of different strategies for drought resistance. The analysis of vegetation water stress for the two main species is aimed at assessing the possibility of tree–grass coexistence due to interannual rainfall variability. Different climatic conditions recorded in the past are analyzed and compared with measured data of canopy coverage to understand the ecosystem response to the climatic forcing.

3. The third case study deals with a shortgrass steppe in north-central Colorado. The large variability in soil characteristics together with the presence of a single dominant grass (*Bouteloua gracilis*, a short C_4 grass) make this ecosystem well-suited for studying the role of soil texture in the water balance and vegetation water stress. A key question is how this role is affected by changes in aridity that accompany rainfall variability at the site. The interplay of rainfall variability and soil type heterogeneity is shown to lead to the so-called "inverse texture effect" (e.g., Noy-Meir, 1973), whereby the optimal soil texture for a given vegetation type varies with annual rainfall amount. It is suggested that the interaction between observed climatic variability and soil texture heterogeneity may explain the dominance of a single species at this site.

4. The analysis of the rainfall statistical characteristics along the Kalahari precipitation gradient shows that the rainfall gradient is mostly due to a decrease in the mean rate of storm arrivals rather than to a change in the mean storm depth. Using this information and typical vegetation and soil parameters, the soil moisture and water stress model previously developed relates the vegetation properties along the transect with those of climate and soil, including the possibility of tree–grass coexistence in the central sector of the Kalahari.

5. Finally, the ecosystem of a blue oak Californian savanna offers an opportunity to investigate the soil moisture and water stress dynamics in a typical Mediterranean climate in which the rainfall input is in counter-phase with the growing season of both trees and grasses. The coexistence in this case seems to be achieved by a temporal separation of water use, in which the shallow-rooted grasses first take advantage of the abundant soil moisture during late winter and then let the deeper-rooted oaks exploit the remaining soil moisture during the early warm season.

5.1 The role of vegetation in the Nylsvley savanna

Since transpiration is the most important loss of moisture in arid and semi-arid ecosystems, vegetation assumes a dominant role in regulating the local soil water balance as well as in determining its own conditions. Because of the relative uniformity of its soil and climate characteristics, the Nylsvley savanna has been chosen to focus on the differences among the water use strategies of

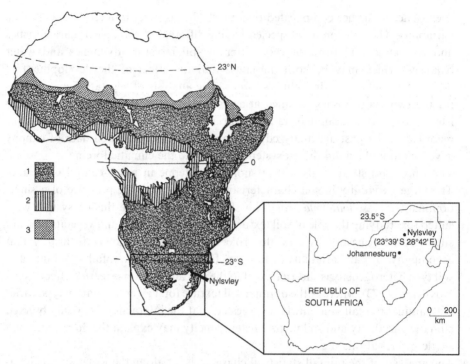

Figure 5.1 The location of the Savanna Biome research site at Nylsvley, South Africa, within the savannas of Africa; (1) areas having > 80% of the vegetation as savanna, (2) transition zones between savannas and forests, and (3) savannas and arid shrublands with 20–80% savanna by area. Adapted from Scholes and Walker (1993).

functionally different vegetation types. The detailed and systematic research carried out in the Nylsvley region under the African Savanna Biome Program (23°39′ S, 28°42′ E) constitutes an important benchmark in savanna studies (see Figure 5.1). All the data for climate, soil, and vegetation used here are taken from the book by Scholes and Walker (1993).

5.1.1 Description of the site

Practically all (98%) the annual precipitation at Nylsvley occurs from September to April, with an average of 60 cm year^{-1}. Mean air temperature ranges from 12.6 ° C in July to 23.0 ° C in January. The net solar radiation goes from 175 W m^{-2} in July to 280 W m^{-2} in January (Scholes and Walker, 1993, page 27). The climate is thus characterized by a warm and dry dormant season followed by a hot and humid growing season. The latter extends approximately from September to April, with a total duration of about 240 days.

Rainfall mostly occurs as convective storms of high intensity and short duration. The rainfall events are well approximated by a Poisson process with arrival rate $\lambda = 0.167$ storms per day and storm depth exponentially distributed with mean $\alpha = 1.5$ cm (Rodríguez-Iturbe et al., 1999c).

The soil in the Nylsvley region is mainly sandy (85–90% of sand) with average porosity of 0.42 and mean saturated hydraulic conductivity $K_s = 109.8$ cm day^{-1}. Effective soil depth for all species is uniform in approximately the range 80–120 cm. The soil water retention curves estimated for the site by Scholes and Walker (1993, page 61) can be used to transform the values of soil water potential to values of soil moisture. Sandy soils generally have a high mean pore size so that adhesion forces acting on water are quite weak, and thus very low soil moisture content can be attained with a not very negative soil matrix potential (Figure 2.6). The relative soil moisture estimated at field capacity, s_{fc}, is 0.29 and the hygroscopic point, s_h, calculated from the retention curves with a matrix potential of -10 MPa, is 0.048.

The plant parameters used in the following analyses are those corresponding to the broad-leafed savannas studied by Scholes and Walker (1993), where the average total canopy coverage of woody plants is 32%, while the total aerial cover of herbaceous species is around 30%, with relevant variations from year to year. The vegetation characteristics are typical of a well-developed savanna ecosystem, with a consistent coverage of trees and grasses and a relevant presence of unvegetated soil. Two types of trees and two types of grasses represent most of the vegetation of this region. As shown in Table 5.1, the Nylsvley broad-leafed savanna is dominated by *Burkea africana* in the tree layer and *Eragostris pallens* in the grass layer, in terms of both canopy (or aerial) cover and biomass (or basal cover).

Table 5.1 *Biomass and canopy cover characteristics for the dominant vegetation in the broad-leafed savanna at Nylsvley (from Scholes and Walker, 1993).*

Woody species	Biomass (% of total woody plants)	Canopy cover (% of total woody plants)
Burkea africana	53.4	40.9
Ochna pulchra	13.1	14.7

Herbaceous species	Basal cover (% of total grasses)	Aerial cover (% of total grasses)
Eragostris pallens	29.0	22.5
Digitaria eriantha	26.6	6.6

Table 5.2 *Parameters characterizing the water use of the four species considered at Nylsvley (data in columns 1–4 from Scholes and Walker, 1993).*

Species	Transpiration reduced below		Permanent wilting point		Maximum daily evapotranspiration
	Ψ_x (kPa)	s^*	Ψ_s (MPa)	s_w	E_{max} (cm day^{-1})
Burkea africana	−71	0.11	−3.1	0.060	0.47
Ochna pulchra	−34	0.14	−3.2	0.058	0.39
Eragostris pallens	−29	0.15	−3.9	0.052	0.61
Digitaria eriantha	−13	0.22	−2.9	0.061	0.47

Table 5.2 shows some parameters characterizing the water use of the four species. The low values of the wilting points reflect a good adaptation of the vegetation to the recurrent drought periods. Interestingly, the permanent wilting point for grasses is, on average, lower than that for trees, while the opposite is true for the soil moisture content below which vegetation experiences water stress, s^*. The maximum daily average evapotranspiration E_{max} is a troublesome parameter to estimate because of the lack of direct measurements. Scholes and Walker (1993, page 69) report typical values of maximum instantaneous evapotranspiration; such values are typical of an eight-hour transpiration period during the peak of the growing season after the occurrence of sizeable rainfall events. These values are corrected in Table 5.2 to obtain a characterization valid for a 24-hour period representative of average conditions during the growing season. After such corrections, the values of E_{max} are consistent with typical values at Nylsvley (0.45 cm day^{-1} for trees and 0.5 cm day^{-1} for grasses; R. Scholes, CSIR-Pretoria, personal communication, 1999). Notice that they also include the rate of soil evaporation, which is normally higher for grassy soils than for shaded soils below woody-plant canopy. Finally, canopy interception at Nylsvley can be modeled as explained in Section 2.1.2 by subtracting a fixed amount to the rainfall depth of each storm event. From measurements at Nylsvley we have used $\Delta_g = 0.1$ cm for grasses and $\Delta_t = 0.2$ cm for trees (Scholes and Walker, 1993, page 66).

5.1.2 *Soil moisture probability density function and water balance*

The data available for the Nylsvley savanna allow the application of the stochastic model for soil moisture dynamics proposed in Chapter 2. The

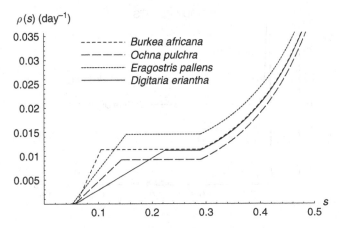

Figure 5.2 Normalized loss functions of the four species considered at Nylsvley, Eq. (2.18). The vegetation parameters are reported in Table 5.2 and the soil texture is characterized by $n = 0.42$, $K_s = 109.8$ cm day^{-1}, $b = 2.25$, $s_h = 0.048$, and $s_{fc} = 0.29$. The soil depth is 100 cm for all the species, the evapotranspiration at the wilting point, E_w, varies slightly for the different species from 0.003 cm day^{-1} to 0.011 cm day^{-1}, due to the differences in s_w. After Laio et al. (2001b).

shape of the normalized loss function (e.g., losses divided by nZ_n) resulting from the data in Table 5.2 is shown in Figure 5.2 for the four considered species. Using the same parameters related to soil properties, it is clear how the different plant characteristics (through s_w, s^*, and E_{max}) significantly affect the soil moisture losses.

The soil moisture steady-state pdf can be calculated for all four species under study using Eq. (2.39). The results are shown in Figure 5.3. Due to the sandy nature of the soil, very low values of s are most probable in all cases. This, however, does not necessarily imply that the available water for plants is proportionally less: in a sandy soil not only the pdf, but also s_w and s^* drop to very low values, thus shifting the entire soil moisture process to drier states of moisture content, as can also be seen from the mean soil moisture values $\langle s \rangle$ reported in Figure 5.3. This, combined with the fact that the soil moisture at the end of the dormant season is usually very low (due to the dry and warm winters), leads to a very fast transient phase at the beginning of the growing season, so that the impact of the soil moisture's initial condition is practically irrelevant. In these conditions, the steady-state pdf of soil moisture is expected to characterize most, if not all, of the growing season.

Figure 5.3 shows the large differences among the pdf's of the different plants. The mode in all cases corresponds to low values of s, especially for *Burkea africana* and *Eragostris pallens*. The major variations stem from the

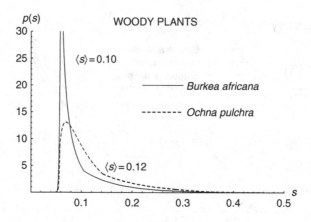

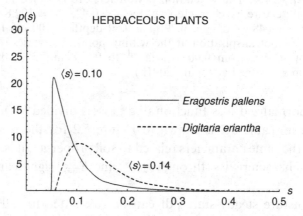

Figure 5.3 Soil moisture probability density functions for the four species analyzed at Nylsvley. The mean depth of a rainfall event is $\alpha = 1.5\,\text{cm}$, the frequency of the rainfall events $\lambda = 0.167\,\text{day}^{-1}$, and canopy interception is $\Delta_t = 0.2\,\text{cm}$ for woody plants and $\Delta_g = 0.1\,\text{cm}$ for herbaceous plants. See Figure 5.2 for other parameters. After Laio et al. (2001b).

differences in E_{max} and s^*, which are responsible for the close link between the slope of $\rho(s)$ in the range $\{s_w, s^*\}$ and the relative importance of the mode of the pdf. In particular, *Burkea africana* has a greater E_{max} and a lower value of s^* than *Ochna pulchra* (Table 5.2) and this is reflected in a higher slope of $\rho(s)$ and greater losses for *Burkea africana* for any value of s (see Figure 5.2), which greatly increases the probability of having very low values of soil moisture. Similar considerations apply to the herbaceous plants: *Eragostris pallens* has much greater evapotranspiration in unstressed conditions than *Digitaria eriantha*, as opposed to the value of s^*, which is much higher for the latter.

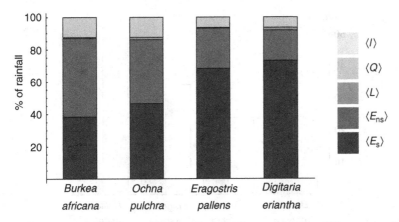

Figure 5.4 Water balance for sites uniformly covered by the different species. The incoming rainfall is partitioned among canopy interception $\langle I \rangle$, runoff $\langle Q \rangle$, leakage $\langle L \rangle$, unstressed evapotranspiration $\langle E_{ns} \rangle$, and evapotranspiration under stress $\langle E_s \rangle$. See Figures 5.2 and 5.3 for the parameter values. After Laio et al. (2001b).

The water balance for each species is shown in Figure 5.4, where the bars respectively show, from bottom to top, the amount of evapotranspiration under stress (e.g., when $s < s^*$), evapotranspiration in unstressed conditions, leakage, runoff, and canopy interception. The sandy soil and gentle topography at Nylsvley result in almost no leakage and negligible runoff (Scholes and Walker, 1993, page 63). Canopy interception is more relevant for woody species than for herbaceous plants, but the key differences in Figure 5.4 arise in the partitioning of evapotranspiration under stressed and unstressed conditions (see Section 2.4). Grasses have a considerably larger percentage of transpiration under stress, because of their higher E_{max} and lower s^* (i.e., they begin the stomatal closure process at higher values of the matrix potential).

The productivity and the capacity for growth and reproduction of a plant are often linked to the amount of water transpired (Hsiao, 1973; Bradford and Hsiao, 1982; Nilsen and Orcutt, 1998), especially under unstressed conditions (e.g., Jones, 1992; Kramer and Boyer, 1995; see also Chapter 6). With this criterion, one would conclude from Figure 5.4 that trees tend to be favored over grasses; for example, *Burkea africana*, thanks to its very low value of s^* (see Table 5.2), appears to be the most suitable plant in terms of growth and reproduction rates, despite its higher probability of undergoing very low values of soil moisture (Figure 5.3). The global picture, however, is much more complex and further insights will be provided by the analysis of soil moisture crossing properties and the resulting water stress.

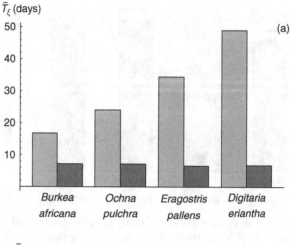

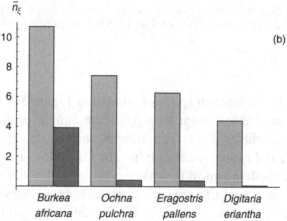

Figure 5.5 Crossing properties of the soil moisture process for the four species analyzed at Nylsvley. (a) Mean duration of an excursion below ζ_{s^*} (light gray) and below ζ_{s_w} (dark gray). (b) Mean number of downcrossings of ζ_{s^*} (light gray) and ζ_{s_w} (dark gray). The duration of the growing season is $T_{seas} = 240$ days. See Figures 5.2 and 5.3 for other parameters. After Laio et al. (2001b).

5.1.3 Soil moisture crossing properties and vegetation water stress

The analysis of the soil moisture crossing properties and plant water stress (Chapters 3 and 4) is now applied to the four main species of the Nylsvley savanna. Figures 5.5a and 5.5b show the mean duration and the mean number of excursions below s^* and s_w. Regarding the threshold s^* (light gray in Figure 5.5), the woody plants, in particular *Burkea africana*, present lower values of $\overline{T}_{s^*}$ but higher frequencies of crossings, while the opposite is true for herbaceous plants, in particular for *Digitaria eriantha*. The differences are quite relevant, with $\overline{T}_{s^*}$ ranging from 17 to 49 days and $\overline{n}_{s^*}$ ranging from 10.7 to 4.5 crossings per growing season. Climate, soil texture, and soil depth being equal, the reason for this behavior is found again in the interplay between E_{max} and s^*. An increase of the former contributes to reduce soil moisture, while at the same time (as the mean is typically below the stress level) a lower s^* reduces the distance $s^* - \langle s \rangle$ (this can be seen by comparing the values of $\langle s \rangle$ in Figure 5.3 with those of s^* in Table 5.2). Thus in the case of *Burkea africana* the trajectory evolves around s^*, with frequent periods of stress of short duration. Differences in E_{max} are also relevant to the explanation of why *Eragostris pallens* has a greater $\overline{T}_{s^*}$ than *Ochna pulchra*, even if their s^* values are similar. A higher E_{max} leads to a soil moisture trajectory further below s^* so that a subsequent rainfall event has a smaller probability of inducing a crossing of s^* (i.e., $\overline{T}_{s^*}$ becomes greater and $\overline{n}_{s^*}$ lower).

The crossing properties of the wilting point (dark gray in Figure 5.5) also show some interesting features. The average excursion below s_w lasts six to seven days for all the plants, while the number of crossings is very different from one case to the other, ranging from four crossings per growing season for *Burkea africana* to 0.07 for *Digitaria eriantha*. Below the wilting point the remaining soil water is lost by evaporation, whose rate is assumed to be independent of plant species (e.g., we do not consider shadowing effects). Since the wilting points are similar in all of the four species, the trajectories below s_w have similar characteristics and this explains the close values of $\overline{T}_{s_w}$. The difference between s_w and $\langle s \rangle$ controls the probability of crossings. *Burkea africana*, with s_w and $\langle s \rangle$ closest to each other, has the largest mean number of crossings, $\overline{n}_{s_w}$.

Figure 5.6 shows the differences in the static water stress $\overline{\zeta'}$ (see Section 4.2) among the various species (light gray for $q = 1$ and dark gray for $q = 3$). Considerations similar to those made for $\overline{T}_{s^*}$ and $\overline{n}_{s^*}$ apply also to $\overline{\zeta'}$: the higher E_{max}, the faster the decay towards s_w (high values of $\overline{\zeta'}$). Analogously, the closer s^* is to s_w, the faster is the approach to high values of plant water stress. As discussed in Chapter 4, lower values of $\overline{\zeta'}$ are generally obtained with higher values of q. However, the effect of changes in q tends to be greater for grasses than for trees, due to the smaller difference between s_w and s^* for trees. It is important to remark again that a higher value of $\overline{\zeta'}$ does not necessarily correspond to a greater average stress during the growing season, but only refers to the fact that, on average, when the plant is under stress the stress is stronger.

A more comprehensive measure of the average vegetation water stress during the growing season is given by the dynamic water stress $\overline{\theta}$ defined in Section 4.3. Figure 5.7 shows the value of $\overline{\theta}$ for the four species considered at

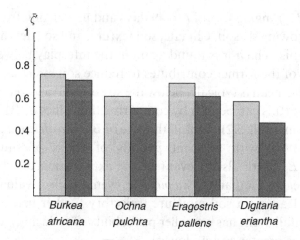

Figure 5.6 Static water stress $\overline{\zeta'}$ of the four species analyzed at Nylsvley (light gray, $q = 1$; dark gray, $q = 3$). See Figures 5.2 and 5.3 for other parameters. After Laio et al. (2001b).

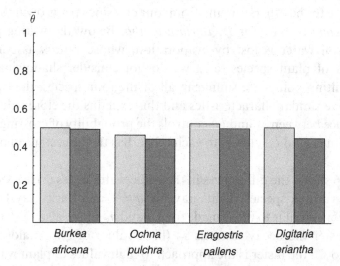

Figure 5.7 Dynamic water stress, $\overline{\theta}$, of the four species analyzed at Nylsvley (light gray, $q = 1$; dark gray, $q = 3$). $T_{seas} = 240$ days, $k = 0.5$. See Figures 5.2 and 5.3 for other parameters. After Laio et al. (2001b).

Nylsvley. The large differences found for $\overline{T}_{s^*}$, $\overline{n}_{s^*}$, and $\overline{\zeta'}$ among the different species are now integrated in the definition of $\overline{\theta}$. The dynamic water stress ranges from 0.46 to 0.52 for $q = 1$ and from 0.44 to 0.49 for $q = 3$. The values of the dynamic stress are very similar among the species with the larger $\overline{\theta}$'s resulting from the larger values of E_{max} (e.g., *Eragostris pallens*). The low variability in $\overline{\theta}$ among the different species may help to explain their coexistence at Nylsvley and show how different approaches to water use by plants can lead to similar conditions of stress. If the dynamic stress conditions had been very different among the four species, only mechanisms other than water stress, such as access to nutrients or the effect of external disturbances like fire and grazing, could have explained the coexistence observed at Nylsvley. Without neglecting the great importance of these mechanisms, the fact that the values of $\bar{\theta}$ are very similar (despite the differences in $\bar{T}_{s^*}$, $\bar{n}_{s^*}$, and $\overline{\zeta'}$) shows that coexistence may be possible at Nylsvley because of specific patterns of water use among the various species. Such an explanation is not linked to the so-called Walter hypothesis (Walter, 1971), which assumes that trees and grasses have access to water in different layers. The rooting depth, Z_r, is not very different for trees and grasses at Nylsvley, and the effect of possible variations of the soil depth in the range reported by Scholes and Walker (1993) does not seem to be very important. In fact similar values of Z_r among the species led Scholes and Walker (1993) to seek explanations other than the Walter hypothesis for the coexistence of trees and grasses at Nylsvley.

It is important to note that Figure 5.7 by itself is not sufficient to explain coexistence. It only states that all four considered species are similarly suited to the soil–climate conditions at Nylsvley. The actual coexistence ultimately depends on interaction and competition among the plants in the use of the water resource, as well as on the evolutionary dynamics of the ecosystem which is linked to rates of growth, reproduction, and colonization. This will be studied in detail in Chapter 11.

5.2 Sensitivity to climate fluctuations in a southern Texas savanna

Savanna ecosystems are particularly sensitive to changes in external factors, such as climatic forcing, grazing, fire, or human interference. Rainfall is often the most relevant one: the manner in how different functional vegetation types respond to interannual rainfall variability may be crucially important for explaining the observed changes in the densities of woody plants and herbaceous vegetation. In this regard, Archer et al. (1988) point out that the Rio Grande Plains of southern Texas are an interesting example of the grassland-to-shrubland conversion process, where "anthropogenic disturbances, reduced fire frequencies, and drought have presumably interacted, perhaps against a backdrop of gradual climatic change, enabling mesquite and other woody plants to increase in stature and density on grasslands" (see Plates 17–18).

This ecosystem has been systematically studied through careful field measurements carried out by the Texas Agriculture Experiment Station in La Copita Research Area (27°40′ N, 98°12′ W). The focus here is to investigate

how different rooting depths and plant physiological characteristics may lead
to different responses when undergoing interannual changes in the frequency
and amount of rainfall during the growing season. The same ecosystem will be
further analyzed in Chapters 7, 8, and 11.

5.2.1 Description of the site

The region has an elevation ranging between 75 m and 90 m above sea level,
with very mild slopes in a flat landscape (Scifres and Koerth, 1987). It is
covered by a diphase tree–grass vegetation with a population of woody plants
consisting mostly of *Prosopis glandulosa* (Honey mesquite) coexisting with C_4
grasses, mainly *Paspalum setaceum* (Scifres and Koerth, 1987; Archer et al.,
1988; Brown and Archer, 1990).

The region has a subtropical climate with hot summers and warm winters,
an average annual temperature of 22.4 ° C, and a mean annual rainfall of about
70 cm. Rain is much more evenly distributed through the year in the Rio
Grande Plains than in the Nylsvley, with approximately 38 cm to 40 cm during
the growing season from May through September. In this respect, the Rio
Grande Plains are not typical savannas, since most savannas have a concentrated wet season (Scholes and Walker, 1993). Much of the Rio Grande Plains
is presently dominated by a subtropical thorn woodland vegetation complex.
Although commonly classified as *Andropogon-Setaria-Prosopis-Acacia*
savanna (e.g., Diamond et al., 1987), its structure seems to belong better to
the tropical thicket, with dense, low-growing formations of woody plants. It is
known that savannas can be induced to develop a thicket structure through
poor management and overgrazing (Archer, 1994).

Figure 5.8 shows the fluctuations in annual rainfall experienced by the
region from 1931 to 1985. One observes large and persistent deviations from
the long-term mean rainfall over the region. The period 1942–60 was characterized by a severe drought, while the period 1961–82 was wet in terms of
total amount as well as in number of years with rainfall higher than average.
Archer et al. (1988) have documented the changes in the woody plant cover in
the region during those periods. The total woody plant coverage increased
from 8% in 1960 to 36% in 1983. In 1941 the woody plant coverage was 13%.
The values of α and λ for the statistical description of rainfall during the
growing season (150 days, from May through September) were obtained from
the rain station at Alice, Texas, from website www.ncdc.noaa.gov/pub/data/
coop-precip/texas.txt. For the dry period 1950–60 the rainfall parameters are
$\lambda_{dry} = 0.166$ day^{-1} and $\alpha_{dry} = 1.342$ cm, while for the wet period 1973–83 they
are $\lambda_{wet} = 0.202$ day^{-1} and $\alpha_{wet} = 1.417$ cm. These values correspond to an

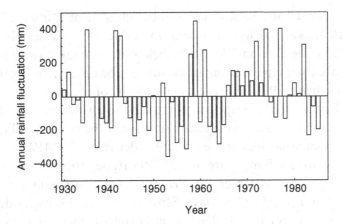

Figure 5.8 Fluctuations around the mean annual rainfall at Alice (Texas) during the period 1931–85. After Archer et al. (1988).

average precipitation per growing season of $\Theta_{dry} = 33.3\,\text{cm}$ and $\Theta_{wet} = 42.9\,\text{cm}$, respectively.

The soil is generally characterized by an A horizon of fine sandy loam with an average porosity of 0.43. Its depth varies from place to place in close relation to the type of vegetation present (see the discussion on rooting depth below). The saturated hydraulic conductivity is log-normally distributed with a geometric mean of 82.2 cm day^{-1} (U.S. Department of Agriculture, 1979; see also Clapp and Hornberger, 1978). The relative soil moisture at field capacity is estimated as $s_{fc} = 0.56$ and the hygroscopic point is $s_h = 0.14$ (see Table 2.1).

The woody vegetation (e.g., *Prosopis glandulosa*) and the C$_4$ grasses (e.g., *Paspalum setaceum*) are characterized by soil matrix potentials at the permanent wilting point of $-3.2\,\text{MPa}$ and $-4.5\,\text{MPa}$, respectively (Haas and Dodd, 1972; Ludlow, 1976). Those soil water potentials correspond to relative soil moistures of 0.18 and 0.167, respectively, where the conversion is made through the retention curves defined in Section 2.1.5. The maximum daily transpiration rate measured for *Prosopis glandulosa* is 3.42 mm day^{-1} (Wan and Sosebee, 1991; Cuomo et al., 1992). For grasses, it is estimated to be 10% more than that of *Prosopis glandulosa*. The evaporation from vegetated soil is taken as 1 mm day^{-1}. The interception losses are assumed equal to those of the Nylsvley case, i.e., 0.2 cm per event for trees and 0.1 cm per event for grasses.

The estimation of the level of relative soil moisture at which transpiration starts being reduced is not simple. *Prosopis glandulosa* is reported to be an "extravagant" user of the soil water resource (Haas and Dodd, 1972), having different possible strategies of drought adaptation. Moreover, it also presents a quite pronounced mechanism of osmotic adjustment (Nilsen et al., 1983, 1984), which makes the use of a constant value of s^* valid only as a first approximation (see Section 7.3). Nevertheless, a reasonable value for the soil matrix potential at incipient stomatal closure can be obtained using the data by Wan and Sosebee (1991). This is estimated to be of the order of -0.12 MPa, which for a loamy sand corresponds to a relative soil water content of 0.35. For the most common C_4 grass (*Paspalum setaceum*), the corresponding value is found to be somewhat higher, i.e., of the order of -0.09 MPa, which corresponds to a relative soil moisture of 0.37 (Rodríguez-Iturbe et al., 1999c).

The rooting depth is larger for *Prosopis glandulosa* than for *Paspalum setaceum*. Below *Prosopis glandulosa*, Stroh et al. (1996) show evidence of an A horizon layer as deep as 1 m. They do not explicitly mention a precise range of values for the rooting depth for grasses, but indicate somewhat shallower soils below grasses. *Prosopis glandulosa* is often reported to be a phreatophyte, which may exploit moisture from the water table by developing very deep tap roots. Although at La Copita *Prosopis glandulosa* does not seem to resort to soil moisture from the water table (as opposed to other semi-arid regions in Arizona and California), the tendency to have deep roots is certainly a typical characteristic of this species. In what follows, a value of $Z_r = 1$ m will be used for *Prosopis glandulosa* and $Z_r = 40$ cm for grasses. Employing slightly different values of Z_r, the results do not qualitatively change, as long as a consistent difference in rooting depth between trees and grasses is maintained.

5.2.2 Soil moisture probability density function and water balance

Figure 5.9 shows a comparison between the normalized loss function, $\rho(s)$, for sites with *Prosopis glandulosa* and *Paspalum setaceum*. Despite the presence of the same type of soil, the two loss functions look quite different, both because of the differences in the plant physiological characteristics and because of the different active soil depths. In particular, the normalization of the loss function by nZ_r makes the effective losses in shallow-rooted sites considerably higher than those in deeper-rooted sites.

The resulting soil moisture pdf's for the two species in the case of wet and dry climates are presented in Figure 5.10. The effect of the different climates is relevant, but quite similar for the two functional vegetation types. Regarding the differences between trees and grasses, it appears that grasses tend to be more frequently at low soil moisture levels. At the same time, however, they also have a fatter right tail in their pdf, due to the more frequent soil moisture excursions at relatively high soil moisture levels. The results are consistent with Figure 2.9, where the trace of soil moisture for deeper soils was found to be

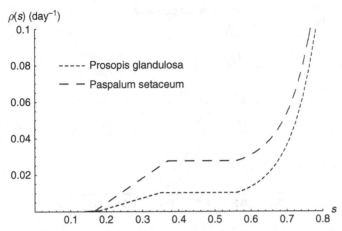

Figure 5.9 Normalized loss functions for the two species considered at La Copita (Texas) (Eq. (2.18)). The soil is a loamy sand with $n = 0.43$, $K_s = 82.2$ cm day^{-1}, $b = 4.9$, $s_h = 0.14$, and $s_{fc} = 0.56$. *Prosopis glandulosa* is characterized by $E_{max} = 0.442$ cm day^{-1}, $s^* = 0.35$, $s_w = 0.18$, $Z_r = 100$ cm, and $E_w = 0.02$ cm day^{-1}. *Paspalum setaceum* has $E_{max} = 0.476$ cm day^{-1}, $s^* = 0.37$, $s_w = 0.167$, $Z_r = 40$ cm, and $E_w = 0.013$ cm day^{-1}. After Laio et al. (2001b).

much less fluctuating than that of shallow soils. The transpiration regime also contributes to enhance the discrepancies in soil moisture dynamics through the values of E_{max} and s^*. As a result, the soil moisture in grassy soils decays more quickly and grasses experience wilting more frequently. Their shallower rooting zone, however, is brought to favorable soil moisture conditions even by very small rainfall events. On the other hand, the greater storage capacity of sites dominated by *Prosopis glandulosa* together with their more parsimonious transpiration regime prevent trees from going under high stress too frequently. As a consequence, mesquites hardly ever experience high levels of soil moisture, and are most often under conditions of water stress, although these are only occasionally extreme.

The previous differences in soil moisture dynamics are also related to the water use efficiency (i.e., total dry matter produced by plants per unit of water used) of trees and grasses. Mesquite is a C$_3$ plant with very low transpiration efficiency (e.g., McGinnes and Arnold, 1939; Nilsen et al., 1983; Wan and Sosebee, 1991), while C$_4$ grasses, such as *Paspalum setaceum*, have high water use efficiency and very quick response to rainfall events (e.g., Kemp and Williams, 1980; Sala and Lauenroth, 1982). In other words, *Prosopis glandulosa* is an extensive soil water user while *Paspalum setaceum* can be classified as an intensive water exploiter. Such physiological characteristics are perfectly suited to the soil water dynamics they contribute to produce: a low water use

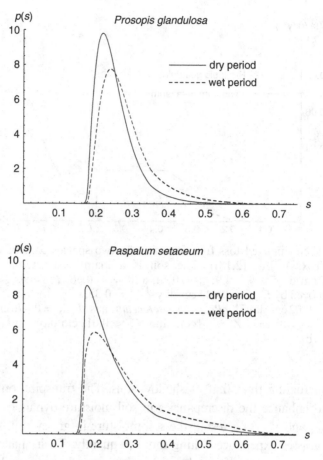

Figure 5.10 Soil moisture probability density function for *Prosopis glandulosa* and *Paspalum setaceum* at La Copita (Texas). For the dry period $\alpha_{dry} = 1.342$ cm, $\lambda_{dry} = 0.166$ day^{-1}; for the wet period $\alpha_{wet} = 1.417$ cm and $\lambda_{wet} = 0.202$ day^{-1}. Canopy interception is $\Delta_t = 0.2$ cm for *Prosopis glandulosa*, $\Delta_g = 0.1$ cm for *Paspalum setaceum*. See Figure 5.9 for the other parameter values. After Laio et al. (2001b).

efficiency for deep-rooted trees subject to slower soil moisture dynamics, and a prompt and efficient response to rapidly varying soil water dynamics for grasses (see Section 7.1). All the above represent an example of opposite strategies of adaptation to water stress (e.g., Grime, 1979). The diversification in the use of the soil water resource may be among the possible mechanisms for tree–grass coexistence in savannas (e.g., Scholes and Archer, 1997).

An indirect example of the differences in water use is provided by the temporal sequence of daily evapotranspiration rates for trees and grasses. As shown by Figure 5.11, grasses transpire at a higher rate soon after each

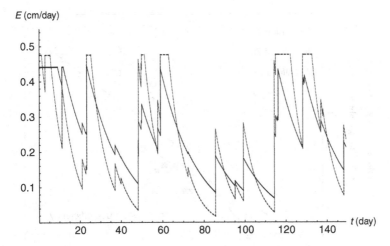

Figure 5.11 Temporal sequence of daily evapotranspiration rates for *Prosopis glandulosa* (continuous line) and *Paspalum setaceum* (dashed line). See Figures 5.9 and 5.10 (wet case) for parameter values. After Laio et al. (2001b).

precipitation event. However, since they quickly lose their available soil moisture, their transpiration rate rapidly drops below that of trees, which instead keep transpiring longer after the rainfall event, thanks to the greater soil moisture storage and slower rate of maximum transpiration.

Further differences among the species may be investigated using empirical relationships between transpiration and productivity. For crop species, Kramer and Boyer (1995) report a linear relationship between total shoot dry mass and amount of water transpired (see Figure 5.12). The greater slope of this relationship (which is precisely the water use efficiency) for C_4 species implies higher productivities than C_3 species for high transpiration rates and lower productivities in the case of low transpiration rates (see Section 2.1.5). Unfortunately, data on the productivity of different functional vegetation types for real environments are scarce and also depend on other factors, such as external disturbances and competition among species. Nevertheless, it is reasonable to believe that similar relationships also hold for trees and grasses in the southern Texan savanna. It is also likely that for natural ecosystems any relationship between productivity and transpiration is more complicated than for agricultural crops. In particular, the distinction between total transpiration and transpiration under unstressed conditions (see Sections 2.1.5 and 2.4) might play an important role in nonirrigated, water-controlled ecosystems. Further analysis on this argument will be given in Section 7.1.

The comparison of the water balance between the dry and wet growing seasons is presented in Figure 5.13. As is evident from the relative values of

Applications to natural ecosystems

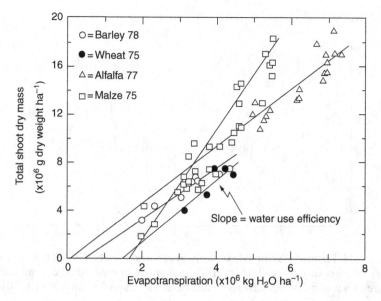

Figure 5.12 Production of aboveground shoot dry matter at various levels of water use in several crops near Logan, Utah. The slope of the linear relation is the water use efficiency, which corresponds to 2.11 g of dry matter per kg of water for barley, 2.50 for wheat, 2.36 for alfalfa, and 4.49 for maize. Note that maize is a C_4 plant and the others are C_3 plants. After Kramer and Boyer (1995).

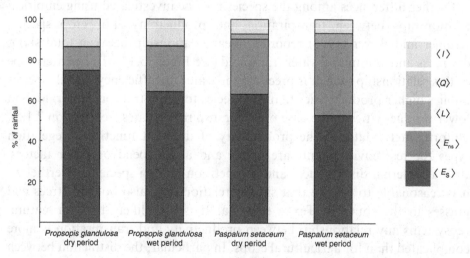

Figure 5.13 Water balance for sites uniformly covered by *Prosopis glandulosa* or *Paspalum setaceum* in dry and in wet growing season periods. Incoming rainfall is partitioned among canopy interception $\langle I \rangle$, runoff $\langle Q \rangle$, leakage $\langle L \rangle$, unstressed evapotranspiration $\langle E_{ns} \rangle$, and evapotranspiration under stress $\langle E_s \rangle$. See Figures 5.9 and 5.10 for the parameter values. After Laio et al. (2001b).

transpiration under stress, in both cases most of the precipitated water is transpired with the stomata partially closed. The increase in average precipitation, although not changing the total fraction of water transpired, produces a consistent increase in the fraction of transpiration under unstressed conditions, especially in the case of mesquite. This higher transpiration is likely related to changes in productivity, which in turn is likely to be responsible for the reported increase in the overall fraction of canopy coverage of mesquite during the wetter period. As far as the other components of the water balance are concerned, runoff production is practically absent; leakage losses are also very small for deep soils, while they become appreciable for shallow soils in the wet climate. Canopy interception is responsible for the loss of 10% to 15% of the incoming rainfall for trees and of 5% to 10% in the case of grasses.

5.2.3 Soil moisture crossing properties and vegetation water stress

Figure 5.14 presents the result of the soil moisture crossing analysis during the growing season for trees and grasses in the case of wet and dry climates. The time of permanence under the wilting point does not appear to be very sensitive to rainfall changes, for trees or for grasses. Differently, the duration of stress periods (i.e., excursions below s^*) displays strong sensitivity to climate. This is particularly evident for mesquite, which in the case of dry climates suffers water stress for more than half of the growing season. The number of crossings of s^* is also different between trees and grasses as well as quite sensitive to different climates. The static water stress (not shown) is somewhat lower for mesquite especially for the wet years; this is because the trace of soil moisture, although being below s^* for most of the time, seldom reaches values as low as the wilting point. The value of the exponent q in the definition of the static stress (Section 4.2) does not qualitatively affect the results of the analysis.

Notwithstanding the remarkable differences in the duration and frequency of periods of water stress (Figure 5.14), the values of dynamic stress shown in Figure 5.15, are relatively close for different species under the same climatic characteristics, although *Paspalum setaceum* has a lower dynamic stress during the dry period. This is very important for the issue of tree–grass coexistence, because it again shows that even with very different temporal dynamics of water stress the two species are similarly suited to this particular environment. The second observation concerns the notable difference in the sensitivity to climatic changes between trees and grasses. The dynamic water stress of the *Prosopis glandulosa* changes considerably for the two different climatic conditions, while that of *Paspalum setaceum* is much less sensitive to climate

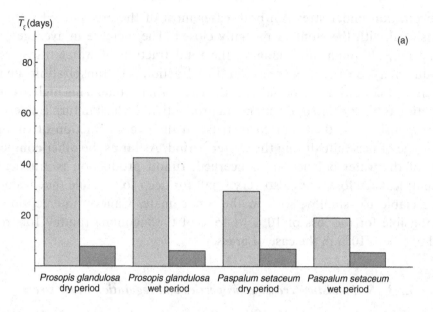

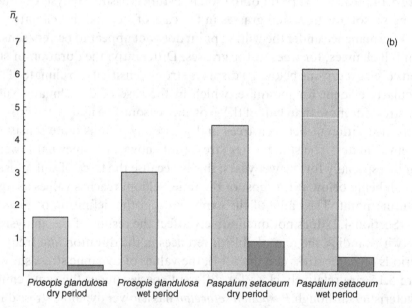

Figure 5.14 Crossing properties of the soil moisture process for the two species at La Copita (Texas) during the dry and wet periods. (a) Mean duration of an excursion below s^* (light gray) and below s_w (dark gray). (b) Mean number of downcrossings of s^* (light gray) and s_w (dark gray). The duration of the growing season is $T_{seas} = 150$ days. See Figures 5.9 and 5.10 for other parameters. After Laio et al. (2001b).

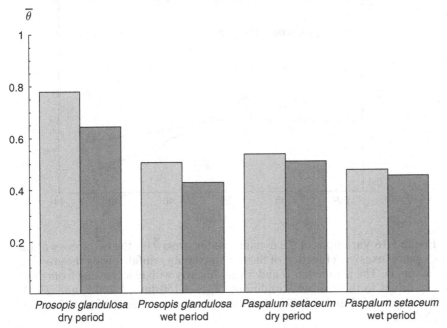

Figure 5.15 Dynamic water stress $\bar{\theta}$ of the two species at La Copita (Texas) corresponding to dry and wet growing season periods (light gray, $q = 1$; dark gray, $q = 3$). $T_{seas} = 150$ days, $k = 0.5$. See Figures 5.9 and 5.10 for other parameters. After Laio et al. (2001b).

fluctuations. As the rainfall increases, there is an inversion in the relative condition of the two species: during the dry period grasses are in better relative condition, but when the rainfall is more abundant the environment becomes more favorable for trees. This is evident in Figure 5.16, where the dynamic water stress is computed for trees and grasses as a function of the amount of rainfall per growing season. The computation is made by linearly increasing the values of α and λ from their dry values to those of wet conditions. Interestingly, for total rainfall above 39 cm trees become less stressed than grasses and the historical long-term average rainfall during the growing season is precisely around that value. Thus the rainfall variability acts as an external forcing that randomly drives the system towards different vegetation conditions: dry periods drive the ecosystem towards a reduction in canopy coverage, while wet periods favor tree encroachment. Due to the interannual persistence of dry and wet conditions, the ecosystem fluctuates over the years between tree-domination and grass-domination. It is important to note that the values of canopy coverage depend on the response time of trees and grasses to favorable or unfavorable conditions as well as on the actual productivity of

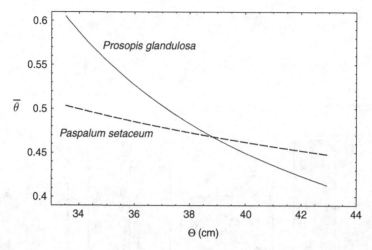

Figure 5.16 Variations of the dynamic water stress $\bar{\theta}$ for the two species at La Copita (Texas) as a function of the total incoming rainfall during the growing season, Θ. The parameters α and λ vary linearly with Θ increasing from their dry values to those of wet conditions. $T_{seas} = 150$ days, $k = 0.5$, and $q = 3$. See Figures 5.9 and 5.10 for the other parameter values. After Laio et al. (2001b).

the species and that coexistence can be stabilized by competition and interaction among species (Scholes and Archer, 1997).

5.3 The role of soil texture in the Colorado shortgrass steppe

Soil texture has been shown to influence patterns of vegetation structure in water-controlled ecosystems through its impact on soil water availability (e.g., Noy-Meir, 1973). The shortgrass steppe in north-central Colorado, with its large variability in soil texture, is an interesting case study for investigating the role of soil texture on soil moisture dynamics and the water stress of its dominant species, *Bouteloua gracilis* (see Plate 16). This is especially important in an area where water availability is the key variable driving the structure and function of the ecosystem. Other studies in this region have focused on the influence of soil texture on the probability of recruitment of *Bouteloua gracilis* (Lauenroth et al., 1994) or on soil water storage patterns within different soil texture sites (Singh et al., 1998). The analytical framework developed in the previous chapters provides a useful tool for assessing the condition of *Bouteloua gracilis* under different climatic and soil texture conditions at a site where drought, together with herbivores, have a long history as selection factors. Noy-Meir (1973) introduced the "inverse soil texture effect" as a diagnostic of arid and semi-arid ecosystems. According to his interpretation, climate and soil texture interact to give rise to different patterns of soil water availability with the result that the same plant can occur at lower rainfall conditions on coarse soils and at higher rainfall on fine soils. The following analysis, taken from Laio et al. (2001b), looks into the occurrence of the inverse soil texture effect as a possible explanation for the dominance over time of *Bouteloua gracilis* at this site. This will be further studied in Chapter 7.

5.3.1 Description of the site

The region of study is the shortgrass steppe Long Term Ecological Research (LTER) site, located on the Central Plains Experimental Range (CPER) in north-central Colorado (40°49′ N, 104°46′ W). Information and data can be found on the project home page: http://sgs.cnr.colostate.edu. The C_4 perennial bunchgrass blue grama (*Bouteloua gracilis*) dominates the shortgrass steppe region of the Great Plains of North America. Other major perennial species at CPER include forbs such as *Sphaeralcea ciccinea*, succulents such as *Opuntia polyacantha*, shrubs like *Artemisia frigida* and other grasses such as *Agropyron smithii* and *Buchloe dactyloides*.

The spatial heterogeneity of soils, geologic substrates, and landforms has been reported to be high within the CPER (Yonker et al., 1988). Although the predominant soil type at the CPER is sandy loam, more than 95% of the area has a sand content greater than 35%, while approximately 70% of the area has a sand content greater than 50% (Burke et al., 1999). To account for the effect of soil texture, Laio et al. (2001b) assigned soil properties to all the textural classes of the USDA soil texture triangle (U.S. Department of Agriculture, 1951; Cosby et al., 1984; see also Figure 7.4). The soil physical parameters such as porosity n, saturated hydraulic conductivity K_s, matrix potential at saturation $\overline{\Psi}_s$ and pore size distribution index b were obtained using the regressions in Cosby et al. (1984) which relate these quantities to soil texture. These regressions were extrapolated to include silty soils.

The climate of the area is semi-arid. Mean monthly temperatures range from −4° C to 22° C. Daily rainfall values from the 1943–94 record at the CPER indicate a mean annual precipitation at this site of 321 mm (SD = 88 mm) with a range from 107 mm to 588 mm. Annual precipitation arrives mainly as large convective storms during the summer months. The high interannual variability of this type of storm results in such a large range of annual precipitation values. Approximately 70% of the mean annual precipitation occurs during the April-to-September growing season (Lauenroth et al., 1978; Sala et al., 1982a). Due to the large range of annual precipitation totals, different climatic scenarios will be superimposed on the soil texture variability. For the purposes of analysis, a year with annual rainfall of one standard deviation below the 321 mm mean is defined as relatively dry, a year with annual rainfall equal to the 321 mm mean is average, and a year with annual rainfall of one standard deviation above the 321 mm mean is relatively wet. Correspondingly, the relatively dry 1974 had mean depth of rainfall events $\alpha_{dry} = 5.76$ mm and mean frequency of storm arrivals $\lambda_{dry} = 0.17$ day^{-1}; the average 1985 had $\alpha_{av} = 4.47$ mm and $\lambda_{av} = 0.29$ day^{-1} and the relatively wet 1957 had $\alpha_{wet} = 6.74$ mm and $\lambda_{wet} = 0.28$ day^{-1}.

The topography of this site is made up of gently rolling hills, low flat-topped terraces and ephemeral stream courses. The uplands make up about 50% of the region, the slopes 28%, while the lowlands are 22% of the area (Yonker et al., 1988). Although there exist clear topographical differences at the landscape scale within the CPER, results by Burke et al. (1999) determined that soil moisture was not significantly different among landscape positions within this site, as required for the application of the stochastic model of soil moisture dynamics presented in Chapter 2.

Bouteloua gracilis does not exceed a height of 3.5 cm for ungrazed conditions at the CPER (Burke et al., 1999). Because of this, rainfall interception for this species is very low. *Bouteloua gracilis* is characterized by a soil water potential at wilting, Ψ_{s,s_w}, of -4 MPa (Lauenroth et al., 1987). This value has been reported as the soil water potential at which new plants of *Bouteloua gracilis* will die. It is known, however, that *Bouteloua gracilis* is able to occasionally withstand much lower soil water potentials, e.g., -10 MPa, and still recover. The onset of water stress is estimated to occur at $\Psi_{s,s^*} = -0.1$ MPa (Sala et al., 1981). The soil water potential at the hygroscopic point is taken as $\Psi_{s,s_h} = -10$ MPa and the evaporation at wilting point is estimated as $E_w = 0.01$ cm day^{-1}. The average daily evapotranspiration rate measured under well-watered conditions is estimated at $E_{max} = 3.7$ mm day^{-1} (Lauenroth and Sims, 1976), while the rooting depth is taken to be $Z_r = 30$ cm, since about 75% of the root biomass in the area has been reported to be in the top 30 cm soil layer (Leetham and Milchunas, 1985; Liang et al., 1989).

5.3.2 *Soil moisture probability density function and water balance*

The study of the soil moisture pdf and water balance was focused on the following three soil classes: sand, clay, and silty loam. Each of these soil classes was characterized by the corresponding percentages of clay, sand, and silt defining the midpoint of the corresponding region in the USDA soil texture triangle (U.S. Department of Agriculture, 1951). These percentages are given

Table 5.3 *Soil and vegetation parameters for Bouteloua gracilis in sand, silty loam, and clay. After Laio et al. (2001b).*

Soil class	% sand	% clay	% silt	s_h	s_w	s^*	s_{fc}	K_s (cm day^{-1})	n	b
Sand	92	3	5	0.04	0.05	0.16	0.44	103	0.37	3.4
Clay	22	58	20	0.44	0.47	0.64	0.78	35	0.46	12.1
Silty loam	17	13	70	0.18	0.16	0.35	0.59	33	0.47	5.0

in Table 5.3. Figure 5.17 shows the shape of the normalized water loss function, $\rho(s)$, for *Bouteloua gracilis* on the three different types of soil being considered. The differences in $\rho(s)$ are due to the differences in the soil and vegetation parameters affected by the soil texture (see Table 5.3).

The steady-state pdf's of soil moisture for average rainfall conditions are shown in Figure 5.18. It is clear that significantly lower values of soil moisture are much more likely for sandy soils than for silty loam soils and clay soils. The pdf shifts to higher soil moisture values when increasing the clay content of the soil. This does not mean, however, that more water is available for plants in a clay or silty loam than in a sand; for water uptake in clay or silty loam, plants have in fact to overcome higher soil water potential as it is reflected by the higher values of s_w and s^*.

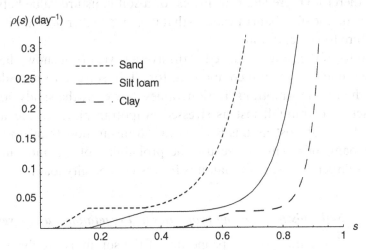

Figure 5.17 Normalized loss functions $\rho(s)$ for *Bouteloua gracilis* corresponding to three soil types considered at the CPER. Soil and vegetation parameters are given in Table 5.3, $Z_r = 30$ cm, $E_w = 0.01$ cm day^{-1}, and $E_{max} = 0.37$ cm day^{-1}. After Laio et al. (2001b).

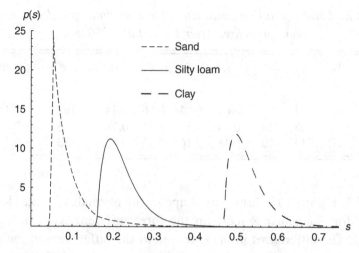

Figure 5.18 Soil moisture probability density function for *Bouteloua gracilis* in three different types of soil considered at the CPER. Rainfall corresponds to the average year with $\alpha = 0.447$ cm and $\lambda = 0.29$ day^{-1}. See Figure 5.17 for the other parameter values. After Laio et al. (2001b).

The mode of the soil moisture pdf is in all cases between s_w and s^* indicating the high likelihood that *Bouteloua gracilis* is under water stress for average rainfall conditions. The mode, however, is more pronounced for sand than for silty loam or clay. The reason for this lies in the slope of $\rho(s)$ in the range $\{s_w, s^*\}$ which is significantly higher for sand than for clay or silty loam. The sand has therefore more vigorous losses for a soil moisture value between s_w and s^* than do the silty loam or clay, so that there is a higher probability for the soil moisture to be near or at s_w.

The results of the water balance for the average year (not shown here) give negligible leakage and runoff components for all the soil classes considered. In terms of the mean evapotranspiration values, however, the sandy soil has a lower fraction of rainfall lost as stressed evapotranspiration and a higher fraction of rainfall lost as unstressed evapotranspiration than do the clay and silty loam soils. This is because the probability of soil moisture being above s^* is larger for the sand than it is for the clay or silty loam.

5.3.3 *Soil moisture crossing properties and vegetation water stress*

Figure 5.19 shows the crossing properties of the soil moisture dynamics for *Bouteloua gracilis* for sand, clay, and silty loam using the average rainfall parameters. These results show the sensitivity of the average time under stress, $\overline{T}_{s^*}$, as well as the average number of crossings of the level of incipient stomatal

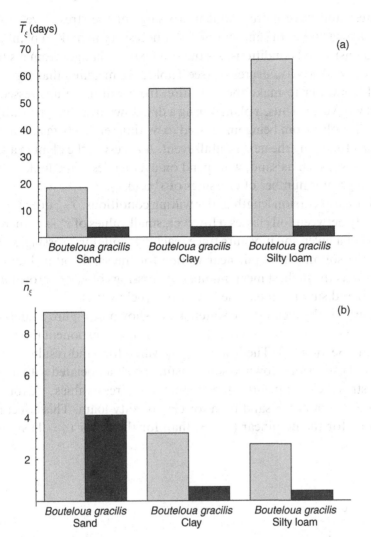

Figure 5.19 Crossing properties of the soil moisture process for *Bouteloua gracilis* in three soil types considered at the CPER. (a) Mean duration of an excursion below s^* (light gray) and below s_w (dark gray). (b) Mean number of downcrossings of s^* (light gray) and s_w (dark gray). The duration of the growing season is $T_{seas} = 183$ days. See Figures 5.17 and 5.18 for values of the other parameters. After Laio et al. (2001b).

closure, $\bar{n}_{s^*}$, to soil texture. These trends can be explained by examining the variation of porosity, n, and the $s^* - s_w$ difference among soil types.

Table 5.3 indicates that as the percentage of sand decreases, so do the porosity and the active soil depth. As explained in Section 3.2, plants with shallower active soil depths experience, on average, shorter periods of time under stress and have more frequent crossings of the stress level. Likewise, $s^* - s_w$ is a measure of the amount of water necessary to make a plant go from wilting to unstressed conditions. As the soil texture changes from a silty loam to a clay or sand, $s^* - s_w$ decreases (see Table 5.3), meaning that a given plant will need less water to make the transition from wilting to unstressed conditions and vice versa. Thus, a plant during a dry-down in a soil type with a small $s^* - s_w$ will evolve from being unstressed to wilting relatively fast, but will also recover fast following the next rainfall event. As a result, the plant in a soil with a small $s^* - s_w$, such as sand, will spend on average less time under stress but will have a greater number of crossings of s^* and s_w.

The average excursion length under wilting conditions, $\overline{T}_{s_w}$, does not change significantly between soil classes. However, small values of $s^* - s_w$ increase the likelihood of a given soil moisture trace reaching, and thus crossing, s_w (see the mode of the soil moisture pdf near wilting for the sand soil in Figure 5.18). Thus sand has the highest mean number of crossings of s_w per growing season (around 4) and silty loam has the lowest one (below 0.5).

The values of the mean static water stress, shown in Figure 5.20, are lower for the $q = 3$ exponent (dark gray) than for the $q = 1$ exponent (light gray) as discussed in Section 4.2. The lower $s^* - s_w$ value for sand results in a higher frequency of excursions down to soil moisture levels associated with high levels of static stress. Consequently, the mean static stress values, $\overline{\zeta'}$, for a given exponent are greater for sand than for clay or silty loam. This effect is more pronounced for the nonlinear ($q = 3$) than for the linear ($q = 1$) static stress

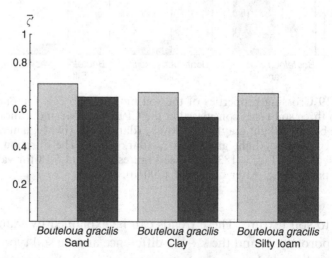

Figure 5.20 Static water stress $\overline{\zeta'}$ of *Bouteloua gracilis* on the three soils under consideration (light gray, $q = 1$; dark gray, $q = 3$). See Figures 5.17 and 5.18 for values of the other parameters. After Laio et al. (2001b).

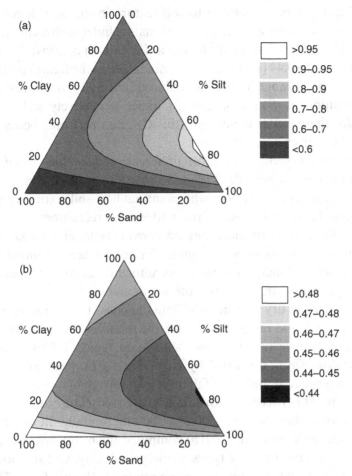

Figure 5.21 Dynamic water stress and the soil texture triangle for *Bouteloua gracilis* (a) under a relatively dry climate, $\alpha = 0.576$ cm and $\lambda = 0.17$ day^{-1}; and (b) under a relatively wet climate, $\alpha = 0.674$ cm and $\lambda = 0.28$ day^{-1}. $T_{\text{seas}} = 183$ days, $q = 3$, and $k = 0.5$. See Figure 5.17 for other parameter values. After Laio et al. (2001b).

formulation due to the larger sensitivity of the nonlinear formulation to soil moisture values near s_{w}.

The mean dynamic water stress, $\bar{\theta}$, combines the information contained in the soil moisture crossing properties and the static water stress to provide a quantitative index of the overall condition of a plant under given edaphic and climatic factors (Section 4.3). Figures 5.21a,b show USDA soil texture triangles illustrating the $\bar{\theta}$ values for *Bouteloua gracilis* during the relatively dry and relatively wet years, respectively. Figure 5.21a shows values of $\bar{\theta}$ in the 0.575 (sand; black in the figure) to 0.95 (silty clay/silty loam; white in the figure) range, indicating large sensitivity to soil texture in the overall condition of *Bouteloua gracilis* under a relatively dry climate. Under such a climate, this C_4 grass does better in a coarse soil than in a fine soil. In contrast, Figure 5.21b shows $\bar{\theta}$ to be in the 0.44 (silty clay/silty loam; black in the figure) to 0.48 (sand; white in the figure) range, indicating reduced sensitivity to soil texture in the overall condition of *Bouteloua gracilis* under a relatively wet climate. In addition, for these wetter conditions, this C_4 grass performs better in a fine soil than in a coarser one.

The above results seem to indicate a preference of *Bouteloua gracilis* for silty loam soils during wet periods (Figure 5.21b). This is in agreement with the results of Lauenroth et al. (1994), who point out how soil texture variability is an important factor determining the patterns of recruitment of *Bouteloua gracilis* at CPER. In particular, they observed silty loam soils as one of the most conducive to seedling establishment for years when recruitment occurs. They also concluded that, on average, recruitment was more probable during years when precipitation was above the site annual mean.

For the relatively dry and intermediate climates, the region of minimum sufferance is the "sand region", while for the relatively wet climate, the most favorable soil texture region is the "silty loam region". This supports the "inverse texture effect" described by Noy-Meir (1973, page 37): "The same vegetation can occur at lower rainfall on coarse soils than it does on fine ones. The balance point between the advantage of coarser texture and its disadvantage occurs somewhere between 300 and 500 mm rainfall." In order to investigate this further, Laio et al. (2001b) computed the mean dynamic stress for *Bouteloua gracilis* on three soil types, namely sand, clay, and silty loam, for a continuously varying total growing season rainfall Θ (Figure 5.22). The values of α and λ were linearly increased from those corresponding to the relatively dry case ($\Theta = 179\,\mathrm{mm}$) to those of the relatively wet year ($\Theta = 345\,\mathrm{mm}$). Figure 5.22 shows how the preferential soil type for this grass differs as a function of Θ. For a relatively dry year, as Figure 5.21a indicated, *Bouteloua gracilis* is fitter in sand than in silty loam or clay. Its better fitness in coarse soils than in fine ones is true for Θ up to approximately 260 mm. As the total growing season rainfall increases above that value, it undergoes a lower mean dynamic stress in fine soils than in coarse soils. Taking into account that the total growing season (April–September) rainfall is approximately 70% of the total annual rainfall for this area (Lauenroth et al., 1978), the point at which coarse soils become more favorable than fine soils for *Bouteloua gracilis*, or vice versa, occurs at an annual rainfall of approximately 370 mm, which is in the range of values indicated by Noy-Meir (1973).

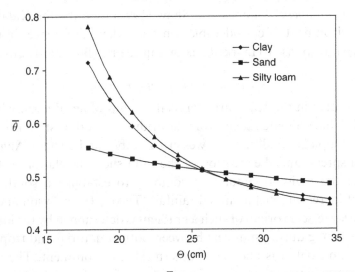

Figure 5.22 Dynamic water stress $\bar{\theta}$ for *Bouteloua gracilis* at CPER as a function of the total incoming rainfall during the growing season, Θ. The parameters α and λ vary linearly with Θ. $T_{seas} = 183$ days, $k = 0.5$, and $q = 3$. See Figure 5.17 for other parameter values. After Laio et al. (2001b).

5.4 Vegetation patterns along the Kalahari precipitation gradient

The Kalahari sand sheet in southern Africa is a 2.5 million kilometer squared area with relatively similar soil but a strong south-to-north increase in rainfall, that provides an excellent basis for gradient studies at the subcontinental scale. For this reason the International Geosphere-Biosphere Programme designated the "Kalahari transect" as one of its "megatransects" to explore continental-scale links between climate, biogeochemistry, and ecosystem structure and function (e.g., Scholes and Parsons, 1997; Annegarn et al., 2001; Dowty et al., 2001; Swap et al., 2001; Scholes et al., 2002).

The focus here is on how the differences in water balance and plant water stress between trees and grasses generate varying preferences for vegetation types along the transect, with deeper-rooted trees favored in the more mesic regions of the northern Kalahari and grasses favored in the drier zones of the southern Kalahari (see Plates 1–13). This section closely follows the paper by Porporato et al. (2003b).

5.4.1 Description of the site

The term Kalahari applies to a loosely defined large area of the interior of central and southern Africa, which has received different types of characterization depending on the subregion of interest. In their beautiful book *The*

Kalahari Environment, Thomas and Shaw (1991) describe the Kalahari as a delicate environment of considerable contrast, where differences in the landscape arise in response to marked changes in the rainfall characteristics.

Climate and precipitation

Climatic control in the Kalahari is crucial to all ecological and hydrological aspects. Kalahari climate ranges from the aridity of southwestern Botswana to the humid tropical conditions of western Zambia and eastern Angola. The latitudinal spread and the location of the sand sheet in relation to the main features of the regional circulation combine to establish a south-to-north gradient of increasing mean annual rainfall (Tyson, 1986; Tyson and Crimp, 1998). The large-scale origin of such a gradient is determined by the interaction (especially strong during summer) between both equatorial and tropical general circulation patterns and the southern African continent. The elevation and interior position of the Kalahari add a dimension of continentality to these atmospheric factors, thus contributing to its aridity (Thomas and Shaw, 1991).

As shown in Figure 5.23, the subsiding components of the southern Hadley cell and southern temperate Ferrel cell are responsible for the nature and position of the subtropical high-pressure belt that so influences the character of Kalahari climate. This average condition of high pressure is modified by two climatic factors that cause important seasonal fluctuations. First, the southern African land mass creates a differential sea–land heating that induces a splitting of the belt into two cells, the South Atlantic and the Indian Ocean anticyclones. Second, these two cells fluctuate with the season and their distinction becomes more marked during summer. Their northwesterly migration during winter months has a significant influence on the seasonal winds and temperature of the Kalahari winter dry season even as far north as the southern half of the Democratic Republic of Congo (Schulze, 1972).

Of particular importance in controlling the rainfall regime is the convective activity of the Inter-Tropical Convergence Zone (ITCZ) or Equatorial Trough (Figure 5.23), which fluctuates in intensity and position throughout the year and from year to year. During summer, in the northern part of the Kalahari, conditions leading to enhanced rainfall associated with the ITCZ are likely to occur in conjunction with the arrival of South Atlantic air. In winter, the southern Kalahari may also occasionally receive rainfall from the depressions that influence the climate farther south (Figure 5.23b).

Variations in rainfall are much more important than temperature in the Kalahari environment. The mean annual precipitation in the Kalahari increases in easterly and northerly directions (Figure 5.24), and there usually exists a strong seasonality in precipitation. The wet season occurs in summer:

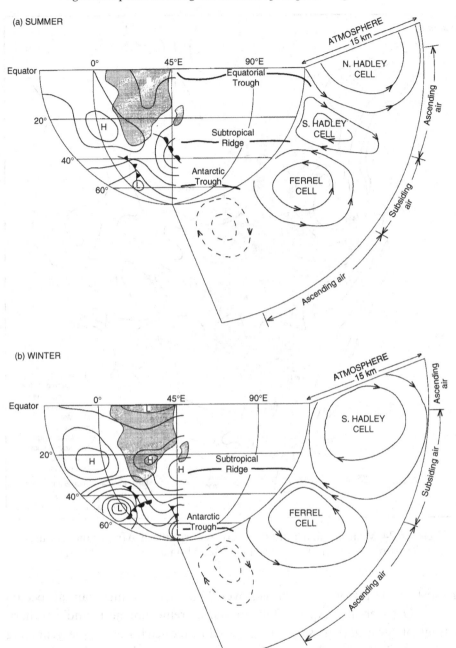

Figure 5.23 Vertical and horizontal components of the atmosphere affecting southern and central Africa in (a) December–February (summer wet season), and (b) June–August (winter dry season). After Tyson (1986), as cited in Thomas and Shaw (1991).

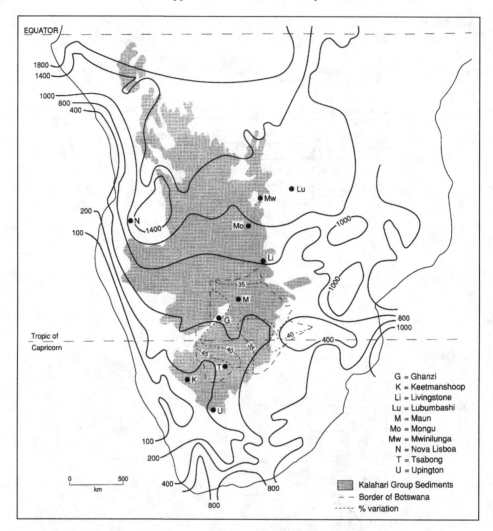

Figure 5.24 Mean annual rainfall (mm) over southern Africa, and annual coefficient of variation over Botswana, after Thomas and Shaw (1991).

for most of the region, on average, over 80% of the annual rainfall occurs between October and April. Only in the extreme northern and southern portions of the Kalahari sands is seasonality reduced, due to the year-long dominance of equatorial convection in the north, and the effect of winter South Atlantic cyclones in the south. Besides the seasonal patterns of precipitation, marked interannual variations are also present. Likewise, the onset and duration of the growing season (wet season) vary considerably from year to year. As the length of the wet season and total precipitation amounts decrease in the south and westerly direction, so the interannual variability increases. As shown in Figure 5.24, the annual coefficient of variation of rainfall exceeds 45% in the driest areas of the southwestern Kalahari.

Most of the rain is in the form of convective thunderstorms, which tend to occur in the late afternoon and early evening. As noted by Thomas and Shaw (1991), however, while most rainfall is derived from high-intensity showers, not every event is a significant one. This is an important aspect for plant water availability, and we will return later to this point when discussing the statistics of the daily amounts of rainfall.

Vegetation distribution

The soils of the Kalahari are relatively homogeneous, mainly made up of sandy sediments. The very low content of organic materials and nutrients makes Kalahari soils highly infertile and places plants under conditions of water and nutrient deficit.

The distribution of vegetation essentially follows the gradient in precipitation and soil moisture. The vegetation type on the sandveld, i.e., the one on deep, well-drained sand, can be broadly defined as a savanna, in the sense of a transition between tropical forest and open grassland (Thomas and Shaw, 1991; Scholes et al., 2002). Schematically, going from the most xeric to the most mesic regions, the vegetation distribution presents the following characteristics. The so-called Kalahari desert of the southern part of the transect largely consists of desert grasslands dominated by annuals such as species of *Aristida*, *Eragostris*, and *Stipagostris*, and only sporadically interrupted by shrubs and few trees. Moving north, trees and shrubs begin to occur in thickets often dominated by stunted *Acacia erioloba*, *Acacia mellifera*, and *Boscia albitrunca*, until one reaches the more mesic sites where trees become increasingly larger and more frequent, including the species *Burkea africana* and *Terminalia sericea*. The deep, soft sands and the greater amounts of rainfall in the northeastern Kalahari often allow trees to develop more abundantly, forming either moist savannas or dry deciduous woodlands. Both of these biomes are dominated by tall (up to 20 m) trees of the species *Baikaea plurijuga*, *Pterocarpus angolensis*, and *Burkea africana*, with an understory of species such as *Baphia obovata* and *Ochna pulchra*. In these northernmost areas, vegetation may either form closed canopy woodlands or more dispersed moist savannas including abundant *Terminalia sericea* and *Combretum hereorense* (Thomas and Shaw, 1991).

The vegetation types form a relatively orderly progression of increasing woody plant cover and height with rainfall gradient, in which the fine-leafed savannas give way to broad-leafed savanna woodlands. This is clearly seen in Figure 5.25, which shows the results of an investigation of ten sites from Zambia to South

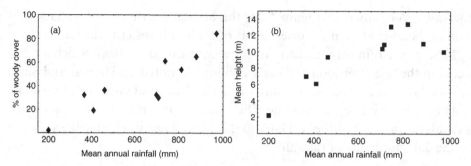

Figure 5.25 Relationships between mean annual rainfall and percentage of woody cover (a), and mean height of the tallest 10% of trees (b). After Scholes et al. (2002).

Africa by Scholes et al. (2002). The overall trend is for the woody biomass, basal area, cover, and height to increase with increasing availability of water to the plants. The increase in woody cover is also associated with an increase in woody plant diversity as well as with a decrease in the relative importance of grasses as contributors to the site biomass. The absolute grass biomass increases with increasing rainfall up to about 600 mm, and then decreases due to competition from woody plants at higher rainfall levels (Scholes et al., 2002).

In addition to climate and rainfall regimes, herbivores (both wild and domestic) and fire are also important disturbances affecting Kalahari vegetation communities. Fires, which can be caused either by lightning or human action, are particularly important and their occurrence and intensity are intimately connected with vegetation and rainfall dynamics (Thomas and Shaw, 1991). Fires are more destructive when rainfall is higher, since plant growth and build-up of fuel materials are higher when the rainfall is more abundant. On the other hand, fires are more likely to occur at the end of the dry season when vegetation is more combustible, and rainfall does not have a direct suppressive effect.

5.4.2 Annual and interannual rainfall variability

Porporato et al. (2003b) examined the historic record of daily rainfall data for a set of stations distributed along the Kalahari transect to assess the seasonal patterns and the statistical characteristics of the rainfall regime, since these factors control the onset and duration of the growing season as well as the soil moisture availability for plant growth.

Four stations, i.e., Mongu, Sesheke, Senanga, and Vastrap, were considered. The first three are along the Zambezi River in western Zambia between the 15° S and the 18° S parallel in the northern part of the transect. The fourth, Vastrap, is located in the southern part of the transect close to the 28° S parallel. Unfortunately, other stations more homogeneously distributed along the transect had to be excluded, either because of the small number of years of observation or because too many gaps were present in the record.

The main statistics of interest are reported in Table 5.4. The mean annual behavior of precipitation, shown in Figure 5.26, reveals a marked seasonality and similar annual regimes for Mongu, Sesheke, and Senanga. For these stations the dry period occurs between May and September, while the wet season goes from October to April; this latter period will be assumed to coincide with the growing season. The months of July and August are on average the driest ones: for all these stations no rainfall was recorded in July during the periods considered, while some occasional events occurred in September.

The rainfall regime at Vastrap requires a separate discussion. As can be seen from Figure 5.26d, the seasonality is almost nonexistent, owing to sporadic winter rainfall associated with mid-latitude disturbances typical of the southernmost Kalahari regions. This anomalous behavior, also noticed by Scholes et al. (2002), will become even more apparent in the subsequent analysis of the statistics of the interstorm periods. It is clear that because of the low total annual precipitation and the lack of any recurrent behavior, Vastrap vegetation must rely on these occasional events randomly distributed along the year rather than on a regular (wet) growing season. Notwithstanding this fact, for the sake of comparison among the stations, the same period from October to April was chosen as a reference growing season for all four sites.

Figure 5.27 presents the normalized histograms of the relative frequency of rainfall depth, h, for rainy days during the growing season for each of the four stations. Both the mean value, α (Table 5.4), and the histograms (Figure 5.27) are quite similar for all the stations, so that a common value of $\alpha = 10\,\text{mm event}^{-1}$ can be used along the entire transect for the probabilistic

Table 5.4 *Rainfall characteristics for the stations considered along the Kalahari transect*

Station	Obs. period[*]	Mean rainfall (mm year^{-1})	α (mm)	λ (day^{-1})	CV_α	CV_λ	$\rho_{\alpha,\lambda}$
Mongu	1935–92 (46)	942	10.1	0.38	0.17	0.16	0.01
Sesheke	1950–92 (39)	715	9.5	0.30	0.20	0.18	0.26
Senanga	1979–92 (14)	737	9.5	0.32	0.16	0.12	0.22
Vastrap	1973–96 (20)	305	10.3	0.09	0.31	0.32	0.21

[*] The number of years considered is in parentheses.

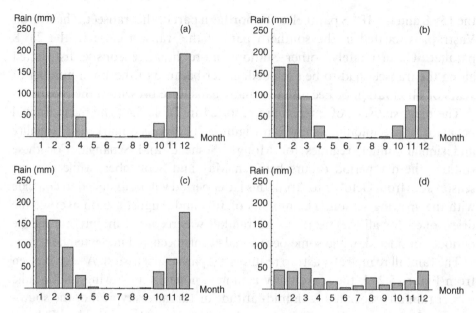

Figure 5.26 Mean rainfall throughout the year for the four stations considered: (a) Mongu, (b) Sesheke, (c) Senanga, and (d) Vastrap. After Porporato et al. (2003b).

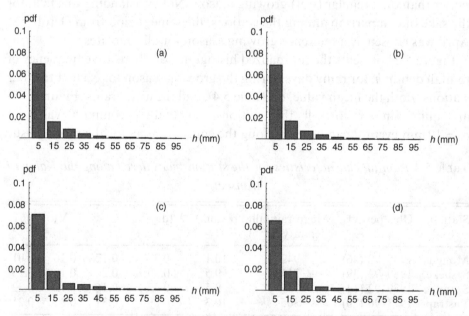

Figure 5.27 Histograms of relative frequency of daily rainfall depth, normalized as pdf's: (a) Mongu, (b) Sesheke, (c) Senanga, and (d) Vastrap.

rainfall model. Notice also that the histograms follow an exponential distribution like the one assumed in the stochastic rainfall model driving the soil moisture dynamics. It is also apparent from Figure 5.27 that most events are of low intensity for all the stations. This is in agreement with the analysis of Botswana data by Pike (1971), where over 80% of rainfall events were found to be of low intensity and rainfall amounts were less than 10 mm for about half of the total number of events. These characteristics have very important implications for vegetation, since interception and surface evaporation may become dominant components of the water balance during such light rainfall events.

The previous analysis implies that the decrease in the mean rainfall amount is mostly a consequence of a reduction in the rate of storm arrivals when moving from north to south. This is confirmed by the mean value of the rate of storm arrivals, λ, which drastically decreases from more than one event every three days in Mongu to less than one event every 11 days in Vastrap in the period from October to April (Table 5.4). The histograms for the relative frequency of the interstorm times, τ, are presented in Figure 5.28. For Mongu, Sesheke, and Senanga, the distribution is well described by an exponential decay, thus agreeing with the Poisson model used for the rainfall arrivals. Only in Vastrap is the distribution of arrival times different, being almost uniform in the range 1–15 days. This confirms the peculiar rainfall regime of this southern

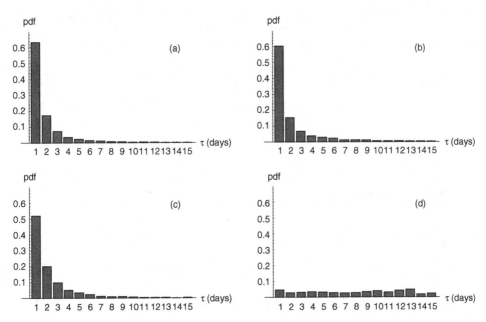

Figure 5.28 Histograms of relative frequency of interstorm times, normalized as pdf's: (a) Mongu, (b) Sesheke, (c) Senanga, and (d) Vastrap.

pre-desertic site, characterized by single, sometimes intense, events separated by long dry periods which may last several weeks.

Table 5.4 also shows the coefficients of variation for α and λ, which exhibit a significant interannual variation around the mean values for each site, especially for Vastrap. No significant correlation, $\rho_{\alpha,\lambda}$, between these two rainfall parameters is present, although both of them appear to vary with mean annual rainfall as shown in Figure 5.29. The values of α and λ during the growing season for each year of observation are shown in Figures 5.30 and 5.31. The

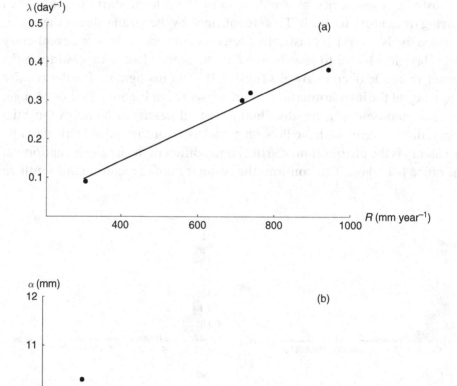

Figure 5.29 Mean frequency (a) and mean depth (b) of rainfall events as a function of mean annual rainfall. After Porporato et al. (2003b).

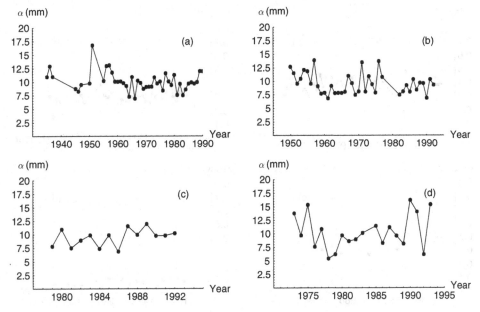

Figure 5.30 Interannual variability of α estimated for each growing season from the available historic record: (a) Mongu, (b) Sesheke, (c) Senanga, and (d) Vastrap.

interannual variability appears to be particularly extreme for the mean rainfall depth, α, in Vastrap (Figure 5.30d).

5.4.3 Soil moisture dynamics and water balance

Assuming a spatially uniform mean rainfall depth per event of $\alpha = 10\,\text{mm}$, Porporato et al. (2003b) varied the mean rate of event arrivals, λ, in the range of $0.5-0.1\,\text{day}^{-1}$ when going from north to south along the transect. The hygroscopic point and the field capacity were assumed to be $s_h = 0.04$ and $s_{fc} = 0.35$, respectively, on account of the uniform sandy soil with low content of soil organic matter, while porosity was set equal to 0.42.

Table 5.5 gives an indication of likely values of the model vegetation parameters for some of the most common species in the Kalahari transect. From these data, it can be seen that trees tend to have higher s_w and lower s^* than grasses. As in the case of the Nylsvley savanna (Section 5.1), this difference can be explained by the reduced drought resistance and high water use efficiency of C_4 grasses. On the other hand, grasses in these regions often have maximum transpiration rates of the order of 10% higher than trees in well-watered conditions (Scholes, 1998, personal communication). Mean typical values for

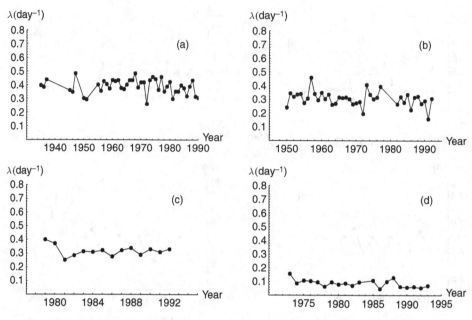

Figure 5.31 Interannual variability of λ estimated for each growing season from the available historic record: (a) Mongu, (b) Sesheke, (c) Senanga, and (d) Vastrap.

trees were assumed to be $s_w = 0.065$, $s^* = 0.12$, and $E_{max} = 0.45$ cm day^{-1}, while for grasses $s_w = 0.05$, $s^* = 0.17$, and $E_{max} = 0.50$ cm day^{-1}. Although all the species are expected to root throughout the upper 1 m of soil, grass roots are typically concentrated closer to the soil surface, while the density of tree

Table 5.5 *Typical vegetation characteristics for the most common plants in the zone of the transect analyzed by Scholes et al. (2002). After Porporato et al. (2003b).*

Species	s_w	Ψ_{s,s_w} (MPa)	s^*	Z_r (cm)	E_{max}(cm day^{-1})
Erythrophleum africanum	0.060	−3.0	0.12	100	0.45
Ochna pulchra	0.060	−3.1	0.14	100	0.39
Pterocarpus angolensis	0.060	−3.0	0.12	100	0.45
Digitaria spp.	0.061	−2.9	0.22	40	0.47
Eragostris spp.	0.052	−3.9	0.15	40	0.61
Brachiaria nigropedata	0.058	−3.2	0.16	40	0.50

roots is more uniform throughout the profile. On account of this, the parameter Z_r (representing the effective rooting depth) was assumed to be 100 cm for trees and 40 cm for grasses (Porporato et al., 2003b).

Figure 5.32 shows the pdf of soil moisture for the case of trees as a function of λ over the range of values encountered along the transect. As λ increases, the shift of the pdf towards higher soil moisture values is evident, progressively moving out of the wilting region and into the region of unstressed conditions. The variance of the distribution also increases with rainfall, mostly because in very arid climates the pdf is bounded from below by the wilting and the hygroscopic points. Grasses show a similar behavior but with higher variance. In the case of trees, the soil moisture attains very high or low values less frequently than grasses. This fact has already been discussed in the framework of the general dependence of the soil moisture pdf as a function of Z_r (Section 2.3).

The computations of the long-term water balance show, as expected, that the water balance over the entire transect is dominated by losses due to evapotranspiration. As for the other main components of the balance, runoff is practically absent while leakage becomes non-negligible only in the more humid northern regions for the case of shallow rooting depths. Since

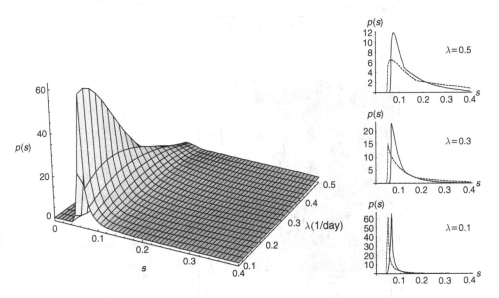

Figure 5.32 Soil moisture pdf as a function of λ for the range of values encountered along the Kalahari transect and using the mean values of the parameters characteristic for trees. The three examples in the insets have mean rainfall rates that are typical of the northern, central, and southern Kalahari, respectively (trees: continuous line; grasses: dashed line). After Porporato et al. (2003b).

interception can be quite important where the canopy cover is denser because of the relatively high fraction of events of low intensity (Figure 5.27), the parameters used were $\Delta_t = 2$ mm for trees and $\Delta_g = 1$ mm for grasses.

Figure 5.33 presents the evapotranspiration and leakage components of the water balance computed as a function of the plant rooting depth, Z_r, and the rate of storm arrivals, λ, using the mean typical values of the evapotranspiration function for grasses and a fixed $\alpha = 10$ mm event^{-1}. From Figure 5.33a it

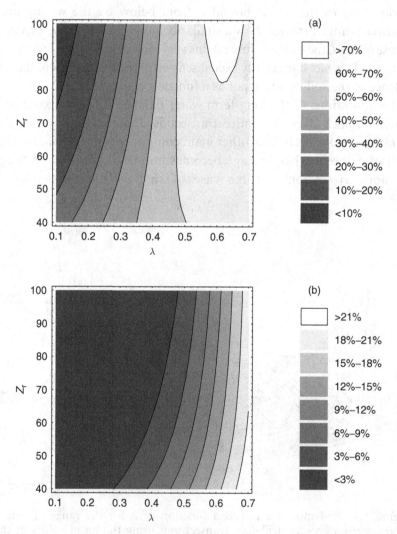

Figure 5.33 Soil water balance using the mean typical daily transpiration function for grasses. (a) Percentage of unstressed evapotranspiration as a function of λ and Z_r. (b) Percentage of leakage losses as a function of λ and Z_r. After Porporato et al. (2003b).

is seen that the percentage of unstressed evapotranspiration can be maximized by adjusting the plant rooting depth. Thus, going from arid to humid climates, the best exploitation of the available soil water is achieved by increasing Z_r. With their 40–60 cm effective rooting depth, grasses are favored in the drier regions, while in the more humid regions the situation is reversed. This is due to more effective water use and decreased leakage losses provided by the different rooting depths. As Figure 5.33b shows, if the rooting depth does not extend below 70 cm, leakage losses may become non-negligible for wet sites, whereas for dry sites the existence of roots below 40–60 cm makes almost no difference. Thus, in the northern regions, deeper roots are important for extracting water that otherwise would percolate below the rooting depth, while in the driest regions, where competition is mostly against evaporative losses, the effort of growing deeper roots becomes pointless in the absence of a permanent source of deep subsurface water.

5.4.4 Plant water stress

The role of the rooting depth in overall plant condition as well as that of the competition between plant transpiration and the other soil water losses becomes more evident when considering the intensity and the temporal statistics of the periods of plant stress through the dynamic water stress index.

Figure 5.34 shows the dynamic water stress computed as a function of the rate of event arrivals for the mean parameter values representative of trees and grasses. As expected, the general behavior is one of progressive increase of

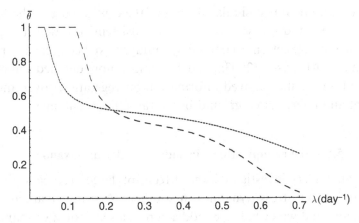

Figure 5.34 Behavior of the dynamic water stress as a function of the mean rate of arrival of rainfall events for trees (dashed line) and grasses (dotted line). $\alpha = 1$ cm, $T_{seas} = 210$ days; see Section 5.4.3 for the values of the other parameters. After Porporato et al. (2003b).

plant stress going from wet to dry climates. The plateau for intermediate values of λ is a consequence of the interplay between the frequency of periods of water stress, which attains its maximum in such a zone, and the duration of the stress periods, which increases with decreasing λ (Figure 4.13). The dependence of the dynamic stress on λ is more marked for trees than for grasses. In particular, for very low rainfall amounts grasses are able to lower water stress more significantly than trees are able to. This is partially due to their lower wilting point, which reduces the effect of water deficit on the plant. More importantly, grasses benefit from their shallow rooting depth, which allows increased access to light rainfall events and reduces the occurrence of long periods of water stress.

The point of equal stress, which could be interpreted as identifying a general region of tree–grass coexistence, is located near $\lambda = 0.2\,\text{day}^{-1}$, which corresponds to a total rainfall of approximately 420 mm for the seven-month period of the wet season (October to April). The fact that the slope of the two curves near the crossing point is fairly mild may contribute to explain the existence of a wide region suitable for tree–grass coexistence at average rainfall rates. The pronounced interannual variability of both rainfall parameters might further enhance the possibility of coexistence by randomly driving the ecosystem from an increase in grasses during dry years to tree encroachment during wet years. Such a mechanism is similar to the one found to facilitate tree–grass coexistence in the savannas of southern Texas (Section 5.2). Moreover, a comparison of Figures 5.33 and 5.34 shows that the region of similar water stress is also the region where the difference in rooting depth is less important in the water balance. This agrees with what is reported for the Nylsvley savanna, where the rainfall regime is similar to that of the central part of the Kalahari transect and trees and grasses both have a consistent root density up to the depth of one meter (Section 5.1). Similar findings have been recently reported by Scanlon and Albertson (2003a, 2003b) using a more detailed soil moisture model based on satellite-derived information on vegetation cover and radiation budget and interpolated ground-based rainfall measurements.

5.5 Tree canopy effects in southern African savannas

Caylor et al. (2004) investigated the effects of large tree canopy on soil moisture and water stress dynamics in southern African savannas, using a coupled energy and water balance model across a series of sites spanning the regional moisture gradient of the Kalahari transect in southern Africa. Their results, which further detail the findings of the previous section, indicate that tree canopies serve to reduce the soil moisture stress of under-canopy vegetation in the middle of the rainfall gradient, while at the dry end of the rainfall gradient, the effect of tree canopies on soil moisture is dependent on the amount of rainfall received in a given growing season. In this manner, as a consequence of the contrasting microclimates existing under and between tree canopies that produce complex patterns of tradeoffs between light and moisture availability, the biophysics of tree and grass canopy performance result in competitive tree–grass interactions in both moist and dry conditions and mutualistic interactions in intermediate conditions.

5.5.1 Modeling framework

The model employed by Caylor et al. (2004) tracks evapotranspiration from five components of the land surface at each site – the tree canopy (denoted $X_{(t,c)}$ when referring to a generic variable X), grass under ($X_{(g,c)}$) and between ($X_{(g,b)}$) tree canopies, and bare soil under ($X_{(s,c)}$) and between ($X_{(s,b)}$) tree canopies.

Tree canopy transpiration is assumed to draw from both the under-canopy and between-canopy soil reservoirs, while the transpiration of under-canopy and between-canopy grasses is localized to s_c and s_b respectively. Loss due to tree canopy transpiration from each of the two reservoirs is constrained by the plant's available moisture in each reservoir, normalized by the total available moisture in both under-canopy and between-canopy conditions. These are called PAM$_b$ and PAM$_c$ and are defined as

$$\text{PAM}_{b,c} = \begin{cases} \frac{s_{b,c}-s_{wt}}{(s_b+s_c)-2s_{wt}} & s_{b,c} \geq s_{wt} \\ 0 & s_{b,c} < s_{wt}, \end{cases} \tag{5.1}$$

where s_{wt} is the wilting point for trees. They represent the partition of the total available soil moisture between under-canopy and between-canopy components.

The total evapotranspiration from the canopy portion of the landscape is given by

$$E_c = \text{LE}_{(s,c)}\lambda_{w(s,c)} + \text{LE}_{(g,c)}\lambda_{w(g,c)} + \text{PAM}_c\text{LE}_{(t,c)}\lambda_{w(t,c)}, \tag{5.2}$$

where $\lambda_{w(\cdot,\cdot)}$ and $\text{LE}_{(\cdot,\cdot)}$ are the latent heat of vaporization of water (function of temperature) and the latent heat flux in each component, respectively. Evapotranspiration from between-canopy areas and total evapotranspiration are found using

$$E_b = \text{LE}_{(s,b)}\lambda_{w(s,b)} + \text{LE}_{(g,b)}\lambda_{w(g,b)} + \text{PAM}_b\text{LE}_{(t,c)}\lambda_{w(t,c)}, \tag{5.3}$$

and

$$E = (1 - f_c)E_b + f_c E_c, \tag{5.4}$$

where f_c is the fractional canopy cover of trees.

The latent heat transfer (W m^{-2}) is computed using the Priestly–Taylor formula (Brutsaert, 1982), i.e.,

$$\mathrm{LE} = \tau(s) \frac{\alpha_{\mathrm{PT}} S}{\gamma_w + S} Q, \tag{5.5}$$

where $\alpha_{\mathrm{PT}} = 1.26$ is the Priestly–Taylor coefficient, $\gamma_w = (p_a c_p)/(0.622\lambda_w)$ is the psychrometric constant, with p_a air pressure and c_p air specific heat capacity, S is the slope of the curve relating saturation vapor pressure to temperature, and Q is the net available energy at the evaporating surface (W m^{-2}). The function $\tau(s)$ accounts for soil moisture limitation of evapotranspiration. It is an exponential function of s for bare soil, $\tau(s) = \exp[-k(1 - s)]$, where k is a limitation coefficient assumed to be equal to 10 by Caylor et al. (2004). As usual, for grass and tree transpiration, $\tau(s)$ is linearly increasing between s_w and s^* and then constant and equal to 1 for $s > s^*$. Full details of the coupled water and energy balance model can be found in Caylor et al. (2004).

Water stress in each vegetation component is characterized using the static moisture stress (see Sections 4.1 and 4.2). In order to assess the potential water stress of juvenile woody vegetation (i.e., sub-canopy trees and shrubs), Caylor et al. (2004) also define $\zeta_{(jt,b)}$ and $\zeta_{(jt,c)}$, which are simply the static stress that would be experienced by juvenile tree vegetation growing in the between-canopy environment or under-canopy environment, respectively. As opposed to the mature trees (i.e., canopy-dominant vegetation) that experience the landscape's average soil moisture, s, the juvenile tree vegetation only experiences the local soil moisture (s_c or s_b). This formulation allows for a comparison of the potential water stress experienced by small trees (i.e., sub-canopy juveniles) growing either under or between large tree canopies (i.e., canopy-dominant adults). Cumulative stress, Φ, is indexed as the number of days during which the static stress is positive for each vegetation component. The ratio of cumulative stress for the under- and between-canopy portions of the landscape, Φ_c/Φ_s, is used as a measure of the different effect of water availability on plant–water relations and stress between these two portions of the landscape.

The rainfall parameters are given in Section 5.4.2, while the necessary structural data are given in Table 5.6 and Figure 5.35. Mean canopy cover,

Table 5.6 *Site locations, description, and vegetation structural parameters used in the coupled energy/water balance model by Caylor et al. (2003).*

Site (latitude–longitude)	Vegetation type	Annual precipitation (mm)	LAI_t	f_c	LAI_g	f_g
Lishuwa Communal Forest, Lukulu, Zambia (14.42 S–23.52 E)	Evergreen woodland	970	5.53	0.844	0.35	.10
Kataba Forest Reserve, Mongu, Zambia (15.44 S–23.25 E)	Kalahari woodland	879	2.47	0.648	0.38	.20
Liangati Forest Reserve, Senanga, Zambia (15.86 S–23.34 E)	Kalahari woodland	811	2.52	0.537	1.1	.30
Maziba Bay Forest, Sioma, Zambia (16.75 S–23.61 E)	Dry Kalahari woodland	737	2.08	0.610	0.81	.40
Sachinga Agricultural Station, Katima Mulilo, Namibia (17.70 S–24.08 E)	Combretum woodland	707	1.16	0.299	2.1	.50
Pandamatenga Agricultural Station, Pandamatenga, Botswana (18.66 S–25.50 E)	Schinziophyton, Baikiaea, Burkea woodland	698	2.04	0.323	1.4	.50
Sandveld Research Station, Gobabis, Namibia (22.02 S–19.17 E)	Acacia-Terminalia woodland	409	0.69	0.191	1.4	.40
Tshane, Tshane, Botswana (24.17 S–21.89 E)	Open Acacia savanna	365	0.61	0.321	0.87	.30
Vastrap Weapons Range, Upington, South Africa (27.75 S–21.42 E)	Open Acacia shrubland	216	0.35	0.058	0.80	.20

grass biomass, grass-specific leaf area and grass leaf area are taken from published field results at the Kalahari transect sites (Caylor et al., 2004). To estimate the site-average leaf area index, the volumetric distribution of woody vegetation leaf area is determined using a combination of field data and allometric relationships. The two-dimensional structure of canopies at each site is revealed using field observations of individual crown dimensions, while

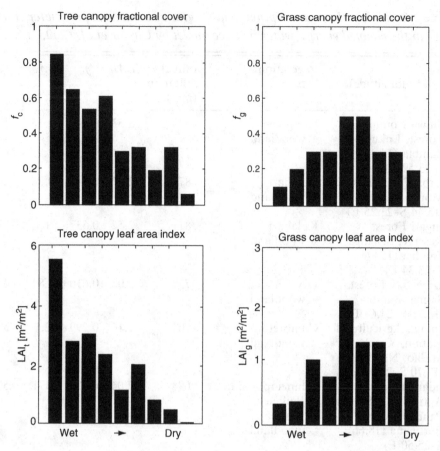

Figure 5.35 Vegetation structural characteristics at each site along the Kalahari transect. Within each graph, the sites proceed from left to right along decreasing latitude and mean annual rainfall. Tree leaf area index (LAI_t) and tree fractional cover (f_c) decrease with rainfall, while grass fractional cover (f_g) and grass leaf area index (LAI_g) reach a peak in the middle of the transect. After Caylor et al. (2003).

the vertical structure of canopies is generated using field observations of canopy height and average canopy depth. Leaf biomass and whole-tree biomass are estimated for each individual using generalized allometric relationships for southern African species (Caylor et al., 2004). Field measurements of specific leaf area are used to calculate leaf area from each tree's allometrically determined leaf biomass. Each individual's leaf area is then distributed evenly throughout the individual's canopy volume, to arrive at a leaf area per unit volume, or leaf area density. Where canopy volumes intersect, leaf area density at intersecting locations is taken to be the sum of the contributing canopies' leaf area densities. Total-site leaf area is the average of leaf area summed vertically. Detailed field observations of canopy structure were used to determine average leaf area, taking into account the relative amount of clumping of canopies, and the overlap between adjacent canopies that may be caused by the aggregated distribution of individuals. A field study examining the distribution of woody vegetation across the Kalahari transect has shown that individuals exhibit rather aggregated distributions at most sites (Caylor et al., 2004).

5.5.2 Model simulations

At each of the nine Kalahari transect sites, Caylor et al. (2004) carried out 1000 yearly simulations at the daily level of the coupled water and energy balance model to determine the changing nature of soil moisture under and between tree canopies across the rainfall gradient. The discussion that follows has been taken literally from Caylor et al. (2004). Simulation of the daily soil moisture under and between tree canopies shows that in the northernmost site, where the mean annual rainfall is ~ 900 mm, daily soil moisture is relatively high, and little relative difference is observed in soil moisture between and under canopies. Only when the interval between rains is longer, can it become somewhat drier between the tree canopies. At intermediate sites (annual rainfall ~ 400 mm), consistent and relatively large differences are seen between the daily soil moisture conditions under and between tree canopies. Locations between tree canopies are almost always drier, usually to a substantial degree. At the southernmost site (Vastrap, ~ 300 mm mean annual rainfall), differences between the soil moisture under and between canopies are negligible. For a few days at the end of the season in Vastrap, the soil moisture under the canopy is lower than the between-canopy value.

Changes in soil moisture under and below the tree canopies affect the distribution of stress days for trees and grasses in the under-canopy and between-canopy portions of the landscape. The accumulated stress index, Φ, is highest for each vegetation component at the southernmost site and negligible at the northernmost site. The difference in soil stress conditions under and between canopies is measured by the difference in stress days. The variation in vegetation stress under and between tree canopies is indexed using the stress thresholds for trees. The overall difference in stress levels at each site can then be determined by calculating the proportion of years during which the total number of stress days in the between-canopy areas exceeds the number of stress days under the canopy.

The leaf area index of tree canopies reduces the amount of energy available under the canopy. Thus at sites where canopy leaf area is higher, the reduction in incoming shortwave radiation available under the tree canopy is greater.

The reduced availability of incoming shortwave radiation directly affects sub-canopy vegetation through energetic controls on transpiration, and indirectly through reductions in bare soil evaporation. For all of the study sites, there are regular patterns in the percentage reduction in shortwave radiation levels below tree canopies at each site and in the likelihood of reduced stress conditions under tree canopies over 1000 simulations (Figure 5.36). At northern

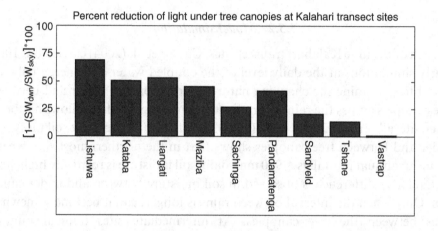

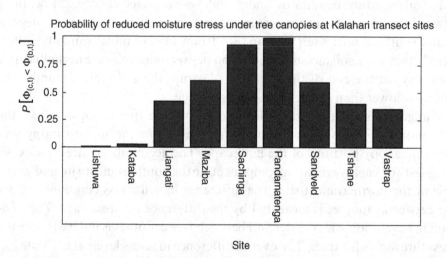

Figure 5.36 Changing light and water stress conditions under large tree canopies across the Kalahari transect. The ratio of incoming shortwave radiation below the canopy to incoming shortwave radiation above and between the tree canopies decreases across the rainfall gradient, while the proportion of simulated years where soil moisture stress is lower under canopies than between canopies is greatest in the middle of the transect. After Caylor et al. (2003).

sites, there is little or no reduction in stress days under tree canopies, but the canopies are seen to reduce incoming shortwave radiation levels up to 75%. At intermediate sites, the under-canopy vegetation almost always experiences fewer stress days than the between-canopy areas. The effect of tree canopies on incoming shortwave radiation attenuation and water stress is reduced at the southern end of the transect.

The pattern in water stress under the tree canopies is such that reduced canopy stress occurs at sites in the middle of the transect (Figure 5.36). At the northern end of the transect, where rainfall is high, the canopy has little effect on soil moisture stress levels, due to the high availability of soil moisture throughout the growing season. As the mean annual rainfall decreases, the proportion of years during which vegetation under the canopy experiences lower total stress days than the vegetation between the canopies increases. Locations where the proportion of years with reduced under-canopy stress is greater than 0.5 indicate sites where the canopy is, on average, less stressed than the between-canopy locations. Across the Kalahari transect, the yearly difference in stress under and between canopies can be expressed as a relationship to the annual rainfall for each simulation. Figure 5.37 displays the probability that the number of stress days under tree canopies will be lower than the number of stress days between canopies for grasses according to the deviation between a given year's rainfall and the long-term mean annual rainfall across the latitudinal gradient. The probability surface is shown for ±2 standard deviations of the mean annual rainfall, which represents 95% of the variation in mean annual precipitation. By linear distance-weighted interpolation between sites, Caylor et al. (2004) obtain a transect-wide determination of how variation in rainfall might affect the relative stress levels between and under tree canopies. At the northernmost, wettest, sites the model results indicate that there are no rainfall amounts within the ±2 standard deviation mean-annual-rainfall envelope, which should lead to differences in water stress levels between and under tree canopies.

A large central portion of the Kalahari transect (from $\sim 16°$ to $\sim 28°$ S) exhibits high probabilities of reduced stress under tree canopies for ±2 standard deviations around the mean annual rainfall (Figure 5.37). At the northern part of this zone, canopy trees reduce stress in years with below-average precipitation ($\sim 16°$ to 18° S) and in the southern parts of the zone ($\sim 20°$ to 28° S) tree canopies only reduce water stress during years with above-average precipitation. As one traverses the rainfall gradient, changes in vegetation structure occur at both the patch and site scale. Mean tree biomass, tree cover, and tree leaf area index are seen to vary directly with rainfall. In contrast to these general trends, the effect of large trees on the light and moisture environment is much more complex (Figure 5.37).

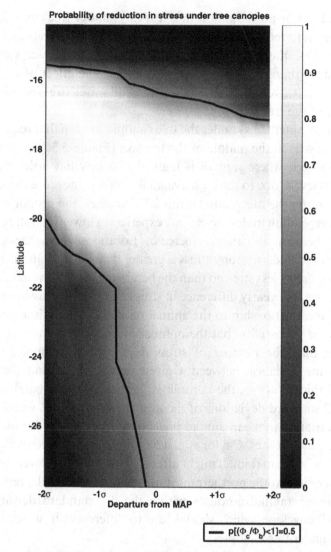

Figure 5.37 Probability of reduced water stress ($\Phi_c/\Phi_b < 1$) under tree canopies across the Kalahari transect as a function of yearly deviation from mean annual precipitation (MAP) (-2σ to $+2\sigma$). Solid lines indicate 50% probability threshold. After Caylor et al. (2003).

5.5.3 *Implications for savanna dynamics*

The consideration of canopy and between-canopy effects adds insight into the ways in which structure and function are coupled in savanna ecosystems (Caylor et al., 2004). The model results indicate that tree canopies can have a significant local effect on the daily distribution of soil moisture across a wide range of rainfall regimes. In addition to pronounced changes in tree–grass interaction across the Kalahari transect rainfall gradient, the results illustrate that the tree–grass interaction in different parts of the Kalahari transect can change in sign and magnitude over time. Given the pattern shown in Figure 5.37, the northern section of the Kalahari should be relatively invariant with neutral effects of trees on the moisture environment beneath them. Further south, the effects of canopy trees on sub-canopy soil moisture are highly likely to be positive for smaller plants growing beneath them. Still further south (below about 20° S), the trees have a positive effect on the soil moisture but in wet years only. Therefore, in the southernmost section of the Kalahari, given the persistence of wet and drought conditions (Tyson and Crimp, 1998), the potential mutualism of larger trees supplying regeneration sites of increased moisture in an overall dry environment turns on and off depending on the climate. Species adapted to regenerating and growing under trees would be advantaged in a wet decade and disadvantaged in another, drier one. Just considering the tree canopy effects on moisture stress and the shading of trees, the patterns of the same suite of life forms interacting along a regular rainfall gradient also should change in space. Therefore, in much of the southern part of the Kalahari transect the rules for interaction could be expected to change in time, even at the same locations (Caylor et al., 2004).

The high degree of control that vegetation structure exerts on the distribution of environmental resources (light and water) necessitates detailed characterization of vegetation structure to assess the potential effects of future and present environmental heterogeneity. In parts of the Kalahari transect (or in different years for much of the transect), if trees successfully establish themselves after an ecological disturbance, there is a diversification of the regeneration niches either for other trees or for other plant functional types. In deforested areas, the absence of trees also implies a reduction in the diversity of regeneration site types (Caylor et al., 2004).

5.6 Soil moisture balance and water stress in a Mediterranean oak savanna

Kiang (2002) and Baldocchi et al. (2004) analyzed the stochastic soil moisture dynamics and the related water stress for a savanna in a Mediterranean climate in California (see Plates 14, 15). The ecosystem differs from the ones previously studied because temperature and precipitation are seasonally out-of-phase and the two functional vegetation types have different growing seasons. The following section is taken from their works.

5.6.1 Description of the ecosystem

The site investigated by Kiang (2002) and Baldocchi et al. (2004) is a California blue oak (*Quercus douglasii*) savanna site with annual grasses in Ione, California, in the low foot-hills of the Sierra Nevada ($38°26'$ N, $120°57'$ $30''$ W). The climate is Mediterranean and semi-arid, with winter rain and summer drought. Mean annual precipitation is 610 mm and mean annual temperature is $16 °C$ (mean maximum $40°$ C and mean minimum $5 °C$). Soil is a rocky loam 0.5–1.0 m deep above greenstone bedrock. Groundwater sources occur at 200–340 m depths and the site is level with maximum slopes in undulations of less than 15%.

In order to account for the clear seasonality in the precipitation, Kiang (2002) specifies three different seasons with distinct precipitation patterns: a winter wet season (November–February), a spring early dry season of initial soil moisture dry-down (March–June), and a late summer dry season (July–October). The precipitation parameters were computed for the growing seasons of grasses and oaks which span two of the previously defined seasons (Table 5.7).

Vegetation at the site is comprised of a scattered canopy of blue oak trees and a small number of grey pines (*Pinus sabiniana*) over an herbaceous understory. The annual grass and herbaceous layer is active during the wet winter to early spring, and the drought-deciduous oaks leaf out 1–2 months before the grasses senesce and remain active through the end of the summer. The density of the blue oak is approximately 200 trees per hectare, with canopy cover of 0.39 and peak leaf area of 0.6. The very minor population of grey pines has a variable density of 3–4 per hectare.

(Kiang, 2002)

Given the deep water table and the drought-deciduous nature of the blue oak the latter ones are not believed to access deep water sources. Their rooting depth varies from 0.5 to 1.5 m, with most of the root biomass located above 0.5 m.

Table 5.7 *Rainfall parameters in the California blue oak savanna. After Kiang (2002).*

Season	λ (day^{-1})	α (cm)
Winter (Nov–Feb)	0.46	1.02
Spring (Mar–Jun)	0.35	0.77
Summer (Jul–Oct)	0.10	0.30
Grass growing season (Nov–Apr)	0.45	0.95
Oak growing season (Mar–Oct)	0.22	0.53

Table 5.8 *Soil and plant characteristics of the California blue oak savanna. After Kiang (2002).*

Soil type	s_h	s_{fc}	K_s (cm day^{-1})	n	b
Ione loam	0.05	0.80	50	0.39	2.57

Species	s_w	s^*	Z_r (cm)	E_{max} (cm day^{-1})	
Grass	0.27	0.65	20	0.25 (winter) 0.40 (spring)	
Blue oak	0.22	0.45	60	0.25	

An average value of 60 cm was used in the computations. The blue oaks have negligible fine root production in the fall and winter and peak production in spring and summer. The annual grass has a peak production in fall and winter with the bulk of fine roots in the upper 0–20 cm (Table 5.8); a weak shift towards deeper layers is observed for grass roots in the early spring slightly before full senescence takes place by the middle of the same period. Oaks and grasses favor different root-depth distributions when their growth is out-of-phase (winter for grass and summer for oaks), while their root-depth distribution is the same when their growth is in-phase (spring). During spring, however, grasses are already senescing and there seems to be little time for competition for soil moisture with the oaks (Kiang, 2002; Baldocchi et al., 2004).

They estimated the values of E_{max} and E_w for grasses and oaks from the maximum and minimum fluxes as measured by eddy flux at the Ione site, while for interception they used $\Delta_{oaks} = 0.2$ cm and $\Delta_{grasses} = 0.1$ cm. The other soil and vegetation parameters are reported in Table 5.8.

5.6.2 Results

Kiang (2002) compared the predictions of the stochastic model of the previous chapters with soil moisture observations during 2001:

The measured soil moisture hovers around field capacity during the winter, then steeply declines in the spring with the end of the winter rains, converging to a minimum value for most of the summer.

The soil moisture pdf's, both predicted and observed, are shown in Figure 5.38 for the full growing season. The agreement is quite good, especially considering that the soil moisture record is rather short, and clear differences

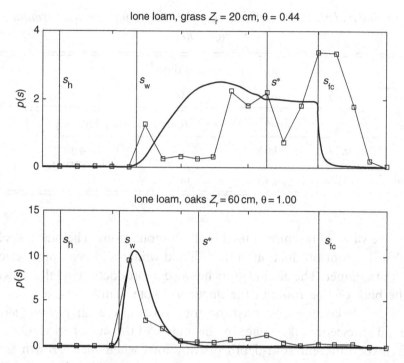

Figure 5.38 Soil moisture pdf's for grass and blue oak. Top: grass full growing season (Nov–Apr); bottom: oak full growing season (Mar–Oct). After Kiang (2002).

are apparent between the two vegetation types during their respective full growing season. The grasses are on average much less stressed than oaks notwithstanding their higher s^*. A more detailed picture of water balance and vegetation conditions is provided by the soil moisture pdf's by season and vegetation type (Figure 5.39) and by the dynamic stress for grasses and oaks subdivided by season (Figure 5.40). Both model and data show the distinct seasonal differences in soil moisture distributions from the wet winter, during which the grasses experience very low stress, to the spring, when grasses are stressed and oaks straddle unstressed and stressed regions, and to the summer when the oaks are near their wilting point. The seasonal breakdown shows why grasses decline in the spring while the oaks take over thanks to an overall lower dynamic stress during this transitional season. The high stress of the oaks during the summer corresponds to their gradual senescence in response to drought during this period. Interestingly, Kiang (2002) also notices that the relative soil moisture at the incipient stress, s^*, for oaks corresponds to the typical transitional soil moisture value between spring and summer.

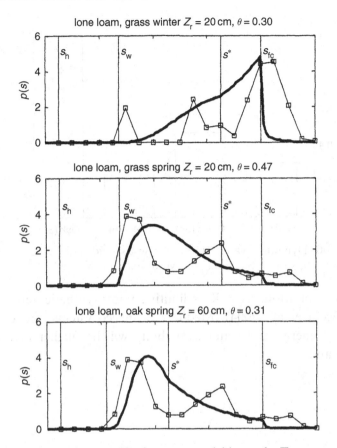

Figure 5.39 Soil moisture pdf's for grass and blue oak. Top: grass winter (Nov–Feb); middle: grass spring (Mar–Jun); bottom: oak spring (Mar–Jun). After Kiang (2002).

The above application to the California blue oak savanna clearly shows the importance of seasonality in temperature and precipitation that, combined with different vegetation types (rooting depth and stress points), in turn produces different soil moisture and evapotranspiration regimes. As Kiang (2002) notices, this becomes even more striking when the present analysis is compared with that of Nylsvley (Section 5.1), where the annual rainfall is the same as in Ione but evenly distributed throughout the warm growing season. Moreover, in the present case of a Mediterranean climate, the stochastic soil moisture model can distinguish stress levels during spring and summer but cannot explain by itself the winter vegetation dynamics, when trees do not share the abundant soil moisture with grasses. The reason seems to lie in the cold winter temperatures that can cause embolism in the oak xylem, thus

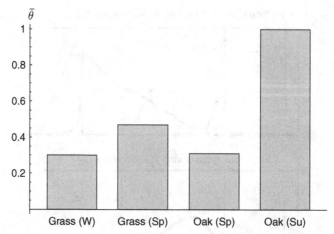

Figure 5.40 Dynamic water stress for grass and blue oak by season (W, winter; Sp, spring; Su, summer). Modified after Kiang (2002).

hindering transpiration, as well as limiting photosynthetic activity (Kiang, 2002; Baldocchi et al., 2004). Mediterranean ecosystems and the issue of seasonality of energy and water availability will be further investigated in Sections 7.1 and 8.1.

6

Coupled dynamics of photosynthesis, transpiration and soil water balance: from hourly to growing-season time scale

This chapter explores the dependence of plant carbon assimilation on soil moisture. Since the soil–plant–atmosphere dynamics involves different time scales, from the hourly dynamics of stomatal control and transpiration, to the daily-to-seasonal time scales of soil moisture and plant growth dynamics, a temporal upscaling is essential to understand how the short time scale processes actually control the long-term dynamics of the water and carbon fluxes and the related evolution of ecosystems.

The analysis starts at the hourly time scale by coupling the equations of the soil–plant–atmosphere continuum (SPAC) to a parsimonious model of the atmospheric boundary layer, to investigate the role of soil, plant, and boundary layer characteristics on the diurnal courses of soil moisture, transpiration, and carbon assimilation during interstorm periods. The temporal evolution of such variables is then integrated at the daily time scale, yielding very similar results to the empirical relationships used in the previous chapters to relate these variables. Finally, the daily leaf carbon assimilation is coupled to the stochastic soil moisture model of Chapter 2, to obtain a probabilistic description of the carbon assimilation during a growing season. The analysis of the duration and frequency of periods with impaired assimilation provides a measure of plant water stress as a function of the soil, vegetation, and climate characteristics that is in good agreement with (and partly complements) the dynamic water stress defined in Chapter 4.

The aim of this chapter is to provide the bases for the temporal upscaling of the transpiration/assimilation dynamics and then proceed with their stochastic analysis at the daily time scale. For this reason the modeling of the SPAC is simplified to only retain its essential dynamical features. It is important to notice, however, that many detailed models of the SPAC have been proposed recently. To such an important branch of ecohydrology, which will not be covered here, we refer elsewhere the reader who is interested in a fine-tuned modeling and prediction of the hourly dynamics of the exchanges in the SPAC (e.g., Williams et al., 1996; Baldocchi and Meyers, 1998; Siqueira et al., 2002) or in the integrated coupling of the soil–plant–atmosphere with the large-scale atmospheric dynamics (e.g., Dickinson et al., 1986; Sellers et al., 1986; Bonan, 1995; Foley et al., 1996; Pollard and Thompson, 1995).

The present chapter closely follows the papers by Daly et al. (2004a, 2004b). After the development of the equations for the water movement through the SPAC, the modeling of the various conductances is discussed. In particular, two stomatal functions are considered according to the approaches of Jarvis (1976) and Leuning (1990, 1995), while their relation to photosynthesis and carbon assimilation is accounted for by means of the classical photosynthesis model of Farquhar et al. (1980). The results of the numerical integration at the hourly time scale give the diurnal patterns of transpiration, assimilation, and leaf stomatal conductance. After upscaling the dynamics to the daily time scale, the results are incorporated into the stochastic model of soil moisture dynamics to analyze the probabilistic structure of plant carbon assimilation under fluctuating climatic conditions during a growing season.

6.1 Transpiration and soil water balance at the hourly time scale

According to the classical framework of the soil–plant–atmosphere continuum (see Section 4.1.1), the water is taken up by the roots, flows through the xylem, and exits through the leaf stomata to the atmosphere, following a path of decreasing water potential, from the soil at Ψ_s, through the leaves at Ψ_l, to the atmosphere at Ψ_a (Figure 4.4). Typically, the water flow inside the plant is assumed to take place as a succession of steady states (i.e., considering the system adjustment to the time-varying equilibrium conditions as practically instantaneous). The exchanges between vegetation and atmosphere are simply modeled using the standard big-leaf schematization.

Under these assumptions, the equation for soil moisture dynamics at the hourly time scale during interstorm periods may be written as (Daly et al., 2004a)

$$nZ_r \frac{ds}{dt} = -E - EV - L, \tag{6.1}$$

where, as usual, s is the relative soil moisture averaged over the root depth Z_r, n is porosity, E, EV, and L are the transpiration, evaporation from the soil, and leakage, respectively, expressed in terms of unit ground area. For reasons of mathematical convenience, EV is assumed to be equal to EV_{max} for s higher than the wilting point s_w and to decrease linearly to zero at the hygroscopic point s_h. Leakage losses are assumed to follow the behavior of the hydraulic conductivity as in Section 2.1.6. In the absence of rain, the solution of Eq. (6.1) represents the decay of soil moisture starting from a given initial condition and provides the boundary condition to the SPAC in terms of water potential.

The flux of water per unit ground area from the roots to the leaves is modeled as proportional to the water potential gradient (van den Honert, 1948) as

$$E = g_{srp}(\Psi_s - \Psi_l),\tag{6.2}$$

where g_{srp} is the soil–root–plant conductance per unit ground area,

$$g_{srp} = \frac{L_{AI}g_{sr}g_p}{g_{sr} + L_{AI}g_p},\tag{6.3}$$

i.e., the series of the soil–root conductance per unit ground area, g_{sr}, and plant conductance in terms of unit leaf area, g_p. Because of this choice of units, g_{srp} depends on both L_{AI} (leaf area index, i.e., the leaf area per unit ground area) and R_{AI} (root area index, i.e., the root area per unit ground area), which express the relative importance of root and xylem conductance, determining which part contributes more to the limitation of water flow (e.g., Sperry et al., 1998, 2002).

The soil–root conductance is assumed to be proportional to the soil hydraulic conductivity divided by the average distance, L_{sr}, traveled by the water from the soil to the root surface. A simplified cylindrical root model allows one to link L_{sr} to the root depth and the root area index as (Katul et al., 2003)

$$g_{sr} = \frac{K\sqrt{R_{AI}}}{\pi g \rho_w Z_r},\tag{6.4}$$

where g is the gravitational acceleration and ρ_w is the density of water. Considering that roots respond continuously to soil moisture dynamics, the plant water availability (and thus soil–root conductance) cannot be judged solely on the basis of soil parameters (Larcher, 1995, page 229). It is therefore reasonable to assume that, as soil dries, the decrease of g_{sr} due to K is in part compensated by the plant through root growth. Thus, for moderate soil-water deficit, the distance L_{sr} decreases when soil water availability decays (e.g., Larcher, 1995). This can be modeled by simply correcting R_{AI} with a multiplicative term that attenuates the reduction due to K, i.e.,

$$R_{AI} = R_{AI}^* s^{-a},\tag{6.5}$$

Table 6.1 *Parameter values used in the model of the hourly dynamics of the SPAC. After Daly et al. (2004a).*

Parameter	Value	Units	Description
a	8	–	Eq. (6.5)
a_1	15	–	Eqs. (6.21) and (6.34)
c	2	–	Eq. (6.6)
c_a	350	$\mu\text{mol mol}^{-1}$	Atmospheric carbon concentration
c_p	1012	$\text{J kg}^{-1}\,\text{K}^{-1}$	Air specific heat at constant pressure
d	2	MPa	Eq. (6.6)
$D_{x,\text{Jarvis}}$	0.0077	kg kg^{-1}	Eq. (6.20)
$D_{x,\text{Leuning}}$	0.0018	kg kg^{-1}	Eq. (6.21)
$g_{p\max}$	11.7	$\mu\text{m MPa}^{-1}\,\text{s}^{-1}$	Maximum plant conductance
g_a	20	mm s^{-1}	Atmospheric conductance
g_b	20	mm s^{-1}	Leaf boundary layer conductance
$g_{s\max}$	25	mm s^{-1}	Maximum stomatal conductance
h_0	50	m	Boundary layer height at night
k_1	0.005	$\text{m}^2\,\text{W}^{-1}$	Eq. (6.18)
k_2	0.0016	K^{-2}	Eq. (6.18)
p_a	$1.013 \cdot 10^5$	Pa	Air pressure
T_{opt}	298	K	Eq. (6.18)
λ_w	$2.5 \cdot 10^6$	J kg^{-1}	Latent heat of water vaporization
ρ	1.2	kg m^{-3}	Air density
$\phi_{\max}$	500	W m^{-2}	Maximum leaf available energy
Ψ_{l1}	−0.05	MPa	Eq. (6.19)
Ψ_{l0}	−4.5	MPa	Eq. (6.19)
Ψ_{l_A1}	−0.5	MPa	Eq. (6.25)
Ψ_{l_A0}	−4.5	MPa	Eq. (6.25)

where R^*_{AI} is the root area index in well-watered conditions and a is a parameter that varies from species to species. Its value (Table 6.1) is chosen to reproduce the typical values of g_{sr} found in the literature (e.g., Nobel, 1999; Daly et al., 2004a).

Plant conductance, g_p, drops when the water potential is too low because of xylem cavitation; this decay is modeled by a vulnerability curve (e.g., Sperry et al., 1998; Katul et al., 2003)

$$g_p = g_{pmax} \exp[-(-\Psi_l/d)^c].$$ (6.6)

The parameters d and c are such that g_p is equal to g_{pmax} for high values of Ψ_l and close to 0 for low Ψ_l (see Table 6.1).

Under steady-state conditions and neglecting cuticular transpiration, the flux inside the plant must equal the transpiration rate, which is proportional to the specific humidity gradient between the leaves and the boundary layer,

$$\rho_w E = g_{sba}\rho[q_l(T_l, \Psi_l) - q(T_a)],$$ (6.7)

where ρ is the air density, ρ_w is the liquid water density, q_l and q are the leaf and air specific humidity, respectively, T_l and T_a are respectively leaf and atmospheric temperatures, and g_{sba} is the equivalent conductance per unit ground area from inside the stomata to the mixed boundary layer. It is modeled as the series of stomatal conductance (per unit leaf area), g_s, leaf boundary layer conductance (per unit leaf area), g_b, and atmospheric conductance (per unit ground area), g_a. The values of the parameters are reported in Table 6.1.

Equation (6.7) is seldom used in this form, since T_l is usually not directly measured or modeled, and the Penman–Monteith combination approach is used instead. To this purpose, Eq. (6.7) is re-written using the relation between the specific humidity and the vapor pressure as

$$E \simeq \frac{0.622 g_{sba}\rho}{p_a \rho_w}[e_l(T_l, \Psi_l) - e_a(T_a)],$$ (6.8)

where p_a is the air pressure. Assuming that the water inside the stomatal pores is freely available, the leaf vapor pressure, $e_l(T_l, \Psi_l)$, can be substituted with its value at saturation, $e_l^* = e_l(T_l, \Psi_l = 0)$. Although this assumption somewhat breaks the SPAC continuum by setting $\Psi_l = 0$, it also simplifies the evaluation of the transpiration rate introducing only a slight overestimation. In fact, the ratio between partial pressure of water vapor, e_l, and its value at saturation, e_l^*, is given by

$$\frac{e_l}{e_l^*} = \exp\left[\frac{\Psi_l V_w}{RT_l}\right],$$ (6.9)

where V_w is the partial molal volume of water and R the universal gas constant; this ratio is only 0.95 at -6.92 MPa at $20\,^\circ$C (Jones, 1992, page 157; see also Section 4.1). Moreover, at low Ψ_l, when $e_l(T_l, \Psi_l) \leq e_l^*(T_l)$ would reduce transpiration, the effective reduction is performed by an earlier stomatal closure.

With the above simplifications, we can first write

$$
\begin{aligned}
e_l^*(T_l) - e_a(T_a) &= [e^*(T_a) - e_a(T_a)] + S(T_l - T_a) \\
&\simeq \frac{p_a}{0.622}[q^*(T_a) - q(T_a)] + S(T_l - T_a) \\
&= \frac{p_a}{0.622}D + S(T_l - T_a),
\end{aligned}
\tag{6.10}
$$

where D is the potential saturation deficit of the air (or water vapor deficit, i.e., the difference between the air relative humidity at saturation, $q^*(T_a)$, and the ambient one, $q(T_a)$) and S is the slope of curve relating saturation vapor pressure to temperature. A simplified leaf energy balance (per unit ground area), can be written as $\phi - H - \lambda_w\rho_w E = 0$, where $H = c_p\rho g_{ba}(T_l - T_a)$ is the sensible heat flux, g_{ba} is the series of leaf boundary layer and atmospheric conductance to sensible heat (and water) flux per unit ground area, $g_{ba} = g_b L_{AI} g_a/(g_b L_{AI} + g_a)$, ϕ is the leaf available energy,[1] λ_w is the latent heat of water vaporization, and c_p air specific heat at constant pressure. This yields

$$
T_l - T_a = \frac{\phi - \rho_w\lambda_w E}{c_p\rho g_{ba}},
\tag{6.11}
$$

which, substituted in Eq. (6.8) and after rearranging the terms, provides the Penman–Monteith equation for transpiration,

$$
E = \frac{\lambda_w\gamma_w g_s L_{AI} g_{ba}\rho D + g_s L_{AI} S\phi}{\rho_w\lambda_w[\gamma_w(g_s L_{AI} + g_{ba}) + g_s L_{AI} S]},
\tag{6.12}
$$

where $\gamma_w = (p_a c_p)/(0.622\lambda_w)$ is the psychrometric constant.

In summary, the system of the four equations

$$
nZ_r\frac{ds(t)}{dt} = -E - EV - L
$$

$$
\Psi_s = \overline{\Psi}_s s^{-b}
\tag{6.13}
$$

$$
E = g_{srp}(\Psi_s - \Psi_l)
$$

$$
E = \frac{\lambda_w\gamma_w g_s L_{AI} g_{ba}\rho D + g_s L_{AI} S\phi}{\rho_w\lambda_w[\gamma_w(g_s L_{AI} + g_{ba}) + g_s L_{AI} S]},
$$

[1] The actual dependence of leaf available energy on L_{AI} is quite complicated (e.g., Jones, 1992). For simplicity, ϕ is assumed to be independent of L_{AI} and equal to the daily effective irradiance on the leaves. See also Section 6.1.1.

in the unknowns s, E, Ψ_s, and Ψ_l, is fully determined once g_s and the time course of the meteorological variables (e.g., radiation, air specific humidity, and air temperature during the day) are given or modeled. The latter ones are computed by solving a suitable model for the boundary layer dynamics, presented in the following section, while the modeling of g_s is discussed in detail in Section 6.2.

6.1.1 Convective boundary layer modeling

Daly et al. (2004a) obtained the specific humidity, q, and potential temperature, ϑ, of the atmosphere by modeling the boundary layer growth according to the scheme of McNaughton and Spriggs (1986).[2] Representing the convective boundary layer as a well-mixed slab of air of thickness h with constant profiles of q and ϑ, the equations for heat and water balance are

$$\rho c_p h \frac{d\vartheta}{dt} = H + \rho c_p (\vartheta_s - \vartheta) \frac{dh}{dt}$$

$$\rho h \frac{dq}{dt} = \rho_w E + \rho (q_s - q) \frac{dh}{dt},$$

(6.14)

where ϑ_s and q_s are the potential temperature and the humidity at height h respectively and H, the sensible heat flux, is equal to $\phi - \rho_w \lambda_w E$. The effect of evaporation is assumed to be negligible compared to transpiration and the boundary layer growth rate is expressed as

$$\frac{dh}{dt} = \frac{H}{\rho c_p h \gamma_\vartheta},$$

(6.15)

γ_ϑ being the gradient of the potential temperature at height h (Lhomme et al., 1998). The temperature and humidity profiles in the overlying undisturbed atmosphere are assumed to be linear (Lhomme et al., 1998),

$$\vartheta_s = \gamma_\vartheta z + \vartheta_{s0} = 4.78z + 293.6$$

$$q_s = \gamma_q z + q_{s0} = -0.00285z + 0.01166,$$

(6.16)

---

[2] We recall that the potential temperature is defined as the temperature that a parcel of dry air would have if brought adiabatically from its initial state to the standard pressure, p_s, of 1000 mbar. It is given by $\vartheta = T\left(\frac{p}{p_s}\right)^{-\frac{R_{dry}}{c_p}}$, where T is the temperature of the parcel in Kelvin, p is the pressure of the parcel, R_{dry} is the gas constant of dry air, $R_{dry} = 287\,\text{J kg}^{-1}\text{K}^{-1}$ ($R_{dry} = R/\mu_{dry}$, where $R = 8314\,\text{J kmol}^{-1}\text{K}^{-1}$ is the gas constant and $\mu_{dry} = 28.98\,\text{kg kmol}^{-1}$ is the apparent molecular weight of dry air), c_p is the specific heat capacity at constant pressure, $c_p = 1012\,\text{J kg}^{-1}\text{K}^{-1}$ (e.g., Peixoto and Oort, 1992; Pielke, 2002).

with z (evaluated at $z = h$) in km, ϑ_s in K and q_s in kg kg^{-1}. For simplicity, in Daly et al. (2004a) the leaf available energy is assumed to be independent of the leaf area index and equal to the effective solar radiation reaching the leaves. This is simulated using the parabolic form

$$\phi(t) = \phi_{\max}\frac{4}{\delta^2}[-t^2 + (\delta + 2t_0)t - t_0(t_0 + \delta)], \qquad (6.17)$$

with maximum daily irradiance $\phi_{\max} = 500$ W m^{-2}, day length $\delta = 12$ hours and $t_0 = 6$ hours (Lhomme et al., 1998).

6.2 Stomatal function

The final step to complete the hourly model described by the system (6.13) is the introduction of a suitable expression for the stomatal conductance. Plants in fact control the opening of the stomata to regulate the water and CO_2 transit during transpiration and photosynthesis (Figure 4.7). This allows them to maintain turgor and reduce dehydration as well as to control leaf temperature and maximize carbon assimilation. The complex mechanisms of stomatal movement depend on both plant physiology and environmental factors (Section 4.1). No complete model for their functioning has been developed so far, so that empirical approaches are usually employed. Daly et al. (2004a) critically analyzed and compared two well-known models for the stomatal functions which are described below.

6.2.1 Jarvis' formulation

Jarvis' empirical formulation (Jarvis, 1976; Jones, 1992; Lhomme et al., 1998; Lhomme, 2001) uses a multiplicative relationship of functions of the main factors affecting stomatal movement, such as solar radiation, ambient temperature, leaf water potential, potential saturation deficit, and CO_2 concentration,

$$g_s = g_{smax}f_\phi(\phi)f_{T_a}(T_a)f_{\Psi_l}(\Psi_l)f_D(D)f_{CO_2}(CO_2), \qquad (6.18)$$

where g_{smax} is the maximum stomatal conductance when none of the factors is limiting. The functions in Eq. (6.18) are obtained from controlled environment studies and account separately for the influence of each variable (Figure 6.1). A brief discussion of these functions is useful for understanding the links of Jarvis' formulation with the more physiological approach described next.

The direct effect of increasing light is usually expressed as an exponential function $f_\phi(\phi) = 1 - \exp[-k_1\phi]$ (Figure 6.1a), where k_1 is equal to

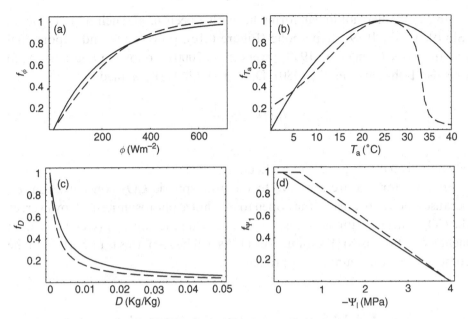

Figure 6.1 Functions controlling stomatal response to the environment: (a) solar irradiance, (b) ambient temperature, (c) water vapor deficit, and (d) leaf water potential, following Jarvis' (continuous line) and Leuning's (dashed line) approaches. After Daly et al. (2004a).

$0.005 \, \mathrm{m^2 \, W^{-1}}$ (Jones, 1992, page 156). In general, stomata open as ambient temperature increases up to an optimum value, T_{opt}, after which they start closing (Figure 6.1b). A simple function commonly used is $f_{T_a} = 1 - k_2(T_a - T_{\mathrm{opt}})^2$, with $k_2 = 0.0016 \, \mathrm{K^{-2}}$ and $T_{\mathrm{opt}} = 298 \, \mathrm{K}$ (Lhomme et al., 1998).

Following Jones (1992), Lhomme et al. (1998), and Lhomme (2001), Daly et al. (2004a) assumed no control of leaf water potential, Ψ_l, up to a certain point Ψ_{l_1} (i.e., well-watered conditions), from which stomatal conductance starts decreasing to zero when Ψ_l drops to Ψ_{l_0} (Figure 6.1d), i.e.,

$$f_{\Psi_l}(\Psi_l) = \begin{cases} 0 & \text{for} & \Psi_l < \Psi_{l_0} \\ \frac{\Psi_l - \Psi_{l_0}}{\Psi_{l_1} - \Psi_{l_0}} & & \Psi_{l_0} \leq \Psi_l \leq \Psi_{l_1} \\ 1 & & \Psi_l > \Psi_{l_1}. \end{cases} \tag{6.19}$$

The parameters Ψ_{l_1} and Ψ_{l_0} are taken here equal to $-0.07 \, \mathrm{MPa}$ and $-4 \, \mathrm{MPa}$, respectively, in agreement with measured values for C_3 plants in semi-arid ecosystems (Scholes and Walker, 1993; Bonan, 2002).

Since in many species an increase in water vapor deficit reduces stomatal conductance independently of Ψ_l, a function $f_D(D)$ describing direct stomatal sensitivity to D is also often adopted. A linear relation is suggested by Jarvis

(1976), but other nonlinear forms have also been used, such as polynomial (Shuttleworth, 1989), hyperbolic (Figure 6.1c), power law, and exponential functions (see Oren et al., 1999; Ewers et al., 2000). Following Leuning (1995) (see also Lohammer et al., 1980), Daly et al. (2004a) assumed

$$f_D(D) = \frac{1}{1 + \frac{D}{D_x}}, \tag{6.20}$$

where D_x is $0.0077 \, \text{kg kg}^{-1}$ (e.g., Leuning, 1995).

Finally, stomata are sensitive to the atmospheric CO_2 concentration, c_a, because when c_a decreases they open to maximize photosynthesis. However, as the CO_2 concentration is almost constant during the day, $f_{CO_2}(CO_2)$ is seldom included in Eq. (6.18) (Lhomme, 2001). As will be seen, this is important in the comparison with Leuning's approach.

6.2.2 Leuning's physiological formulation

The model introduced by Ball et al. (1987) and improved by Leuning (1990, 1995) is based on the observed linear dependence of CO_2 stomatal conductance, g_{s,CO_2}, on the net assimilation rate, A_n, for normal environmental conditions. Leuning's formulation is

$$g_{s,CO_2} = a_1 A_n \frac{f_D(D)}{c_s - \Gamma^*}, \tag{6.21}$$

where A_n is net carbon assimilation per unit leaf area, a_1 is an empirical constant, whose typical value is around 15 (Leuning, 1995), Γ^* is the CO_2 compensation point (see Eq. (6.33) below), and c_s the carbon concentration at the leaf surface. The latter is related to the CO_2 atmospheric concentration (see Figure 6.2 and Eq. (6.22)). The same Eq. (6.20) is used in Eq. (6.21) for $f_D(D)$, but the value of D_x used in this second approach is $0.0018 \, \text{kg kg}^{-1}$ (Leuning, 1995), which gives similar values of transpiration and assimilation in wellwatered conditions with both approaches. The residual (i.e., cuticular) value of conductance is assumed to be negligible in Eq. (6.21), which is equivalent to setting the cuticular conductance to zero and assuming that there is no carbon flux when the stomata are closed (see next section).

Differently from Jarvis' approach, which completely determines g_s, Eq. (6.21) requires a model for the net carbon assimilation A_n and thus for leaf photosynthesis and respiration. Such a model, which is described in the next section, will also suggest some interesting connections with Jarvis' formulation.

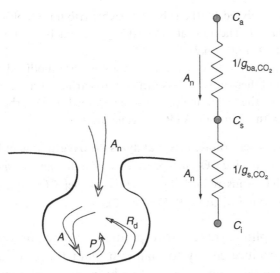

Figure 6.2 Scheme of the gas balance within stomata. After Daly et al. (2004a).

6.3 Leaf carbon assimilation and photosynthesis

Photosynthesis is the process by which light energy is absorbed by green plants and used to produce carbohydrates from carbon dioxide and water. It occurs in the chloroplasts of the leaves through two subsequent processes. The first process involves light directly and is thus called "light reactions" whilst the second is called "dark reactions" as light is not directly involved. We schematically review its basic functioning while defining both notation and terminology.

1. *Light reactions*: light energy is first absorbed by pigment molecules (mostly chlorophyll) and transferred to reaction centers where a series of biochemical reactions are initiated to create reducing power in the form of NADPH (reduced nicotinamide adenine dinucleotide phosphate) and chemical energy in the form of ATP (adenosine triphosphate). Water is necessary in this phase as it is oxidized into two protons, two electrons and $\frac{1}{2}$ oxygen (photolysis). For this reason, leaf cells need to be well hydrated and the leaf water potential Ψ_l must be high, otherwise photosynthesis is reduced. This is important in water-controlled ecosystems (e.g., Larcher, 1995; Bonan, 2002) but is often neglected in the models. The water abundance in the leaf cells is also the cause of the great amount of water lost by evaporation when the stomata are opened to allow CO_2 influx.

2. *Dark reactions*: NADPH and ATP are used to convert the CO_2 (taken up from the atmosphere through the stomata) into carbohydrates through the Calvin cycle. This consists of three phases:

   (a) *Carboxylation*: the five-carbon sugar RuBP (ribulose-1,5-biphosphate) combines with CO_2 and water to form three-carbon compounds (C_3 plants). This reaction is catalyzed by the rubisco enzyme (ribulose biphosphate carboxylase-oxygenase). The rate at which CO_2 is fixed by carboxylation will be indicated as A (Figure 6.2).

(b) *Reduction*: next the three-carbon compounds are modified using ATP and NADPH and then in part transformed into carbohydrates.

(c) *Regeneration*: the remaining part of the modified three-carbon compounds is combined with additional ATP to regenerate RuBP.

During the Calvin cycle, rubisco also catalyzes the oxygenation of RuBP, consuming oxygen and producing CO_2, in a process called either *photorespiration* or *RuBP oxygenation*, which thus reduces by 30–50% the net CO_2 uptake. The rate of photorespiration is indicated by P (Figure 6.2).

The opposite of photosynthesis is respiration by which organic compounds are oxidized to produce energy to maintain plant functions and grow new tissues. Respiration takes place day and night in almost all the plant parts; in the leaf cells at daytime it occurs simultaneously during photosynthesis through the stomata. This type of respiration is called either *dark respiration* or *daytime respiration* (e.g., Farquhar et al., 1980; Campbell and Norman, 1998) and is distinct from photorespiration. The rate of dark respiration is indicated by R_d (Figure 6.2).

6.3.1 Modeling net assimilation

During the day, as a result of photosynthesis and respiration (and simultaneously with transpiration), plants exchange CO_2 with the environment as a diffusion process through the stomata. Under steady-state conditions, the net CO_2 flux per unit leaf area (i.e., net assimilation) is

$$A_n = g_{sba,CO_2}(c_a - c_i), \qquad (6.22)$$

where c_a and c_i are the CO_2 concentrations in the atmosphere and leaf pores, respectively (Figure 6.2), and g_{sba,CO_2} is given by the series of the CO_2 stomatal, leaf boundary layer, and atmospheric conductance. It is usually assumed that $g_s = 1.6g_{s,CO_2}$ (e.g., Jones, 1992), $g_{b,CO_2} = g_b/1.37$ (Bonan, 2002), and $g_{a,CO_2} = g_a$ (Jones, 1992), where the conductances are expressed in mol_{H_2O} m^{-2} s^{-1} and mol_{CO_2} m^{-2} s^{-1}, respectively.

As the adaptation of CO_2 concentration is much faster than the time scale of stomatal adjustment, the CO_2 balance inside the stomatal pores may be described by the steady-state condition (Figure 6.2)

$$A_n = A - P - R_d. \qquad (6.23)$$

The daytime respiration, R_d, depends on leaf temperature and is reduced at low leaf water potentials (e.g. Larcher, 1995, page 114). However, since R_d is typically a small fraction of assimilation, it will be neglected for the sake of simplicity, as was also done by Dewar (1995) and Katul et al. (2003).

The rate of CO_2 fixation by carboxylation, A, and the rate of photorespiration, P, primarily depend on ϕ, c_i, T_l, and Ψ_l (nutrient status and internal oxygen concentration are factors of secondary importance that are not considered here). As already mentioned, the limitation of A and P by low leaf water potential is seldom considered in modeling studies. However, such a dependence becomes crucial for studies focusing on water-stressed conditions. For this reason, Daly et al. (2004a) assumed an independent linear reduction of assimilation from an unrestricted rate at $\Psi_{l_{A1}}$ to zero at $\Psi_{l_{A0}}$, that is

$$A_n \simeq A - P = A_{\Psi_l} \cdot A_{\phi, c_i, T_l}(\phi, c_i, T_l), \qquad (6.24)$$

where

$$A_{\Psi_l} = \begin{cases} 0 & \text{for} & \Psi_l \leq \Psi_{l_{A0}} \\ \frac{\Psi_l - \Psi_{l_{A0}}}{\Psi_{l_{A1}} - \Psi_{l_{A0}}} & & \Psi_{l_{A0}} \leq \Psi_l \leq \Psi_{l_{A1}} \\ 1 & & \Psi_l > \Psi_{l_{A1}}. \end{cases} \qquad (6.25)$$

Such behavior is similar to those reported by Larcher (1995, pages 113–21), Tezara et al. (1999), and Bonan (2002, page 303) and gives a dependence of g_s on Ψ_l very similar to that used for Jarvis' model. Since the assimilation reduction caused by chemical action presumably starts for a leaf water potential level lower than that corresponding to incipient stomata closure (Larcher, 1995), Ψ_{l_1}, $\Psi_{l_{A1}}$ is assumed to be equal to $-0.5\,\text{MPa}$. For simplicity, Daly et al. (2004a) chose $\Psi_{l_{A0}}$ equal to Ψ_{l_0} (Eq. (6.19)), since assimilation does not occur when stomata are completely closed.

The dependence of $(A - P)$ on ϕ, c_i, and T_l is given by the model of Farquhar et al. (1980; see also Collatz et al., 1991; Leuning, 1995; Foley et al., 1996). Although in this chapter we only consider C_3 plants, the model can easily be extended to C_4 plants following Collatz et al. (1991; see also Foley et al., 1996). When the leaf water potential is not limiting, the rate of CO_2 fixation is expressed as a function of three potential capacities to fix carbon, i.e.,

$$A_{\phi, c_i, T_l} = f(A_c, A_q, A_s), \qquad (6.26)$$

where A_c is the assimilation rate limited by rubisco activity (i.e., limited by CO_2 concentration), which depends on c_i, T_l, and atmospheric oxygen concentration, o_i; A_q is the assimilation rate when the photosynthetic electron transport limits RuPB regeneration (light limitation), which depends on ϕ, c_i, and T_l;

and A_s is the assimilation rate in conditions of high c_i and ϕ, when photosynthesis is limited by T_l.

Following Farquhar et al. (1980) and the modifications by Collatz et al. (1991), Leuning (1995), and Foley et al. (1996), the rubisco-limited rate of photosynthesis is given by

$$A_c = V_{c,max} \frac{c_i - \Gamma^*}{c_i + K_c(1 + c_i/K_o)}, \tag{6.27}$$

where $V_{c,max}$ is the maximum carboxylation rate, K_c and K_o are the Michaelis–Menten coefficients for CO_2 and O_2 respectively, and Γ^* is the CO_2 compensation point. These are given by Eq. (6.33) below, while the oxygen concentration is assumed to be constant (Collatz et al., 1991).

The assimilation rate when the photosynthetic electron transport limits RuPB regeneration (light limitation) is

$$A_q = \frac{J}{4} \frac{c_i - \Gamma^*}{(c_i + 2\Gamma^*)}, \tag{6.28}$$

where J is the electron transport rate, given by the lower root of the equation

$$\kappa_1 J^2 - (\kappa_2 Q + J_{max})J + \kappa_2 Q J_{max} = 0, \tag{6.29}$$

where Q (mol photons m^{-2} s^{-1}) is the absorbed photon irradiance (proportional to the leaf available energy, ϕ) and

$$J_{max} = J_{max0} \frac{\exp\left[\frac{H_{vJ}}{RT_0}\left(1 - \frac{T_0}{T_l}\right)\right]}{1 + \exp\left(\frac{S_v T_l - H_{dJ}}{RT_l}\right)}, \tag{6.30}$$

with the parameters as in Table 6.2.

A_s is approximately the maximum value of net assimilation in saturating conditions of carbon concentration and irradiance, modeled as (Collatz et al., 1991)

$$A_s = \frac{V_{c,max}}{2}. \tag{6.31}$$

To account for the gradual transition among the three carbon assimilation rates and to obtain A_{ϕ,c_i,T_l} in Eq. (6.26), Collatz et al. (1991) proposed using the roots of the following set of equations

$$\begin{cases} \beta_1 A_p^2 - (A_c + A_q)A_p + A_c\,A_q = 0 \\ \beta_2 A_{\phi,c_i,T_l}^2 - (A_p + A_s)A_{\phi,c_i,T_l} + A_p\,A_s = 0, \end{cases} \tag{6.32}$$

Table 6.2 *Parameters for the model of* C_3 *photosynthesis. After Daly et al.* (2004a).

Parameter	Value	Units	Description
H_{K_c}	59 430	J mol^{-1}	Activation energy for K_c
H_{K_o}	36 000	J mol^{-1}	Activation energy for K_o
H_{vV}	116 300	J mol^{-1}	Activation energy for $V_{c,max}$
H_{dV}	202 900	J mol^{-1}	Deactivation energy for $V_{c,max}$
H_{vJ}	79 500	J mol^{-1}	Activation energy for J_{max}
H_{dJ}	201 000	J mol^{-1}	Deactivation energy for J_{max}
J_{max0}	17.6	W m^{-2}	Eq. (6.30)
K_{c0}	302	μmol mol^{-1}	Michaelis constant for CO_2 at T_0
K_{o0}	256	mmol mol^{-1}	Michaelis constant for O_2 at T_0
o_i	0.209	mol mol^{-1}	Oxygen concentration
R	8 314	J kmol^{-1}K^{-1}	Gas constant
S_v	650	J mol^{-1}	Entropy term
T_0	293.2	K	Reference temperature
$V_{c,max0}$	50	μmol m^{-2} s^{-1}	Eq. (6.33)
κ_1	0.95	–	Eq. (6.29)
κ_2	0.20	(mol electrons)/ (mol photons)	Eq. (6.29)
β_1	0.9	–	Eq. (6.32)
β_2	0.9	–	Eq. (6.32)
γ_0	34.6	μmol mol^{-1}	CO_2 compensation point at T_0
γ_1	0.0451	K^{-1}	Eq. (6.33)
γ_2	0.000347	K^{-2}	Eq. (6.33)

where A_p is an intermediate value that gives the minimum between A_c and A_q, and β_1 and β_2 are empirical constants (see Table 6.2).

The Michaelis–Menten coefficients, the maximum carboxylation rate, and the T_l dependence of the CO_2 carboxylation point are modeled using the following empirical functions (Leuning, 1995)

$$K_x = K_{x0} \exp\left[\frac{H_{K_x}}{RT_0}\left(1 - \frac{T_0}{T_l}\right)\right]$$

$$\Gamma^* = \gamma_0[1 + \gamma_1(T_l - T_0) + \gamma_2(T_l - T_0)^2] \tag{6.33}$$

$$V_{c,max} = V_{c,max0}\frac{\exp\left[\frac{H_{vV}}{RT_0}\left(1 - \frac{T_0}{T_l}\right)\right]}{1 + \exp\left(\frac{S_v T_l - H_{dV}}{RT_l}\right)},$$

where x stands for c or o and the terms are reported in Table 6.2.

With the above model of photosynthesis, Eqs. (6.22) and (6.26) allow either the direct evaluation of assimilation with Jarvis' approach once the other environmental variables are determined or the computation of the stomatal conductance, assimilation, and transpiration with Leuning's approach, when coupled to Eq. (6.21).

The stomatal function for Leuning's approach is thus

$$g_s = 1.6 \, a_1 A_{\Psi_l} A_{\phi, c_i, T_l}(\phi, c_i, T_l) \frac{f_D(D)}{c_s - \Gamma_*}, \tag{6.34}$$

which now has many similarities with Jarvis' Eq. (6.18). As shown by Figure 6.1, Eqs. (6.34) and (6.18) behave quite similarly to a function of D, ϕ, and Ψ_l. However, an important difference between the two approaches is in the dependence of g_s on CO_2 concentration. As discussed in Section 6.2.1, Jarvis' formula does not usually include a CO_2 control, consistent with the observation that stomata tend to remain open for very low CO_2 concentrations to favor assimilation, unless the other factors (i.e., ϕ, T_a, Ψ_l, D) become limiting. In Eq. (6.34), instead, a double dependence on CO_2 concentration is present: one inherent in A_n and the other in the corrective term $c_s - \Gamma^*$ introduced by Leuning (1995; see also Jarvis et al., 1999) to account for the fact that, when external CO_2 decreases towards the compensation point, the stomata tend to open, the other external parameters remaining the same. In this way, when c_s tends to Γ^*, both net assimilation and $c_s - \Gamma^*$ tend to zero, leading to a finite (and higher than the usual values) limit for g_s.

6.4 Hourly dynamics

The system (6.13), coupled to the stomatal function, Eq. (6.18) or (6.34), and to the convective boundary layer model (Section 6.1.1) was solved numerically by Daly et al. (2004a) using the parameters given in Tables 6.1 and 6.2 as well as the suitable initial conditions of soil moisture and boundary layer height. Assimilation is given by Eq. (6.24). The plant parameters are typical of a C_3 woody plant adapted to semi-arid conditions (e.g. Scholes and Walker, 1993).

Figure 6.3 shows the temporal dynamics of some of the main variables characterizing the SPAC during a dry-down of several days. It is apparent how soil moisture decays more slowly during the night for lack of transpiration. Leaf water potential is equal to soil water potential during the night, while during the day it is much lower because of transpiration (Figure 6.3b). Through the energy balance, transpiration also influences the boundary layer, affecting leaf temperature (not shown) and potential saturation deficit (Figure 6.3c). D and Ψ_l affect stomatal behavior and, in turn, cause a reduction in

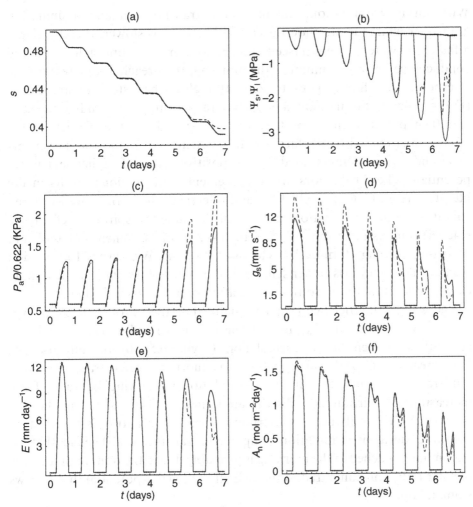

Figure 6.3 Results at hourly time scale using Jarvis' approach (continuous line) and Leuning's formulation (dashed line), with an initial value of s equal to 0.45. (a) Relative soil moisture, (b) soil and leaf water potential, (c) water vapor pressure deficit, (d) stomatal conductance, (e) transpiration, and (f) net leaf carbon assimilation. $R_{AI} = 5.6$, $L_{AI} = 1.4$, and $Z_r = 60$ cm; soil is a loam. Plant parameters as in Tables 6.1 and 6.2. After Daly et al. (2004a).

assimilation (Figure 6.3f). Both the daily maximum assimilation and minimum leaf water potential decrease with time, while the maximum D increases. The carbon concentration inside the stomata (not shown), driven by the assimilation rate, is lower during the day (at night c_i is imposed equal to c_a).

Figure 6.3 also highlights some differences due to the two stomatal functions. The higher leaf water potential given by Leuning leads to higher assimilation even in the afternoon, in spite of the faster stomata closure (Figure 6.3d).

With Jarvis' formulation, the maximum transpiration and assimilation are higher while the afternoon decay of E and A_n is slower, thereby giving a faster soil water depletion (see Figure 6.3a). For both approaches stomatal conductance has a maximum in the morning, then reaches a plateau in the afternoon, and finally goes to zero at night, when stomata are closed (Figure 6.4a, b). Stomata quickly open in the morning when radiation starts and then partially close for the effect of D and Ψ_l; the plateau in the afternoon is related to the fact that the reduction of ϕ and the increments in D and T_a are compensated by the positive effect of higher leaf water potentials. These behaviors are in agreement with previous results in the literature (e.g., Collatz et al., 1992 and references therein). In well-watered conditions, both formulations give about the same transpiration (cf. Figure 6.4c, d) and assimilation (cf. Figure 6.4e, f) rates. When soil moisture decreases, Leuning's formula leads to lower E and A_n. With Jarvis' formulation the daily course of E is rather symmetrical, while with Leuning's one it more closely follows that of g_s so that in the afternoon it tends to be lower.

The water use efficiency (WUE) – defined as the ratio of A_n/A_{max} over E/E_{max}, where A_{max} and E_{max} refer to daily rates under well-watered conditions – also differs between the two formulations (Figure 6.5a, b). In both cases the WUE reaches its peak value in the early morning, when transpiration is hindered by the moister air and assimilation is growing because of the increasing solar irradiance. As the air humidity declines during the day so does WUE, until the late afternoon when the situation is reversed by the slower decay of assimilation compared to transpiration. The fact that the maximum is reached in the morning is probably due to the dominant increase of solar radiation compared to the reduction of water potential that follows stomatal opening.

Some differences between the two approaches in the evaluation of WUE are shown in Figure 6.6. With Jarvis' formulation the early-morning high values of WUE tend to increase at the beginning of the dry-down, but they dramatically decrease when the soil moisture level becomes lower than a certain value (as will be seen, the value of s at which WUE starts diminishing approximately corresponds to the soil moisture level s^*, below which a plant can be considered to be under water stress conditions). A steeper drop is evident for the WUE evaluated at midday. A similar pattern also holds for Leuning's approach, which leads to lower early-morning values of WUE under well-watered conditions and gives higher WUE in conditions of low soil moisture. This difference is due to the dependence of stomatal conductance on net assimilation in Leuning's formulation, which, differently from Jarvis' one, couples A_n to E.

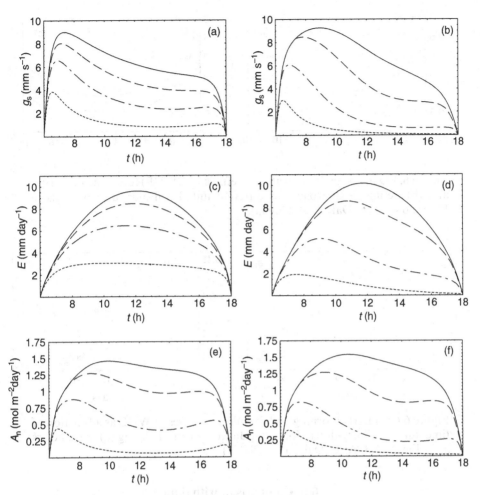

Figure 6.4 Results at hourly time scale following Jarvis' (left column) and Leuning's (right column) formulation for different initial soil moisture levels: $s = 1$ (continuous line), 0.45 (dashed line), 0.4 (chain line) and 0.35 (dotted line). Parameters as in Figure 6.3. After Daly et al. (2004a).

This last analysis of the results at the hourly time scale partly anticipates the results of the scaling at the daily time scale (see next sections). In fact, it already shows the existence of two different regimes in the assimilation dynamics of the plant: an unstressed one, in which soil moisture deficit does not affect WUE, and a subsequent one of progressive worsening of plant water stress with decreasing WUE.

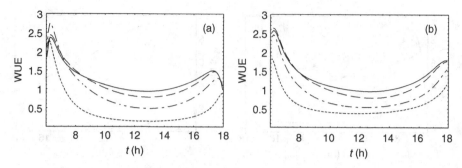

Figure 6.5 Daily behavior of water use efficiency (WUE) following (a) Jarvis' and (b) Leuning's approach for different initial soil moisture levels, as in Figure 6.4. After Daly et al. (2004a).

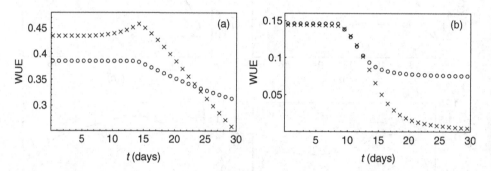

Figure 6.6 (a) Early morning and (b) midday values of WUE as a function of time during soil dry-down, following Jarvis ($\times$) and Leuning ($\circ$); parameters as in Figure 6.3. After Daly et al. (2004a).

6.5 Comparison with data

A close inspection of the diurnal patterns of transpiration and assimilation (Figures 6.3 and 6.4) actually shows that they are modulated mostly by the changes in soil water potential. The trends of their mean values are thus expected to be more regular, allowing simple and parsimonious modeling of daily transpiration, assimilation, and soil moisture dynamics. Examples and detailed discussion of the upscaling at the daily level will be given in Section 6.6. Here we only analyze and compare with measured data the functional dependence of leaf conductance on relative soil moisture at the daily level. Figure 6.7 compares the model results with the measured leaf conductance of the C_3 plant *Nerium oleander* (Gollan et al., 1985) as a function of the so-called extractable soil moisture. As the measurements were conducted under constant irradiance, the model results during a phase of dry-down are sampled at a given hour of the day to ensure constant irradiance. The time of the sampling is

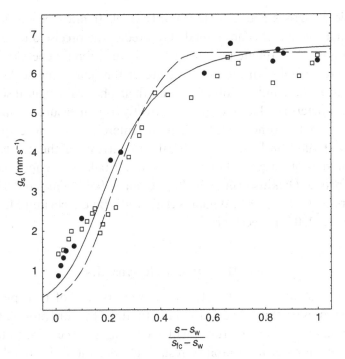

Figure 6.7 Stomatal conductance g_s as a function of the extractable soil moisture for *Nerium oleander* measured maintaining constant irradiance and two different levels of vapor pressure deficit (solid circles and open squares; after Gollan et al., 1985). Comparison with model results of stomatal conductance sampled every day at a given time for Jarvis' (continuous line) and Leuning's (dashed line) formulation. The time of the sampling is chosen so as to ensure the same stomatal conductances as the measured data under well-watered conditions. The extractable soil moisture is defined using typical values for loamy soil and a C_3 plant, e.g., $s_w = 0.20$ and $s_{fc} = 0.55$. See also Figure 2.4a.

chosen so that the modeled stomatal conductance matched the measurements at high soil water content. The general behavior is very well reproduced for both approaches with a plateau for well-watered conditions and a regular (almost linear) decay at low soil moisture values. The results of the model can thus be used with confidence to obtain the soil moisture dependence of transpiration and assimilation at the daily time scale, as will be seen in the next section.

In summary, the above comparison shows that, even if some important simplifications have been made, the model reliably reproduces the basic functioning of the coupled dynamics of photosynthesis, transpiration, and soil water balance at the hourly time scale, and thus fulfills our goal of obtaining a physical interpretation of the daily time scale from an understanding of the basic processes acting at the hourly level. It must be borne in mind, however, that the model is not intended to predict the precise diurnal behavior or to quantitatively estimate parameters of the functional dependence of transpiration and assimilation on soil moisture at the daily time scale. Once the typical form of these links is obtained and their physical origin discussed, the relevant parameters need to be estimated directly from measured data (e.g., as in Chapter 5). On the other hand, when the interest is in more quantitative predictions at the hourly level, a critical comparison of the results of more complete models of integrated dynamics of water balance, ecophysiology, and meteorology (e.g. Dickinson et al., 1986; Bonan, 1995; Pollard and Thompson, 1995; Sellers et al., 1986, 1997; Williams et al., 1996; Baldocchi and Meyers, 1998; Siqueira et al., 2002) is necessary.

6.6 Daily time scale dynamics

The dynamics of plant water potential as well as that of transpiration and assimilation rates have marked diurnal fluctuations that are modulated by soil moisture dynamics whose features evolve on a longer (i.e., daily) time scale. Actually, the short-term variability resulting from the diurnal solar forcing mostly impacts plant conditions through its daily-averaged behavior and its link with the soil water balance. To pinpoint the essential interactions between hydrological processes and vegetation it is thus convenient to bypass the diurnal fluctuations and focus on the behavior of their daily means, whose values are expected to be more regular, allowing simple and parsimonious modeling of daily transpiration, assimilation, and soil moisture dynamics (Daly et al., 2004b).

Examples of the daily processes obtained from the integration of the results at the hourly time scale are shown in Figures 6.8 and 6.9. The mean daily values of soil moisture during the drying phase are observed to follow a very similar pattern to those at the hourly time scale. The behavior of transpiration, shown in Figures 6.8b and 6.9a, closely resembles the empirical relationship often employed before (Chapter 2), where transpiration is approximated as a piecewise function of relative soil moisture, s, constant above a certain s^* and linearly decreasing to zero at the wilting point, s_w. The integrated results of Figure 6.8a are important for various reasons. First of all, they offer a sounder justification of the simplified model of soil moisture dynamics employed up to now: the empirical components in the hourly time scale model are more specific from a physical point of view and, therefore, easier to justify. Second, as discussed in detail in the next section, the three parameters of the transpiration function, s_w, s^*, and E_{max}, are now connected by the temporal

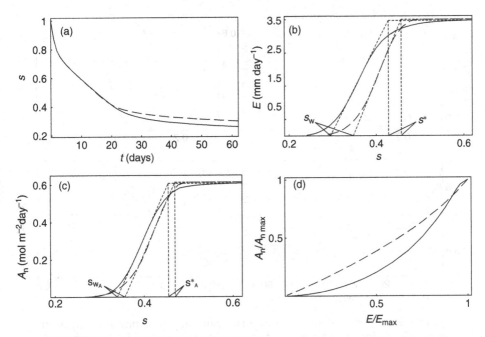

Figure 6.8 Results at the daily time scale for Jarvis' approach (continuous line) and Leuning's formulation (dashed line): (a) relative soil moisture during a drying period, (b) and (c) daily transpiration and daily net assimilation, respectively, as a function of relative soil moisture (shown by the dotted lines are the approximating piecewise functions, see text for details), (d) relation between assimilation and transpiration. Loamy soil, $R_{AI} = 5.6$, $L_{AI} = 1.4$, and soil depth $Z_r = 60$ cm. Vegetation and soil parameters refer to Tables 6.1 and 6.2. After Daly et al. (2004b).

scaling to the plant, soil, and climate characteristics. Finally, because of the links of s_w and s^* with the plant water stress (Chapter 4), the parameters of the climate–soil–vegetation system modeled at the daily scale also include the most important feedbacks between the hydrological processes and plant conditions.

A behavior similar to that of daily transpiration is also found for the dependence of daily carbon assimilation on soil moisture (Figures 6.8c and 6.9b). This dependence may be functionally approximated as (Daly et al., 2004b)

$$A_n(s) = \begin{cases} 0 & s \leq s_{W_A} \\ \frac{s - s_{W_A}}{s^*_A - s_{W_A}} A_{max} & s_{W_A} \leq s \leq s^*_A \\ A_{max} & s^*_A \leq s \leq 1. \end{cases} \quad (6.35)$$

As will be seen in the next section, this simple description of assimilation as a function of soil moisture at the daily time scale is key to linking photosynthesis and assimilation to the probabilistic description of soil moisture dynamics. Owing to the direct effects of low leaf water potential on the chemical reactions

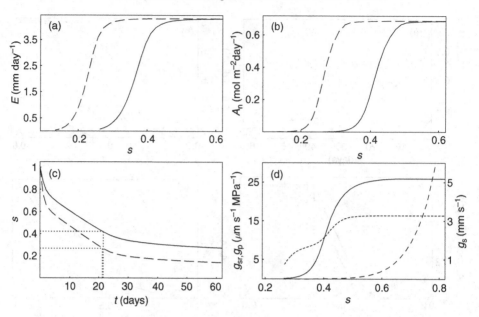

Figure 6.9 Influence of soil type on (a) mean daily transpiration, (b) mean daily assimilation, and (c) mean daily soil moisture using Jarvis' approach: loam (continuous line) and loamy sand (dashed line). Dotted lines in (c) show the times after which soil moisture reaches s^* with the two considered soil types. In (d) comparison of different mean daily conductances for a loam with $Z_r = 60$ cm: stomatal conductance (continuous line), soil–root conductance (dashed line), and plant conductance (dotted line). See caption of Figure 6.8 for other parameters. After Daly et al. (2004b).

of photosynthesis, the values of s_{w_A} and s_A^* in Eq. (6.35) are slightly higher than the corresponding points for transpiration. The relation between E and A_n (Figure 6.8d) shows a nonlinear behavior linked to the faster decrease of A_n compared to that of E at low soil moisture. This difference is less marked with Leuning's formulation because in that case E is related to A_n through g_s.

Jarvis' and Leuning's approaches give almost the same s_{w_A} and s_A^*, while for Leuning's formulation the assimilation rate under stressed conditions is lower (Figure 6.8c) and s_w and s^* are higher. This is probably an artifact in Leuning's formulation that could be solved by making the parameter a_1 of Eqs. (6.21) and (6.34) a function of soil moisture. Daly et al. (2004b) only consider Jarvis' formulation for its simplicity, but the qualitative similarity of the results of the two approaches gives confidence about the robustness of the hourly time scale model, whose integrated response at the daily time scale proves to be independent of the details at the hourly time scale (Figure 6.7) and the specific formulation of stomatal conductance (e.g., Jarvis' or Leuning's). It is important to remark, however, that the quantitative differences in the values of

s^* and s_w may become important when the model is applied to specific ecosystems. In this case a fine-tuned calibration of the parameters of the hourly time scale model or a direct field estimation of the parameters of the daily time scale model are required.

6.7 Physical interpretation of the parameters

Daly et al. (2004b) analyzed in detail the dependence of the parameters of the transpiration and assimilation functions on soil and vegetation characteristics. Figures 6.9a–c show the impact of two different soil types on soil moisture dynamics. As expected, soil properties, especially hydraulic conductivity and soil texture, control transpiration through their influence on both soil–root conductance and soil water potential. Plants in soils with higher saturated hydraulic conductivity, K_s, tend to have lower s^*, because of the lower soil resistance to water uptake and transpiration, but reach s^* approximately at the same time during the drying phase because of the higher leakage (Figure 6.9c). The soil type has negligible influence on the maximum transpiration and assimilation rates, which essentially only depend on the leaf–plant conductances that are dominating under well-watered conditions.

The plant type may play an important role in the value of s^*, through the root and leaf area index, the maximum leaf–plant conductances, and the thresholds controlling stomatal functioning and photosynthesis. The root area index (R_{AI}) has some impact on the resulting value of s^*, because a decrease in R_{AI} reduces the soil–root conductance and thus s^* increases. The leaf area index (L_{AI}) has a greater influence on s^* for its direct control on transpiration. A higher L_{AI} causes a faster soil drying accompanied by higher values of s^* (Figure 6.10) so that, when plants allocate more biomass to leaves than to roots, their sensitivity to water deficit is increased. On the other hand, since an increase in L_{AI} implies higher transpiration rates and an earlier onset of water stress, plants maintaining a constant R_{AI}/L_{AI} have lower s^* variability even with a high leaf area index and, therefore, are probably less sensitive to water stress. The strategy of allocating more biomass in the leaves in humid conditions in order to achieve higher assimilation rates and of preferentially allocating biomass in the roots in arid environments for a better soil water use (Larcher, 1995, page 143) supports these observations. The value of s^* is also conditioned by Ψ_{l_1}, i.e., the leaf water potential level at which stomata start closing in Jarvis' formulation. Low values of Ψ_{l_1}, which are typical of plants with high resistance to water stress, correspond to low values of s^*, when all the other parameters are the same.

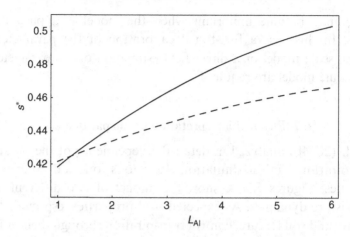

Figure 6.10 Influence of root area index and leaf area index on s^*. Continuous line represents constant $R_{AI} = 5.6$, while the dashed line is for constant $R_{AI}/L_{AI} = 4$. Soil type is a loam with $Z_r = 60$ cm; plant parameters are typical of a C_3 plant adapted to semi-arid conditions (see Tables 6.1 and 6.2). After Daly et al. (2004b).

If s^* appears to be significantly affected by R_{AI} and L_{AI}, these two indexes do not seem to influence s_w very much, which instead mainly depends on K_s through g_{sr} which is the first conductance to become limiting at low soil moisture levels (Figure 6.9d). The plant influence on s_w occurs through Ψ_{l_0}, the leaf water potential corresponding to complete stomata closure (i.e., $g_s = 0$) and to widespread xylem cavitation (i.e., $g_p \sim 0$). In a similar way, s_{w_A} is strongly influenced by the combined effect of Ψ_{l_1} and $\Psi_{l_{A_1}}$, i.e., the values at which the leaf water potential starts affecting respectively stomata movement and net assimilation. A reduction in $\Psi_{l_{A_1}}$ lowers the water potential at which the chemical reactions of photosynthesis start being impaired. This in turn increases the steepness of the curve A_n versus s between s_{w_A} and s_A^*. The maximum transpiration and assimilation rates, E_{max} and A_{max}, essentially depend on both vegetation type, through the physiological parameters g_{smax} and g_{pmax}, and the atmospheric boundary layer conditions (Daly et al., 2004b).

6.8 Probabilistic dynamics of carbon assimilation

The previous description of the daily dynamics of soil moisture, transpiration, and carbon assimilation during periods without rainfall was extended to include stochastic rainfall fluctuations (Daly et al., 2004b). Using the framework of Chapter 2 for the rainfall model and the resulting infiltration, the soil moisture dynamics at the daily time scale obtained in the previous section remains valid during randomly varying inter-storm periods, provided the initial conditions are properly updated according to the stochastic infiltration dynamics.

With the only proviso of accounting separately for evaporation and transpiration, the equation for the vertically averaged soil moisture is the same as in Section 2.1, i.e.,

$$nZ_r \frac{ds(t)}{dt} = I[s(t), t] - E[s(t)] - EV[s(t)] - L[s(t)], \tag{6.36}$$

and so is the steady-state probability density function (pdf) of soil moisture,

$$p(s) = \frac{C}{\rho(s)} e^{-\gamma s + \lambda' \int_s^{s_w} \frac{du}{\rho(u)}}, \tag{6.37}$$

where C is a normalization constant and $\rho(s)$ is the normalized sum of the losses, i.e., $\rho(s) = \{E[s(t)] + EV[s(t)] + L[s(t)]\}/nZ_r$.

The probabilistic structure of the daily assimilation can be easily obtained as a derived distribution of $p(s)$ (Daly et al., 2004b). The pdf of A_n has an atom of probability in zero, $P_{A_0} = \int_0^{s_{wA}} p(s)\, ds$, one in A_{max} equal to $P_{A_{max}} = \int_{s_A^*}^1 p(s)ds$, and is continuous in between. The two atoms of probability represent, respectively, the fraction of time during a growing season in which soil moisture is too low for a plant to perform photosynthesis and the fraction of time when the plant achieves maximum assimilation. The probability $P_{A_{st}} = \int_{s_{wA}}^{s_A^*} p(s)ds$ is the average fraction of time in a growing season in which assimilation takes place in nonoptimum (i.e., stressed) conditions. These statistics, through the soil moisture dynamics, synthesize the action of climate, soil, and vegetation characteristics on plant photosynthesis and carbon assimilation. Figure 6.11 shows an example of pdf's of s and A_n for two climatic conditions acting on a loamy soil and plant characteristics as used before in this chapter. In arid climates, the mode of soil moisture is located at low values of s with a high P_{A_0} value and a fast decrease of $p(A_n)$ for high A_n values, while in wetter climates $P_{A_{max}}$ is higher and the decrease of $p(A_n)$ with A_n is slower.

The impact of the rainfall regime, both in terms of total amounts per growing season and as a function of the frequency and amount of rainfall per event, may be studied by varying the parameters λ and α. Figure 6.12a shows the effect of increasing the total rainfall by varying λ while keeping α constant. As expected, a greater water availability increases the probability of maximum assimilation rate, $P_{A_{max}}$ and lowers the total time of zero assimilation, P_{A_0}. Interestingly, the maximum of assimilation under stressed conditions, $P_{A_{st}}$, roughly corresponds to a rainfall regime for which previous

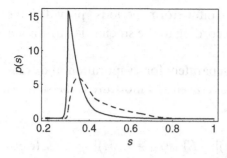
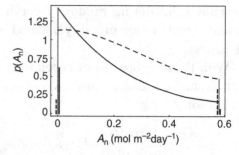

Figure 6.11 Probability density functions of soil moisture and assimilation as a function of frequency of rainfall events ($\lambda = 0.15$, continuous line; $\lambda = 0.3$, dashed line). The vertical bars are the atoms of probability at zero and A_{max}. Soil type is a loam with $Z_r = 60$ cm; $E_{max} = 3.5$ mm day^{-1}, $L_{AI} = 1.4$, $R_{AI} = 5.6$, $A_{max} = 0.58$ mol m^{-2} day^{-1}, $s_w = 0.3$, $s^* = 0.43$, $s_{w_A} = 0.34$, and $s^*_A = 0.45$. After Daly et al. (2004b).

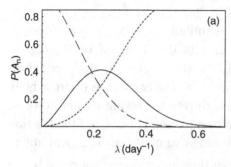
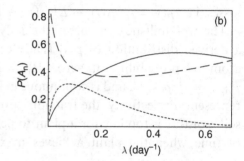

Figure 6.12 Probabilities of assimilation rate, P_0 (dashed line), $P_{A_{st}}$ (continuous line), and $P_{A_{max}}$ (dotted line), as a function of frequency of rainfall, λ, (a) for a given rainfall depth ($\alpha = 1.5$ cm) and (b) for constant mean total rainfall ($\Theta = 60$ cm) during a growing season ($T_{seas} = 200$ days, $\Delta = 0.2$ cm). Soil type is a loam with $Z_r = 60$ cm; $E_{max} = 3.5$ mm day^{-1}, $L_{AI} = 1.4$, $R_{AI} = 5.6$, $A_{max} = 0.58$ mol m^{-2} day^{-1}, $s_w = 0.3$, $s^* = 0.43$, $s_{w_A} = 0.34$, and $s^*_A = 0.45$. After Daly et al. (2004b).

analysis based on the crossing analysis of soil moisture (e.g., Chapter 4) found the transition between stressed and unstressed plant conditions.

Figure 6.12b shows the result of the probability of assimilation when the total rainfall is kept constant but the frequency and the mean depth of rainfall events are changed. A maximum in the probability of both stressed and unstressed assimilation is present. In particular, a maximum in $P_{A_{max}}$ for the same total rainfall implies a maximum in the efficiency of the rainfall regime for assimilation and photosynthesis. In such conditions runoff and interception losses are low compared to transpiration and, at the same time, the soil moisture is at a level that allows the plant to maintain relatively high water potential and hence sufficient levels of tissue turgor, hydration, and photosynthesis.

6.9 Mean carbon assimilation and plant water stress

Although A_n does not take into account the carbon lost by plant respiration (see Section 6.3.1), the net carbon assimilation at the leaf level can be considered an index of plant productivity. In this respect, the analysis of the mean assimilation during a growing season, $\langle A_n \rangle = \int_0^{A_{max}} A_n p(A_n) dA_n$, is of particular interest (Daly et al., 2004b).

Figure 6.13 shows the interplay between the timing and the amount of rainfall and $\langle A_n \rangle$. Similarly to the results in Figure 6.12b, net assimilation reaches a maximum at intermediate values of λ. Increasing the mean total rainfall Θ induces higher $\langle A_n \rangle$, but also shifts the maximum towards higher values of λ. For a given type of soil and vegetation, a maximum in $\langle A_n \rangle$ is linked to the efficiency of the rainfall regime and recalls the results of Chapters 2 and 4, which reported the existence of optimal plant conditions, in terms of transpiration and water stress, for average conditions of frequency and amounts of rainfall events.

In order to fully describe the link between assimilation and plant conditions, the analysis needs to be complemented by suitable statistics accounting for the

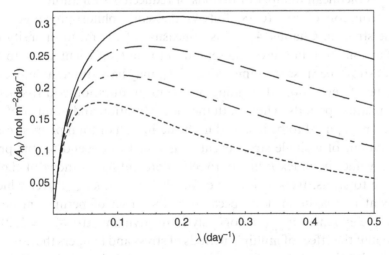

Figure 6.13 Mean assimilation rate as a function of frequency of rainfall events for different total rainfall in a growing season ($\Theta = 80$ cm continuous line, 70 cm dashed line, 60 cm chain line, and 50 cm dotted line). Soil type is a loam with $Z_r = 60$ cm; $E_{max} = 3.5$ mm day^{-1}, $L_{AI} = 1.4$, $R_{AI} = 5.6$, $A_{max} = 0.58$ mol m^{-2} day^{-1}, $s_w = 0.3$, $s^* = 0.43$, $s_{w_A} = 0.34$, and $s_A^* = 0.45$. After Daly et al. (2004b).

temporal dynamics of assimilation. One possibility is to proceed by analyzing the crossing properties of assimilation levels, analogously to the methodology of Chapter 4, where the dynamic water stress was used to describe plant response to drought conditions. Along this line, Daly et al. (2004b) linked the plant water stress to the statistics of reduction of carbon assimilation by soil moisture deficit. For this purpose, the problem is considerably simplified by the one-to-one correspondence of the crossing statistics of levels of s and A_n that follows from Eq. (6.35). In fact, it is readily found that when the soil moisture level passes below a threshold ξ, net assimilation rate falls below the value $A_n(\xi)$; thus the mean duration and frequency of soil moisture below a certain level ξ is equal to the mean duration and frequency of assimilation below $A_n(\xi)$ (Daly et al., 2004b).

Considering now that the worst plant conditions in terms of plant productivity occur when assimilation is zero, a new measure of plant stress may be defined as

$$
\bar{\theta}_A = \begin{cases} \left(\dfrac{P_{A_0}\bar{T}_{s_{w_A}}}{kT_{\text{seas}}}\right)^{\bar{n}_{s_{w_A}}^{-r}} & \text{if } \ P_{A_0}\bar{T}_{s_{w_A}} < kT_{\text{seas}} \\ 1 & \text{otherwise,} \end{cases} \tag{6.38}
$$

where P_{A_0} represents the average fraction of a growing season in which photosynthesis is zero, $\bar{T}_{s_{w_A}}$ is the mean duration of periods of no assimilation, and $\bar{n}_{s_{w_A}}$ is the mean number of periods of reduced assimilation.

The definition of Eq. (6.38) follows the same philosophy used for the dynamic stress in Chapter 4. P_{A_0} is a measure of the mean intensity of the lack of assimilation in a growing season but is not sufficient alone to define the plant stress, because the same value of P_{A_0} may have different impacts on a given type of vegetation, depending on the mean duration and frequency of no-assimilation periods. Thus, both the mean duration and number of excursions below s_{w_A} need to be included in the definition of total plant stress. The mean amount of a single stress event is assumed to depend on the product $P_{A_0}\bar{T}_{s_{w_A}}$, while the index k in Eq. (6.38), representing an index of the plant resistance to stress, fixes a threshold for plant stress, kT_{seas}, over which the stress is at its maximum level because of the onset of permanent damage. Finally, the exponent $\bar{n}_{s_{w_A}}^{-r}$ (analogously to the dynamic stress, $r = 1/2$) takes into account the effect of multiple periods of stress and tempers the importance of very high values of $\bar{n}_{s_{w_A}}$, thus avoiding erroneously high values of $\bar{\theta}_A$ in the case of very short, but frequent, stress periods. The behavior of the plant stress based on net assimilation as a function of $(P_{A_0}\bar{T}_{s_{w_A}})/(kT_{\text{seas}})$ and $\bar{n}_{s_{w_A}}$ is similar to that of the dynamic water stress in Figure 4.12. When $P_{A_0}\bar{T}_{s_{w_A}}$ exceeds the threshold value kT_{seas}, the resulting plant stress is equal to 1, independently of

$\bar{n}_{s_{w_A}}$, while when $(P_{A_0}\bar{T}_{s_{w_A}})/(kT_{\text{seas}}) < 1$, the total plant stress increases in a way that depends on the value of $\bar{n}_{s_{w_A}}$, whose importance becomes negligible when $(P_{A_0}\bar{T}_{s_{w_A}})/(kT_{\text{seas}})$ tends to 1. In fact, when the stress of a single excursion below s_{w_A} is very high, the total plant stress is already so high that having more than one period of stress becomes irrelevant.

Figure 6.14 shows a comparison of the mean assimilation rate and the two stress measures, $\bar{\theta}$ directly based on water availability (see Eq. (4.21)) and $\bar{\theta}_A$ defined through the assimilation rate (Eq. (6.38)). Interestingly, the clear maximum reached by the net assimilation, evident in both low and high total rainfall conditions, does not correspond to a sharp minimum in $\bar{\theta}$ and $\bar{\theta}_A$. On the contrary and differently from $\langle A_n \rangle$, the two stresses have a wide minimum at slightly higher values of λ. This suggests that the consideration of the temporal dynamics of assimilation makes more realistic the description of plant responses to hydrologic fluctuations. Thus, favorable conditions for plants are not expected to be concentrated around a narrow range of parameter values, but rather to change slowly from optimal to nonoptimal ones.

The analysis of plant carbon assimilation under stochastic rainfall conditions provides an interesting synthesis that can be useful for improving the understanding of the impact of hydrologic processes on plant conditions. A recent analysis by Porporato et al. (2004) using the minimalistic soil moisture model with linear losses (Section 2.6.2) and the simple representation of assimilation of Eq. (6.35) assessed the effects of alterations of rainfall regime due to climate change on ecosystem productivity. Figure 6.15 compares the theoretical results with the recent findings of a four-year manipulative

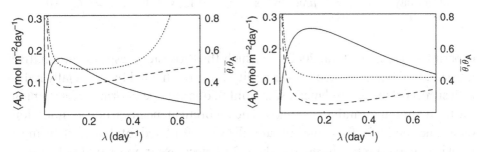

Figure 6.14 Comparison of mean assimilation $\langle A_n \rangle$ (continuous line), dynamic plant water stress $\bar{\theta}$ (dotted line), and assimilation-based plant water stress $\bar{\theta}_A$ (dashed line) as a function of frequency of rainfall events for different total rainfall, $\Theta = 50$ cm (on the left) and 70 cm (on the right). Soil type is a loam with $Z_r = 60$ cm, $T_{\text{seas}} = 200$ days, $k = 0.5$, $q = 3$, and interception $\Delta = 0.2$ cm; $E_{\text{max}} = 3.5$ mm day^{-1}, $L_{\text{AI}} = 1.4$, $R_{\text{AI}} = 5.6$, $A_{\text{max}} = 0.58$ mol m^{-2} day^{-1}, $s_w = 0.3$, $s^* = 0.43$, $s_{w_A} = 0.34$, and $s_A^* = 0.45$. After Daly et al. (2004b).

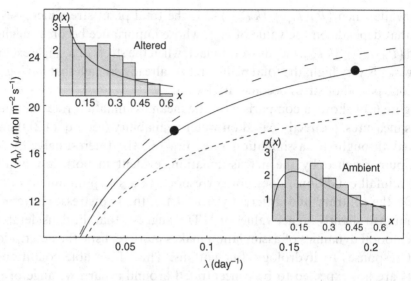

Figure 6.15 Mean daily carbon assimilation rate as a function of the frequency of rainfall events for constant total amount of precipitation during a growing season. The lines are the theoretical curves derived from the soil moisture pdf, while the two points are field data published by Knapp et al. (2002), who compared the response of a mesic grassland to ambient rainfall pattern versus an artificially increased rainfall variability. The point on the right corresponds to the ambient conditions and the one on the left to artificially modified conditions while keeping the total rainfall the same. The continuous line is for mean total rainfall during a growing season of 500 mm, the dashed line for 700 mm, and the dotted line for 300 mm. The two insets show observed and theoretical soil moisture pdf's for ambient and altered conditions. The parameters used are (Knapp et al., 2002): $\lambda_{ambient} = 0.19$ day^{-1}, $\lambda_{altered} = 0.08$ day^{-1}, $n = 0.38$, $Z_r = 30$ cm, $E_{max} = 0.6$ cm day^{-1}, $A_{max} = 28$ μmol m^{-2} s^{-1}, $s_w = 0.12$, $s^* = 0.30$, $s_1 = 0.65$. After Porporato et al. (2004).

experiment (Knapp et al., 2002), in which the ecosystem response of a native grassland to increased rainfall variability was investigated by artificially reducing storm frequency and increasing rainfall quantity per storm, while keeping the total annual rainfall unchanged. The theoretical mean carbon assimilation (as a function of the frequency of rainfall events for fixed total rainfall during a growing season) reproduces well the ~20% decrease in measured net assimilation for the altered rainfall pattern, going from a mean net assimilation of 23 μmol m^{-2} s^{-1} in natural conditions to ~18.4 μmol m^{-2} s^{-1} when total rainfall was the same but concentrated in fewer events. As shown by the effective relative soil moisture pdf's, the dramatic shift in the rainfall frequency changes the grassland water balance from an intermediate type to a dry one (cf. Figures 6.15 and 2.17). The analysis also shows that, in the grassland ecosystem studied by Knapp et al. (2002), the impact on carbon assimilation of a decrease in total rainfall is more pronounced when such a decrease is accompanied by a reduction in the frequency of rainfall events. If the mean total rainfall is kept constant, then the sensitivity of mean assimilation to the frequency of rainfall events becomes much more pronounced for dry periods.

7

Plant strategies and water use

In water-controlled ecosystems, water demand by plants is generally higher than water availability. To efficiently cope with water stress, plants have thus developed different strategies, which become more sophisticated the more intense and unpredictable the water deficit is. As we have already seen in the previous chapters, low soil moisture periods bring about an abatement of transpiration and cell turgor which with time may initiate a chain of damages of increasing seriousness. On the other hand, given the external conditions in terms of soil and climate, plants have the possibility to act on some of the components of the water balance to reduce their water stress level. Many species combine a number of complementary measures to do so, the most extreme of which include permanent forms of adaptation, such as changes in resource allocation, specialized growth of roots (e.g., cactuses build a dense network of roots to capture the light rainfall events, while some desert shrubs – the so-called phreatotypes – develop deep roots to tap the water table when present), specialized photosynthetic pathways (e.g., the CAM pathway), short and intense life cycles during favorable periods, dormancy, drought deciduousness, specialized metabolism and leaf structure to reduce water losses (high cuticular resistance, protection and changes in the dimension and density of the stomata), etc. (Jones, 1992; Larcher, 1995).

We will not deal with all this gamut of strategies, but rather focus on some of the less extreme ways of adaptation that are more commonly adopted in semiarid climates. These include the development of different photosynthetic pathways and root distributions, the adaptation of water use according to climate and soil characteristics, as well as temporary modifications of transpiration through stomatal control and osmotic adjustment. Since our focus will be on the hydrologic role of such strategies, the approach will be, again, a simplified one and will make use of the analytical tools previously developed. The first application follows the analysis of Rodríguez-Iturbe et al. (2001b) and concentrates on the difference between extensive and intensive use of soil moisture by plants that, respectively, adopt C_3 and C_4 photosynthetic pathways and different rooting depths. The second case, taken from Fernandez-Illescas et al. (2001), is an extension of the analysis of inverse texture effect (already introduced in Section 5.3) to assess its robustness to interannual rainfall fluctuations and its role in producing patterns of water stress which may favor the coexistence of different species. The last application concerns the role of the transpiration function and the hydrologic significance of plant osmoregulatory measures in providing optimal water use in conditions of stochastic water availability.

7.1 Extensive and intensive users of soil moisture

In ecosystems with typically dry growing seasons following a wet winter season (e.g., Mediterranean climates), some plants rely on the dependable winter recharge that is stored deeply in the soil, whereas others, not being able to tap such a resource, quickly respond to the intermittent and uncertain rainfall during the hot growing season. These kinds of plants are called extensive and intensive users, respectively. Intensive users develop a dense network of shallow roots to absorb moisture originating from rainfall during the growing season before it evaporates. Typical examples are shallow-rooted grasses with a C_4 photosynthetic pathway and a fine-tuned system of stomatal control that allows a rapid response to the intermittent rainfall. In contrast, extensive users are well adapted to low temperatures and have root systems that penetrate larger and deeper volumes of soils and a C_3 photosynthetic pathway. They extract water from both shallow and deep soil layers and are favored by winter rains that infiltrate deep into the soil (Burgess, 1995).

Figures 7.1a and 7.1b show typical daily traces of soil moisture for extensive and intensive users during the growing season. As previously discussed in Chapter 3, the effects of the winter moisture storage, namely the initial soil moisture condition at the start of the growing season, are much more important for deep-rooted plants whose transpiration dynamics and larger soil reservoir lead to a smoother soil moisture evolution. Intensive users respond quickly even to light and brief rainfall events, leading to a succession of water deficit periods substantially different from that impacting extensive users.

In the presence of an initial transient period, the mean time $\overline{T}_{s^*}(s_0)$ to reach the threshold of water stress, s^*, from an arbitrary soil water content, s_0, at the start of the growing season becomes crucial in the vegetation strategy to cope with stochastic water availability. The analytical derivation of $\overline{T}_{s^*}(s_0)$ is given in Section 3.1.1 (see also Section 3.3), where it was seen that, except for very

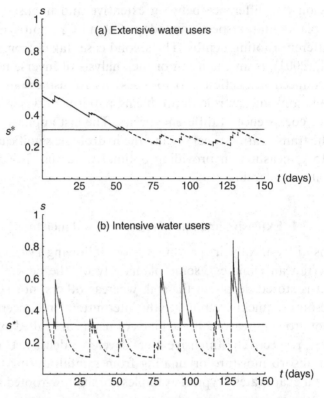

Figure 7.1 Examples of relative soil moisture traces modeled at a daily time scale for extensive (a) and intensive (b) water users. In (a) the mean rooting depth is $Z_r = 180$ cm, while in (b) $Z_r = 20$ cm. The dashed parts of the curves represent the periods of plant water stress. The values of $s^* = 0.31$ and $s_0 = s_{fc} = 0.51$ considered in the examples are typical for plants in loamy sand soils. The mean maximum daily evapotranspiration under well-watered conditions is $E_{max} = 0.45$ cm day^{-1}, the average frequency of rainfall events $\lambda = 0.167$ day^{-1}, and the average depth per event $\alpha = 2$ cm. After Rodríguez-Iturbe et al. (2001b).

dry conditions, the impact of winter storage increases dramatically with the amount of the growing season rainfall for effective soil depths larger than 60 cm. To account for the possible water storage at the start of the growing season, Rodríguez-Iturbe et al. (2001b) extended the previous definition of dynamic water stress, $\bar{\theta}$ (see Section 4.3), assuming that the water stress is zero at the beginning of the growing season, until $s(t)$ reaches s^*, and then equal to $\bar{\theta}$, i.e.,

$$\bar{\theta}' = \frac{T_{seas} - \overline{T}_{s^*}(s_0)}{T_{seas}} \bar{\theta}, \tag{7.1}$$

where T_{seas} is the duration of the growing season.[1] Despite its simplicity, this new normalized water stress provides an effective synthesis of the action of the soil water balance on plant conditions in ecosystems when the transient at the beginning of the growing season is important (Rodríguez-Iturbe et al., 2001b).

7.1.1 Applications

Eastern Amazonian forest

The evergreen forests of Eastern Amazonia are subject to prolonged dry seasons (typically five months from July to November) during which they are able to maintain evapotranspiration by taking up water from depths of six to eight meters or more (Nepstad et al., 1994). Deforestation transforms such forests into pastures with much shallower roots, which are able to withstand the dry season by responding more efficiently to the scarce precipitation (intensive users). Using the extended definition of stress to account for the high soil moisture at the start of the dry season (Eq. (7.1)), Rodríguez-Iturbe et al. (2001b) showed that the pastures may be a low-stress alternative to the deep-rooted forests. As appears from Figure 7.2, because of the scarcity of precipitation during the dry season, the rooting depth required to sustain a strategy of extensive soil water use needs to be larger than two meters. Moreover, the inter-annual coefficient of variation (CV ≈ 0.25), characteristic of the total rainfall during the wet season, imposes a large degree of uncertainty on the soil moisture storage s_0 at the start of the dry season. The likely occurrence of s_0 values smaller than the soil field capacity, s_{fc}, makes roots considerably deeper than the four meters necessary in order for extensive users to be competitive.

Southern Texas savanna

A different type of ecosystem is the subtropical savanna of southern Texas. As already seen in Section 5.2, this ecosystem has been reported to be very sensitive to rainfall fluctuations during previous decades, with tree density increasing after a sequence of wet years which followed a drought that decimated both trees and grasses in the region (Archer, 1989). The previous analyses (Section 5.2) considered steady-state conditions during the growing season and showed that in terms of water stress *Prosopis glandulosa* is favored for growing season rainfall above 400 mm, compared to the dominant grass, *Paspalum setaceum*. To refine the analysis, Rodríguez-Iturbe et al. (2001b) investigated how the two different

---

[1] When the steady-state conditions are very dry and the mean time to reach steady state from s^* is very long, the lower threshold s^* in Eq. (7.1) could be replaced by a more representative level, such as the mean or the mode of the steady-state distribution.

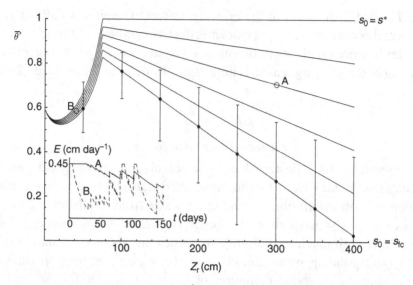

Figure 7.2 Mean total water stress $\overline{\theta}'$ during the dry season as a function of the average rooting depth Z_r for the case of the Eastern Amazonian forest. From top to bottom, the different curves refer to values of s_0 linearly increasing from s^* to s_{fc}. The climate parameters during the dry season are $\alpha = 1.7$ cm and $\lambda = 0.1$ day^{-1}, the soil is a clay-loam, $E_{max} = 0.45$ cm day^{-1} and $s^* = 0.5$ (Nepstad et al., 1994). The error bars on the bottom curve show that the effect on the total water stress of variations of E_{max} and s^* in a range of 20% around their mean values does not change the pattern of favorable conditions for extensive or intensive water users. The inset shows the dramatic difference in the transpiration patterns of an extensive (A) and an intensive water user (B) for such climate and soil conditions. After Rodríguez-Iturbe et al. (2001b).

functional vegetation types are affected by changes both in the winter soil moisture recharge and in the rainfall regime during the growing season. As Figure 7.3 shows, the point of equal water stress (see also Figure 5.16) moves towards lower rainfall as the winter recharge is increased. Soil moisture storage ($s_0 > s^*$) favors *Prosopis glandulosa* which is a phreatophyte (i.e., with deeper roots). On the other hand, if the recharge is not enough to overcome a particularly dry growing season, the more superficial roots ($Z_r \approx 40$ cm) and the more efficient transpiration system of *Paspalum setaceum* become the best way to exploit the uncertain and fleeting summer rainfall. Throughout the years, the ecosystem is thus in a situation of unstable equilibrium, evolving between tree encroachment and grass domination depending on both summer and winter rainfall amounts.

Colorado shortgrass steppe

An interesting situation is that occurring for two typical grasses of the short-grass steppe of Colorado, the C$_4$ blue grama (*Bouteloua gracilis*) and the C$_3$

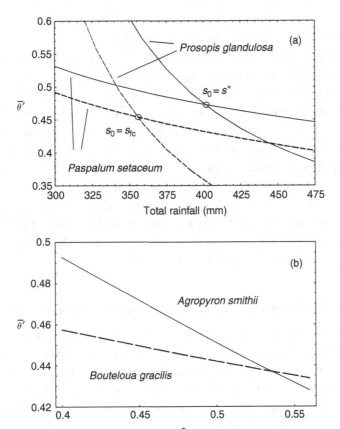

Figure 7.3 Mean total water stress, $\bar{\theta}'$, during the growing season: (a) as a function of the total rainfall for the southern Texas savanna and (b) as a function of the soil moisture value at the beginning of the growing season, s_0, for the Colorado shortgrass steppe. In (a) the continuous lines are for $s_0 = 0.37$, while the dashed lines are for $s_0 = 0.56$. λ and α vary linearly between the driest and wettest historic conditions as recorded in Alice, Texas, USA. The soil is a sandy loam, $Z_r = 100$ cm, $E_{max} = 0.4$ cm day^{-1}, $s^* = 0.37$ for *Prosopis glandulosa* and $Z_r = 40$ cm, $E_{max} = 0.44$ cm day^{-1}, $s^* = 0.35$ for *Paspalum setaceum* (Wan and Sosebee, 1991). In (b) the soil is a sandy loam, $\lambda = 0.28$ day^{-1}, $\alpha = 0.65$ cm, $E_{max} = 0.35$ cm day^{-1}, $Z_r = 65$ cm and $s^* = 0.40$ for *Agropyron smithii* and $E_{max} = 0.37$ cm day^{-1}, $Z_r = 30$ cm and $s^* = 0.36$ for *Bouteloua gracilis* (Sala et al., 1982b; Lauenroth et al., 1994, and references therein). After Rodríguez-Iturbe et al. (2001b).

western wheatgrass (*Agropyron smithii*). The ecosystem has already been analyzed in Section 5.3, where it was shown how, depending on the soil texture, the dominant *Bouteloua gracilis* may have different stress conditions during the growing season. Following Rodríguez-Iturbe et al. (2001b), here we analyze the possible effects of the transient of soil moisture at the beginning of the growing season in altering such a dominance.

The shallow-rooted *Bouteloua gracilis* ($Z_r \approx 30$ cm) is able to utilize the light summer rainfall by transpiring even during the early morning at very low soil water potential, while the deeper-rooted *Agropyron smithii* ($Z_r \approx 65$ cm) prefers high soil moisture levels at the beginning of the growing season. Figure 7.3b shows that, only if the winter recharge is sufficient to bring s_0 close to field capacity, $s_{fc} = 0.57$, *Agropyron smithii* has a lower water stress; otherwise, *Bouteloua gracilis* is better suited to the intermittent and light rainfall of the arid growing season. The different strategies of these two species help them to select their temporal and spatial niches for water use, thus reducing the pressure of competition for the limiting resource. Only very rarely, however, the winter recharge is such to invert the situation and make the more extensive user, *Agropyron smithii*, the superior competitor.

7.2 The inverse texture effect in the southern Texas savanna

This section follows Fernandez-Illescas et al. (2001) and analyzes in greater detail the inverse texture effect, already introduced in Section 5.3, and its dependence on plant type and interannual rainfall variability. The analysis focuses on the two most common vegetation types coexisting in La Copita, Texas, namely the herbaceous C_4 *Paspalum setaceum* and the woody C_3 *Prosopis glandulosa* (see Section 5.2).

7.2.1 Soil texture

Fernandez-Illescas et al. (2001) focused on the 12 textural classes within the USDA soil texture classification (Figure 7.4), which assumes that all soil particle sizes can be sorted into three convenient size ranges called separates or textural fractions, namely sand, silt, and clay. Conventionally, the overall textural classification of a soil is determined as a function of the mass ratios of these three textural separates. Soil texture is also used as a descriptor of soil physical properties such as porosity, n, saturated hydraulic conductivity, K_s, soil matrix potential at saturation, $\overline{\Psi}_s$, and pore size distribution index, b (Cosby et al., 1984). Although other descriptors such as horizon and structural size certainly influence the hydraulic parameters of soils, Cosby et al. (1984) carried out a two-way analysis of variance of nine descriptors to conclude that soil texture alone can account for most of the discernible patterns in K_s, $\overline{\Psi}_s$, b, and n. As a result, under given climatic and vegetation conditions, those soil-texture-dependent physical properties determine the soil wetness values which

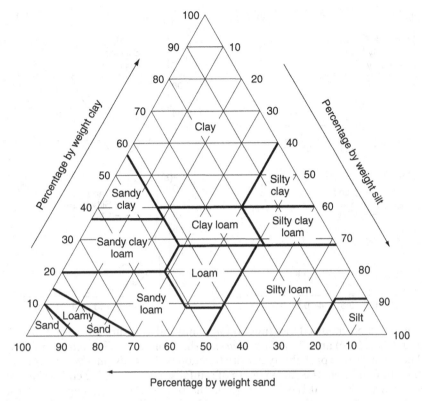

Figure 7.4 The USDA soil textural triangle.

in turn establish the water condition of the plant. Sand, for instance, has a lower n than clay and, as a result, is able to store less water than clay for a given soil depth. Clay, on the other hand, with a smaller K_s than sand, produces less amount of drainage under saturated conditions in the soil column and, because of the smaller size of the individual pores which increase the adsorptive forces between grain and water, leads to more negative values of soil water potentials for a given soil moisture value. This makes water extraction by plants more difficult and may reduce transpiration rates.

Fernandez-Illescas et al. (2001) obtained n, K_s, $\overline{\Psi}_s$, and b from the univariate regression equations defined by Table 5 in Cosby et al. (1984). These univariate regressions were deemed to be sufficient to describe most of the variability in hydraulic parameters (Cosby et al., 1984) and were extrapolated to include silty soils.

7.2.2 Soil moisture probability density function and soil texture

The daily average dependence during the growing season of the normalized evapotranspiration and leakage losses on relative soil moisture,

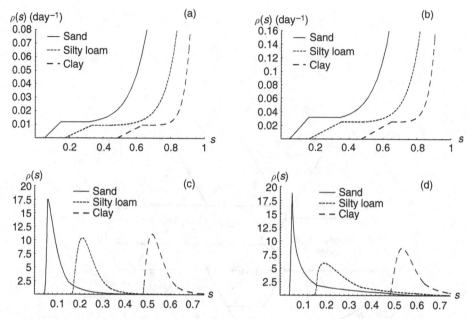

Figure 7.5 Normalized loss function $\rho(s)$ and soil moisture pdf's for *Prosopis glandulosa* (a, c) and *Paspalum setaceum* (b, d) on sand, silty loam, and clay. The soil moisture probability density function of *Prosopis glandulosa* is for dry climatic conditions (i.e., mean depth of rainfall events $\alpha = 1.342$ cm and rate of storm arrivals $\lambda = 0.166$ day^{-1}), while the one of *Paspalum setaceum* is for wetter climatic conditions (i.e., mean depth of rainfall events $\alpha = 1.417$ cm and rate of storm arrivals $\lambda = 0.202$ day^{-1}). After Fernandez-Illescas et al. (2001).

$\rho(s) = \chi(s)/nZ_r$, is shown in Figures 7.5a, b for *Prosopis glandulosa* and *Paspalum setaceum* for sand, silty loam, and clay. Figure 7.5c shows the behavior of the growing season steady-state pdf of soil moisture for *Prosopis glandulosa* on these three soil textural classes under the dry climatic conditions observed in the region during the period 1950–60, while Figure 7.5d shows the same for *Paspalum setaceum* under the wet climatic conditions observed during 1973–83. Figure 7.6 shows the steady-state mean soil moisture, $\langle s \rangle$, corresponding to the different categories of the USDA soil textural triangle for *Prosopis glandulosa* under the dry climatic conditions. It is clear that as the percentage of clay increases, the pdf and the loss function shift to higher soil moisture values. These shifts occur because of the adsorptive forces which permit higher water retention at any given suction. It is also clear that these soil texture related patterns are robust to vegetation and climatic characteristics.

While the pdf's of soil moisture shown in Figures 7.5c, d are quite different, all of them have their mode between the corresponding s_w and s^* indicating that both species are in water stress conditions at this site. The mode, however,

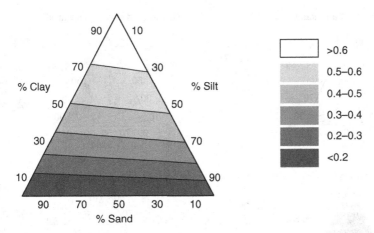

Figure 7.6 Mean soil moisture for *Prosopis glandulosa* under dry climatic conditions on the soil textural triangle. After Fernandez-Illescas et al. (2001).

is more pronounced for sand than for silty loam or clay because of the higher probability of soil moisture being at or near s_w for sandy soils resulting from the larger slope of the loss function, $\rho(s)$, in the $\{s_w, s^*\}$ range.

7.2.3 Water balance components

Figure 7.7 shows the components of the water balance, expressed as a percentage of growing season rainfall for *Prosopis glandulosa* under dry climatic conditions. Interception is independent of soil texture and is therefore not shown. All the components show considerable variability within the soil texture triangle.

From Figures 7.7a and 7.7b one observes that sandy soils have the lowest mean levels of stressed evapotranspiration while unstressed evapotranspiration is highest for the silty loam region of the triangle. The manner in which average percentages of stressed and unstressed evapotranspiration depend on soil texture results from the dependence on soil properties of the probability of soil moisture being below s^*, $P(s^*)$, and the probability of soil moisture being above s^*.

Figure 7.7c shows that runoff constitutes a negligible portion of the water balance at this site and its variability within the soil triangle appears to follow gradients in mean soil moisture, $\langle s \rangle$ (Figure 7.6). Clayish soils, with higher mean values of soil moisture, are associated with larger amounts of runoff. Mean leakage levels are also quite small (Figure 7.7d) and increase as the probability of s being above s_{fc} increases. As a consequence, for *Prosopis glandulosa* under dry climatic conditions, clayish soils tend to exhibit a greater volume of leakage than sandy soils despite having a much lower K_s value, although in both cases the amounts are minimal.

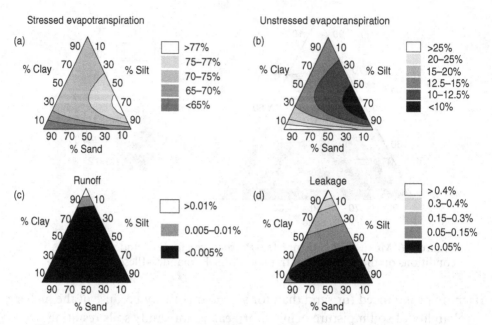

Figure 7.7 Mean components of the water balance for *Prosopis glandulosa* under dry climatic conditions at La Copita Research Area for the USDA soil textural triangle. After Fernandez-Illescas et al. (2001).

For dry climatic conditions, stressed evapotranspiration is the most important component of the water balance in terms of its magnitude for both *Prosopis glandulosa* and *Paspalum setaceum*. This suggests that under such climatic conditions and for any soil texture conditions, both species are water stressed most of the time. Leakage and runoff are negligible for *Prosopis glandulosa* and only slightly larger for *Paspalum setaceum*. As the climate becomes wetter, the magnitude of the unstressed evapotranspiration significantly increases for both species and the occurrence of leakage becomes greater for the shallow-rooted plant. Differences in soil water dynamics due to soil texture are large enough to suggest a pronounced influence on vegetation water stress regimes, which are the focus of the next section.

7.2.4 *Sensitivity of soil moisture crossing properties and water stress to soil texture*

Figures 7.8a–c show the mean duration of an excursion below s^*, $\overline{T}_{s^*}$, the mean number of crossing of s^*, $\overline{n}_{s^*}$, and the mean static water stress, $\overline{\zeta'}$, for *Prosopis glandulosa* under dry climatic conditions as a function of soil texture.

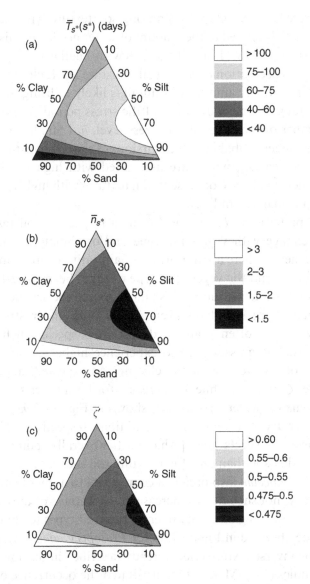

Figure 7.8 Crossing properties and static water stress for *Prosopis glandulosa* under dry climatic conditions at La Copita. (a) Mean duration of an excursion below s^*, (b) mean number of downcrossings of s^*, (c) mean static water stress using $q = 3$. The duration of the growing season is $T_{seas} = 153$ days. After Fernandez-Illescas et al. (2001).

The $\overline{T}_{s^*}$ values show a minimum for sandy soils while they show a maximum for the silty loam soils. In contrast, Figures 7.8b and 7.8c indicate that the $\overline{n}_{s^*}$ and $\overline{\zeta'}$ values both show the opposite pattern, i.e., a minimum for the silty loam soils and a maximum for the sandiest soils. These patterns appear to mimic the variability of the quantity $n(s^* - s_w)$ among soil textures (Fernandez-Illescas et al., 2001), which is related to the amount of water needed to raise the relative soil moisture from s_w to s^*. For soil textures with small $n(s^* - s_w)$, wilting and therefore water conservation occur at soil moisture levels close to the onset of stress s^*. As a result, any amount of rainfall is likely to bring the plant out of stress, diminishing the mean duration of the stress period, $\overline{T}_{s^*}$, but increasing the mean number of crossings of the stress level, $\overline{n}_{s^*}$. Small $n(s^* - s_w)$ values also result in a higher probability of excursions of the soil moisture trace to soil moisture values near s_w, which are associated with higher values of static stress, i.e., higher $\overline{\zeta'}$ values. Conversely, soil textures with high $n(s^* - s_w)$ result in large $\overline{T}_{s^*}$ and small $\overline{n}_{s^*}$ and $\overline{\zeta'}$.

The above patterns of $\overline{T}_{s^*}, \overline{n}_{s^*}$, and $\overline{\zeta'}$ with respect to soil texture do not qualitatively change with vegetation type and/or climate in the case of La Copita. Nevertheless, there are important quantitative changes in those measures as the climate and the vegetation type change (Fernandez-Illescas et al., 2001). For both dry and wet climates, shallow-rooted species such as *Paspalum setaceum* (not shown) spend on average less time under stress and cross the stress level more often than do deep-rooted species such as *Prosopis glandulosa*. Also, the grasses tend to experience higher $\overline{\zeta'}$ values than trees independently of the wetness of the climate. As expected, a wetter climate results in lower $\overline{\zeta'}$ and $\overline{T}_{s^*}$, while $\overline{n}_{s^*}$ increases for both species.

Values of dynamic water stress, $\overline{\theta}$, are shown in Figures 7.9a, b for *Prosopis glandulosa* under dry and wet climatic conditions, respectively. Under the dry climatic scenario, this vegetation type has a better overall condition (i.e., lower $\overline{\theta}$ values) on a sandier soil than on a finer texture soil such as a silty loam (Figure 7.9a). *Paspalum setaceum* also prefers a coarser soil to a finer texture soil when the climate is dry (not shown). An increase in the wetness of the climate results in a shift of the low $\overline{\theta}$ region to the finer texture soil types for both vegetation types (as can be observed in Figure 7.9b for *Prosopis glandulosa*), revealing the existence of the inverse texture effect already described in Section 5.3.

Previous studies (Noy-Meir, 1973) attributed the occurrence of the inverse texture effect to the impact of climate and soil texture on plant water availability and the manner in which this impact changes with the degree of climate aridity. Water loss in humid climates is greatly influenced by leakage losses, favoring plant growth in soils with high water holding capacities such as silty loams. In contrast, surface evaporation is a significant loss mechanism in more arid climates, favoring sandy soils that are capable of quickly draining water away from the surface and avoiding high levels of water loss due to direct soil evaporation. Since our vertically integrated model of soil moisture does not differentiate near-surface water available for direct soil evaporation from

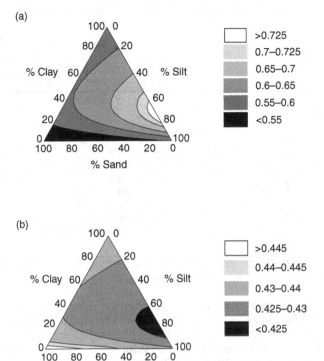

Figure 7.9 Dynamic water stress $\bar{\theta}$ for *Prosopis glandulosa* on the USDA soil textural triangle under (a) dry climatic conditions and (b) wet climatic conditions. $T_{\text{seas}} = 153$ days, $q = 3$ and $k = 0.5$. After Fernandez-Illescas et al. (2001).

water available for transpiration within Z_r, within our modeling framework such an effect is driven by the climate and soil sensitivity of the transition between stressed and unstressed conditions. In dry climates, vegetation favors soil textural types where a minimum amount of water is required to raise soil moisture levels above s^* and thus even weak storm events are capable of taking plants out of water stress. Consequently, vegetation will experience lower levels of dynamic water stress for coarse-textured soils with small $n(s^* - s_w)$. In contrast, wetter climates favor fine-textured soils with large $n(s^* - s_w)$ so that, if s^* is reached during a drying period, the soil moisture trace does not fall to levels associated with large static stress values. Thus, the finer textured soils have the lowest dynamic water stress values under wet climatic conditions.

7.2.5 The interaction of climate and soil and their impact on coexistence patterns

Figure 7.10 shows the relationship between the total growing season rainfall, Θ, and the difference in dynamic water stress between *Paspalum setaceum* and

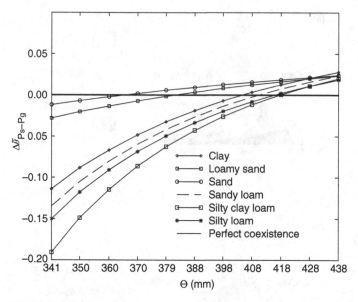

Figure 7.10 Difference in the dynamic water stress between *Paspalum setaceum* and *Prosopis glandulosa*, $\Delta\bar{\theta}_{Ps-Pg}$, as a function of the total incoming rainfall during the growing season, Θ. The parameters α and λ vary linearly with Θ increasing from their dry to wet values. $T_{seas} = 153$ days, $q = 3$ and $k = 0.5$. After Fernandez-Illescas et al. (2001).

Prosopis glandulosa, $\Delta\bar{\theta}_{Ps-Pg}$, for six soil textural classes. The total growing season rainfall was computed by linearly increasing, in a joint manner, the mean depth of rainfall events, α, and the mean frequency of storm arrivals, λ, from their dry to their wet period values (see Section 5.2 for a detailed description of the climate characteristics in La Copita, Texas). Figure 7.10 shows a decreasing sensitivity of the difference in the overall condition of the two species, $\Delta\bar{\theta}_{Ps-Pg}$, to total rainfall as the climate gets wetter. The figure also shows a transition in the sign of $\Delta\bar{\theta}_{Ps-Pg}$ with increasing growing season rainfall for all six soil textures. For dry periods grasses tend to dominate (values of $\Delta\bar{\theta}_{Ps-Pg} < 0$) while the opposite is true for wet periods (values of $\Delta\bar{\theta}_{Ps-Pg} > 0$). In addition to what is reported in Section 5.2, one observes that the interannual rainfall variability present at the site drives the system towards different states where soil texture plays a key role in determining the dominant vegetation type. Moreover, a sandy soil in this site leads to much smaller differences in the overall condition of the two species under strong interannual rainfall variability than those occurring, for example, within a silty clay loam. Wet periods are more conducive to *Prosopis glandulosa* but the differences in fitness between the two species are significantly smaller than those occurring in dry periods in favor of the grasses.

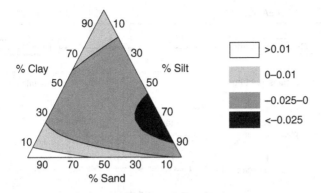

Figure 7.11 Difference in the dynamic water stress between *Prosopis glandulosa* and *Paspalum setaceum*, $\Delta\bar{\theta}_{Ps-Pg}$, for an average climate, $\alpha = 1.387$ cm and $\lambda = 0.188$ day^{-1}, for the USDA soil textural triangle. $T_{seas} = 153$ days, $q = 3$ and $k = 0.5$. After Fernandez-Illescas et al. (2001).

Fernandez-Illescas et al. (2001) selected a growing season rainfall of 40 cm, equal to the 1950–85 mean value, and assigned the mean depth per rainfall event, α, and the mean frequency of arrival of storms, λ, from the previously described linearly varying values to match the mean rainfall (i.e., $\alpha = 1.387$ cm and $\lambda = 0.188$ day^{-1}, respectively). The corresponding $\Delta\bar{\theta}_{Ps-Pg}$ values for the USDA soil texture triangle are shown in Figure 7.11, where it is seen that the transition between grass and tree domination can occur at this site at almost any soil texture and even more so within the sandy loam region which is the predominant soil in the area.

7.3 Stochastic water availability and adaptation of transpiration characteristics

We turn now to analyze the role of the transpiration function under conditions of stochastic soil water availability. In Chapter 6 we justified the simple piecewise form of the transpiration function and discussed the physical origin of its parameters, s_w, s^*, and E_{max}. Possible adjustments of the transpiration function may be performed by the plant through stomatal control, which manifests itself in changes of the parameter s^*. As discussed in Chapters 4 and 6, by means of stomatal control plants both minimize internal and soil water losses and maximize the photosynthetic carbon gain (at the same time they also try to maintain the cooling effect of evaporation). Such a regulation is made possible by osmoregulatory measures (i.e., osmotic adjustment), which allow changes of turgor of the guard cells (Larcher, 1995; Nilsen and Orcutt, 1998).

In conditions of marked water deficit, osmotic adjustment also involves other vital activities with varied consequences for the growth and survival of

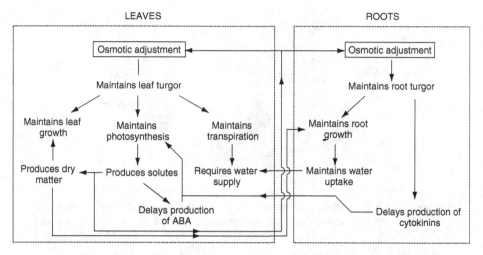

Figure 7.12 Effects produced by osmotic adjustment in roots and leaves. Cytokinins are plant phytohormones that, along with the abscisic acid (ABA), are involved in the stomatal regulatory system and in the processes of growth and development. Redrawn after Turner (1986).

plants (Figure 7.12). As water potential declines, live cells adjust their water status by accumulating osmotically active compounds. This increases their osmotic potential and thus helps maintain their turgor (Section 4.1.2, Figure 4.5). Under such conditions, stomata remain open longer, leaving more time for carbon assimilation, and plants continue to acquire water from the soil at low water potentials (Larcher, 1995; Lambers et al., 1998). Turgor maintenance through osmotic adjustment in response to water stress is widespread, occurring in leaves, roots, and the reproductive organs of many species (Turner and Jones, 1980). Even a partial turgor maintenance can be advantageous and may also be accomplished by a passive concentration effect as cell water decreases.

From the hydrological point of view, the most important consequence of osmotic adjustment is to facilitate water uptake at low soil moisture levels to avoid dehydration and keep performing photosynthesis. However, as the soil water reserve is very limited in such conditions, the osmotic adjustment is somewhat in contrast with the requirement of reducing soil water use (as a parsimonious strategy of water use would suggest). This, in addition to the existence of other soil water losses and the unavoidable cost of osmotic adjustment, clearly indicates that an efficient transpiration strategy may only be realized by taking into account the actual environmental variability of the soil water balance. Moreover, the unpredictability of future rainfall forces the plants to seek a sort of stochastic optimization of such clashing exigencies: a parsimonious water use that could imply low productivity and competitiveness versus an unnecessary and costly consumption of soil water that successively could end up in intense water stress. Using Jones' terminology (Jones, 1992), two such extremes correspond to the so-called pessimistic and optimistic water use behaviors. The balance between these two strategies depends on the intrinsic resistance and productivity potential under intense stress of each species (or plant), as well as on the existing environmental conditions.

As a detailed treatment of the physiological aspects of transpiration and osmotic adjustment is outside the scope of this book, here we will only analyze its ecohydrological consequences for which our stochastic soil moisture approach is particularly suited. Paraphrasing Cowan (1986), our analysis does not deal with the mechanistic description of the processes in terms of intrinsic plant metabolic characteristics, but tackles the problem in terms of its environmental dependence considering the statistical properties of the rainfall regime, leaf environment, and competition with other water losses. The importance of rainfall stochasticity in the assessment of optimal patterns of water use and plant conditions was previously pointed out by Cowan (1986) and Jones (1992), who stressed how optimal stomatal behavior depends on all the environmental and plant life aspects.

7.3.1 The role of s^* in the soil water balance and plant stress

It is assumed here that plants adjust their transpiration function by changing the point of incipient stomatal closure, s^*, and that osmotic adjustment results in a reduction of s^*. The effect of changes in s^* on soil moisture dynamics and water balance is analyzed in Figures 7.13 and 7.14. All the other parameters remaining constant, a reduction of s^* enhances transpiration at low soil moisture contents. Since in arid and semi-arid environments low soil moisture conditions are the most common ones, the consequences are noticeable. As the value of s^* is reduced, the most immediate result is the reduction of the soil moisture content in the soil (Figure 7.13) along with an increase of the transpiration under unstressed conditions in the water balance (Figure 7.14). The tradeoff between maintaining assimilation and parsimony in water use (notice that the water left in the soil is also prone to other losses and competition from neighboring plants) is clear.

Such contrasting aspects are even more evident when a single phase of drying is analyzed (Figure 7.15). With lower values of s^* soil moisture is depleted faster, but at the same time the appearance of static stress is delayed (Section 4.2). With high s^*, plants save soil water (provided no

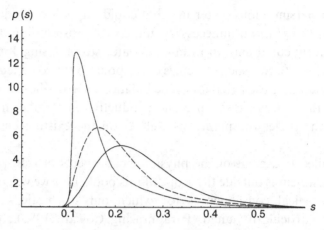

Figure 7.13 Examples of soil moisture pdf's for different values of s^*. Continuous line: $s^* = 0.4$; dashed line: $s^* = 0.3$; chain line: $s^* = 0.2$. Other parameters common to all curves: $\alpha = 1\,\text{cm}$, $\lambda = 0.3\,\text{day}^{-1}$, $E_{\max} = 0.45\,\text{cm day}^{-1}$, $E_w = 0.02\,\text{cm day}^{-1}$, $s_h = 0.08$, $s_w = 0.11$, $Z_r = 60\,\text{cm}$, $\Delta = 0.2\,\text{cm}$. The soil is a loam (see Table 2.1).

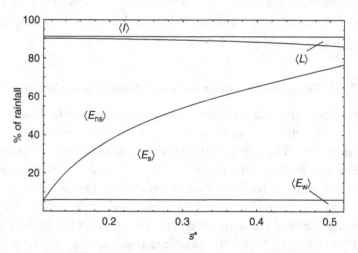

Figure 7.14 Water balance components as a function of s^* varying between the wilting point and the field capacity (parameters are the same as in Figure 7.13).

other losses are present) but also start experiencing water stress earlier than plants that perform osmotic adjustment. Obviously, osmotic adjustment cannot prevent the appearance of stress, and the situation is inverted after some time. The general behavior shown in Figure 7.15b is maintained also when different types of nonlinearities are used in the definition of the static stress.

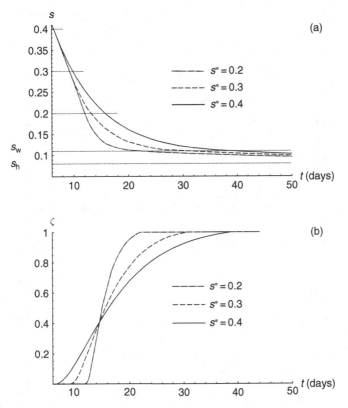

Figure 7.15 (a) Temporal evolution of soil moisture and (b) instantaneous static water stress (with $q = 2$) as a function of s^* (soil, climate, and vegetation parameters are the same as in Figure 7.13).

As mentioned before, the actual advantage of reducing s^* crucially depends on the stochastic aspects of precipitation, which control the mean duration of the stress periods as well as their mean frequency and intensity (Chapters 3 and 4). The case shown in Figure 7.16a refers to a climate with relatively infrequent storms which is typical of water-controlled ecosystems. Presumably, in such conditions, the static water stress ζ is often above the inversion point of Figure 7.15b, so that the average static stress $\overline{\zeta'}$ is lower with higher s^* (however, the difference in stress is high only at very low s^*, where also the cost of the osmotic adjustment is expected to be higher). The advantage of low values of s^* becomes clear when considering the duration of such periods of stress. Although the mean static water stress is somewhat more intense (Figure 7.16a) and more frequent (Figure 7.16b) with lower values of s^*, the average duration of stress periods is drastically reduced (e.g., 10 versus 40 days; Figure 7.16b).

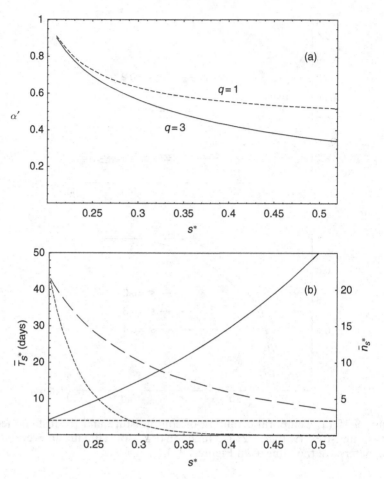

Figure 7.16 (a) Static water stress as a function of s^* for two different values of the nonlinear exponent q (Section 4.2). (b) Crossing properties of soil moisture levels s^* and s_w as a function of s^*. Continuous line: $\overline{T}_{s^*}$; dashed line: $\overline{n}_{s^*}$; dotted lines refer to $\overline{T}_{s_w}$ and $\overline{n}_{s_w}$ (parameters are the same as in Figure 7.13).

7.3.2 Transpiration control and minimum water stress

Figures 7.16a and 7.16b show the two opposite hydrological consequences of lowering s^*: an increase of intensity of water stress but also a reduction of its duration. The combination of these two conditions is synthesized by the dynamic water stress $\overline{\theta}$ (Section 4.3). The results are shown in Figure 7.17. When varying the value of s^* there is a minimum of dynamic water stress that corresponds to the most advantageous level of osmotic adjustment which in turn provides the optimum point at which to begin stomatal control to minimize water stress.

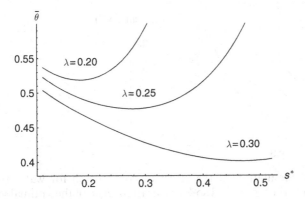

Figure 7.17 Dynamic water stress as a function of s^* for different values of the mean rate of storm arrival (other parameters are the same as in Figure 7.13).

Figure 7.17 also shows how the optimal transpiration function depends on environmental conditions. The drier the climate and the longer the interstorm period, the lower is the required value of s^* to reduce the average global stress level. One should notice however that, in the case of very arid conditions, the physiological cost of osmotic adjustment may be unbearable if not associated with other remedies. Moreover, since the dynamic water stress does not consider directly either the relationship between transpiration and productivity or the cost of osmotic adjustment at very low soil moisture values, the actual advantage of osmotic adjustment may be underestimated by the present analysis. The exact location of the minimum is also sensitive to the specific value of the nonlinear exponent q used to define the static water stress.

The sensitivity of the optimal transpiration function to changes in the other losses in the water balance provides interesting insights. By increasing the value of E_w, one may account for the effects of an increase in both soil evaporation and competition for moisture from neighboring plants. The behavior of the dynamic water stress as a function of s^* for different values of E_w is shown in Figure 7.18. Due to the more intense competition for water from the other losses, the optimal values of s^* must be reduced in order to ensure a more efficient (and aggressive) water use strategy. This agrees with the results by Cowan (1986, Figure 5.13 therein), who found steeper curves of optimal daily rate of assimilation (which is itself closely related to transpiration) in the case of higher values of water deficit due to increased water losses by neighboring plants.

Finally, we investigate the optimal transpiration conditions for the dominant herbaceous and woody species in the Nylsvley savanna already considered in Section 5.1. Figure 7.19 shows the comparison between the field measurements of s^* (see Table 5.2) and the theoretical values of s^* obtained

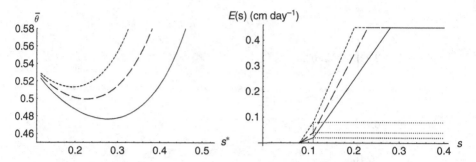

Figure 7.18 Simulation of the effect of competition for water from other plants as well as of increased soil evaporation (E_w) on the optimal value of s^*. (a) Minimization of dynamic water stress and (b) corresponding transpiration function. Continuous line: $E_w = 0.02$ cm day^{-1}; dashed line: $E_w = 0.05$ cm day^{-1}; dotted line: $E_w = 0.08$ cm day^{-1} (other parameters as in Figure 7.13).

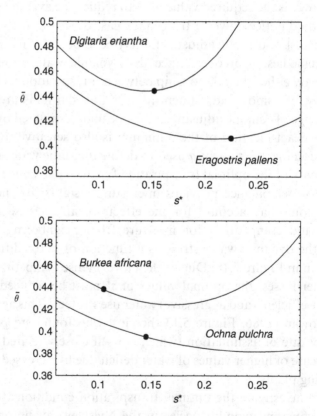

Figure 7.19 Comparison of measured values of s^* for plants of the Nylsvley savanna (Section 5.1) with those computed by minimizing the dynamic water stress. See Section 5.1 for climate, soil, and vegetation parameters.

by minimizing the dynamic water stress. The values of s^* for grasses are in very good agreement with the field measurements of Scholes and Walker (1993); the measured values for trees, however, are overestimated. Competition for water by neighboring plants (see Figure 7.18) may be one of the factors responsible for the difference (the value of E_w used in Figure 7.19 is too low to account for competition). The many simplifications in our analysis also reduce our ability to make more precise conclusions on such delicate aspects. The general findings, however, are encouraging and seem to provide a reasonable explanation for the underlying different transpiration strategies adopted by plants in water-controlled ecosystems.

8
Seasonal and interannual fluctuations in soil moisture dynamics

Apart from the notable exception of the analysis of the transient dynamics caused by the initial condition of soil moisture at the start of the growing season (Sections 3.3 and 7.1), until now only statistically steady conditions during the growing season have been considered. In the context of this book, however, two other sources of temporal dynamics may also be important: the seasonality of both the rainfall regime and the evapotranspiration demand, and the interannual rainfall fluctuations. The former may be especially relevant in Mediterranean climates or at mid latitudes where the rainfall regime may have multiple maxima (e.g., spring and fall) and the passage between the growing season and the dormant season is more gradual. In some cases, however, when the seasonality is marked but the regimes of the two seasons are distinct, the steady-state analysis of the growing season often proves to be satisfactory, provided that the initial soil moisture conditions are properly accounted for (see Chapters 5 and 7). The second form of variability, i.e., the interannual one, is in many aspects even more important, and tends to be more intense in arid and semi-arid regions. Its origin stems from the fact that each growing season is characterized by a rainfall regime in part controlled by general circulation and global climate characteristics, which are affected by year-to-year variability. This brings about interannual fluctuations in the growing season rainfall regime that are more intense than what would simply result from different realizations of precipitation obtained using the same stochastic process with constant coefficients (e.g., α and λ).

Through the action of the long-term water balance, both types of fluctuations may be key factors for the existence of temporal niches of soil moisture availability, which in turn may affect the coexistence of species and the development of different strategies of soil water use. The analysis of the seasonal variability follows the paper by Laio et al. (2002), while the analysis of the impact of interannual fluctuations is summarized after the work by D'Odorico et al. (2000a) and Ridolfi et al. (2000b). The role played by the interannual variability as a driving mechanism for the long-term evolution of ecosystems will be discussed in Chapter 11.

8.1 Seasonal mean soil moisture dynamics

The analysis of seasonal variability in soil moisture dynamics is extremely important not only for ecohydrology (e.g., Rodríguez-Iturbe et al., 2001a; Porporato and Rodríguez-Iturbe, 2002) but also for a number of other issues, such as the evaluation of the most probable timing of floods (e.g., Ettrick et al., 1987; Black and Werritty, 1997) and the determination of the possible feedbacks between the soil moisture and climate dynamics (e.g., Eltahir, 1998; Porporato et al., 2000).

The vertically integrated soil moisture balance equation (see Section 2.1) with seasonal components in the rainfall and evapotranspiration regime assumes the form

$$nZ_r\frac{ds}{dt} = R(t) - I(t) - Q(s) - E(s, t) - L(s),\qquad(8.1)$$

where the rainfall rate, $R(t)$, and the evapotranspiration rate, $E(s,t)$, now explicitly include a temporal (typically periodic) dependence. Equation (8.1) is a nonlinear first-order stochastic differential equation driven by a non-Gaussian noise and time-varying coefficients. As such it is quite difficult to approach analytically.

Two different avenues to deal with seasonal water balance have been pursued in the literature: (i) to disregard the stochastic component of the water balance and thus deal with a fully deterministic problem (e.g., Milly, 1994b); (ii) to tackle the problem in its entire complexity, either using real data or by means of numerical simulations of stochastic models (e.g., Cordova and Bras, 1981; Milly, 1994a; Simmons and Meyer, 2000). However, both approaches have serious drawbacks. The first one is an evident oversimplification that was shown not to be accurate even for the description of the mean soil moisture (Laio et al., 2002), while in the second approach the use of real data is strongly limited by the scarcity of consistent soil moisture databases and the numerical simulations hamper the possible generalization of the results.

Laio et al. (2002) followed an intermediate approach, extending the model of Chapter 2 to include seasonal components and considering an approximate equation for the first moment of the soil moisture process. In this way, they obtained analytical solutions that describe well the behavior of the full solution in most of the environmental conditions encountered in water-controlled ecosystems.

Laio et al. (2002) idealize the occurrence of rainfall as a series of point events in continuous time, arising in a Poisson process with time-varying rate λ_t so that the number of events in the interval $[0,t]$ follows a Poisson distribution with parameter $\int_0^t \lambda_u du$ (e.g., Ross, 1996). Likewise, each rain event is assumed to carry a random amount of rainfall h extracted from an exponential distribution with mean α_t. Interception is included defining a reduced value of the rainfall rate, $\lambda_t' = \lambda_t e^{-\frac{\Delta_t}{\alpha_t}}$, where Δ_t is the maximum depth of rainfall intercepted by the vegetation canopy during a single rain event. The evapotranspiration regime also follows a seasonal trend, through an explicit dependence of the maximum evapotranspiration rate on time, i.e., $E_{\max,t}$. In the above quantities and in the rest of this section, the t subscript underlines the time dependence of the corresponding parameters.

8.1.1 The macroscopic equation

With the previous hypotheses, the evolution of the soil moisture pdf is given by the Master equation (see Section 2.2)

$$\frac{\partial}{\partial t}p(s,t) = \frac{\partial}{\partial s}[p(s,t)\rho_t(s)] - \lambda_t'p(s,t) + \lambda_t' \int_0^s p(u,t)\gamma_t e^{-\gamma_t(s-u)}du$$

$$+ \delta(s-1)\lambda_t' \int_0^s p(u,t)e^{-\gamma_t(1-u)}du,$$

(8.2)

where $\rho_t(s) = \frac{E(s,t)+L(s)}{nZ_r}$ and $\gamma_t = \frac{nZ_r}{\alpha_t}$. Multiplication by s and integration over the whole range of s yields

$$\frac{d\langle s\rangle_t}{dt} = \int_0^1 s\frac{\partial}{\partial s}[p(s,t)\rho_t(s)]ds - \lambda_t'\langle s\rangle_t$$

$$+ \lambda_t' \int_0^1 \gamma_t s e^{-\gamma_t s} \int_0^s p(u,t)e^{\gamma_t u}duds$$

(8.3)

$$+ \lambda_t' \int_0^1 p(u,t)e^{-\gamma_t(1-u)}du,$$

where $\langle s\rangle_t = \int_0^1 sp(s,t)ds$ is the time-dependent ensemble mean of soil moisture. Integration by parts of the first and third terms on the r.h.s. and reorganization of the terms gives

$$\frac{d\langle s\rangle_t}{dt} = \frac{\lambda_t'}{\gamma_t} - \int_0^1 \rho_t(u)p(u,t)du - \frac{\lambda_t'}{\gamma_t} \int_0^1 e^{-\gamma_t(1-u)}p(u,t)du.$$

(8.4)

An equation of this type is often referred to as the macroscopic equation of the stochastic process, because it represents the macroscopic approximation of the Master equation in the case that the fluctuations around the mean are negligible (Van Kampen, 1992).

From the physical viewpoint, the macroscopic Eq. (8.4) (multiplied by nZ_r) can be interpreted as an average water balance (e.g., Laio et al., 2002), with

$$\frac{\lambda'_t}{\gamma_t} = \frac{\langle R(t) - I(t) \rangle}{nZ_r}, \tag{8.5}$$

$$\int_0^1 \rho_t(u)p(u,t)\mathrm{d}u = \langle \rho_t(s) \rangle = \frac{\langle E(s,t) + L(s) \rangle}{nZ_r}, \tag{8.6}$$

and

$$\frac{\lambda'_t}{\gamma_t} \int_0^1 \mathrm{e}^{-\gamma_t(1-u)}p(u,t)\mathrm{d}u = \frac{\langle Q(t) \rangle}{nZ_r}. \tag{8.7}$$

The closure problem

The major difficulty for the solution of the differential Eq. (8.4) derives from the presence of the last two terms, whose evaluation would require knowledge of $p(s,t)$ which, instead, is unknown. Some simplification is thus called for.

Unless the soil is very shallow, in water-controlled ecosystems the runoff term is usually negligible compared to the other terms and can be canceled from Eq. (8.4). This does not mean however that surface runoff is nonexistent, but only that its episodic nature allows us to neglect its contribution to the mean water balance and thus to the mean soil moisture evolution. In special cases when the runoff is of certain importance for the long-term water balance, it can be approximated using the first term of the Taylor expansion as

$$\frac{\lambda'_t}{\gamma_t}\mathrm{e}^{-\gamma_t\left(1-\langle s \rangle_t\right)}. \tag{8.8}$$

Much more important and difficult to deal with is the term related to the deterministic losses, $\int_0^1 \rho_t(u)p(u,t)\mathrm{d}u = \langle \rho_t(s) \rangle$. Only in the special case when $\rho_t(s)$ is linear (i.e., $\rho_t(s) = B_t s$) does one have $\langle \rho_t(s) \rangle = B_t \langle s \rangle_t$ and Eq. (8.4) is easily solvable. For nonlinear losses, as in our case, one could think of taking the average of the Taylor expansion of $\rho_t(s)$ around $\langle s \rangle_t$,

$$\langle \rho_t(s) \rangle = \rho_t(\langle s \rangle_t) + \frac{1}{2} \langle (s - \langle s \rangle_t)^2 \rangle \frac{d^2 \rho_t(s)}{ds^2} \bigg|_{s=\langle s \rangle_t} + \cdots \quad (8.9)$$

However, some problems occur in the application of Eq. (8.9). First, the condition of continuity of the derivatives of the expanded function is not respected because of the piecewise form of $\rho_t(s)$, which should be replaced by a suitable function (e.g., a sigmoidal function plus an exponential). In any case, the strong nonlinearities of the loss function would reduce the interval of convergence of the Taylor expansion to a narrow range around $\langle s \rangle_t$, and this contrasts with the presence of noise that gives rise to single realizations very far from the mean. Second, any expansion to an order n higher than the first introduces terms in the macroscopic Eq. (8.4) that depend on moments of all orders up to the nth. This implies that the evolution of the mean is influenced by all the higher-order moments so that the solution of Eq. (8.4) is subordinated to that of the system of n differential equations for the moments obtained multiplying Eq. (8.2) by $s, s^2, \ldots, s^n$ (Van Kampen, 1992).

Moreover, the differential equation for the mth moment includes a term of the form $\int_0^1 s^m \rho_t(s) p(s, t) ds = \langle s^m \rho_t(s) \rangle$; when the Taylor expansion of $\rho_t(s)$ is extended to the nth order as before, this term introduces the dependence on all the moments up to the order $m + n$ in the equation for the mth moment. All the higher-order moments thus affect the evolution of the mean, and closing the system of differential equations without suitable hypotheses becomes impossible. The usual "closure" procedure, which consists of truncating the Taylor expansion to the order $n - m$ in the equation for the mth moment, requires special conditions for the coefficients of the Taylor expansion (Van Kampen, 1992, pages 125–6) that are not fulfilled by the highly nonlinear form of $\rho_t(s)$.

Under a more general viewpoint, the presence of both strong nonlinearities in the loss function and large jumps in the soil moisture trajectory not only prevents the use of a Taylor expansion of $\rho_t(s)$ in the macroscopic equation, but also inhibits the derivation of the Fokker–Planck equation for the process as an approximation of the Master equation (8.2) (Van Kampen, 1992, page 198). For similar reasons, the more widely applicable "Van Kampen's system size expansion" of the Master equation (Van Kampen, 1992, page 244) is not feasible here either.

Approximate closure model

The approach to overcome these problems followed by Laio et al. (2002) was to keep only the first term of the Taylor expansion (Eq. (8.9)), $\rho_t(\langle s \rangle_t)$, and to model the remaining terms with a correction factor $g(\langle s \rangle_t)$ calculated on the basis of numerical simulations, namely $\langle \rho_t(s) \rangle = \rho_t(\langle s \rangle_t) + g(\langle s \rangle_t)$. The solution is approximate, because it is impossible to exactly reproduce terms depending on the higher-order moments with a function of the mean soil moisture only, but it was shown to be accurate enough from a hydrological perspective throughout a wide range of environmental conditions.

When the correction factor is introduced in the macroscopic Eq. (8.4), the latter assumes the form

$$\frac{d\langle s \rangle_t}{dt} = \frac{\lambda_t'}{\gamma_t} - \rho_t(\langle s \rangle_t) - g(\langle s \rangle_t) = \frac{\lambda_t'}{\gamma_t} - f(\langle s \rangle_t), \tag{8.10}$$

where $f(\langle s \rangle_t)$ is the overall, effective loss function for the mean soil moisture process. The correction function $g(\langle s \rangle_t)$ can be estimated by numerically simulating the time-dependent mean soil moisture for different initial values and climatic conditions. The value of $\frac{d\langle s \rangle_t}{dt}$ is then calculated as the difference between two subsequent values of $\langle s \rangle_t$, and finally the correction function is evaluated as $g(\langle s \rangle_t) = \frac{\lambda_t'}{\gamma_t} - \rho_t(\langle s \rangle_t) - \frac{d\langle s \rangle_t}{dt}$.

A good balance between the mathematical tractability of the equation and the accuracy of approximation is achieved with a correction function of the form

$$g(\langle s \rangle_t) = \frac{\lambda_t'}{\gamma_t} \epsilon(\langle s \rangle_t), \tag{8.11}$$

where $\epsilon(\langle s \rangle_t)$ is the linear stepwise function shown in Figure 8.1a (notice that Z_r must be in centimeters). The effects of the correction on the original loss function $\rho_t(s)$ are shown in Figure 8.1b: under dry conditions ($\frac{\lambda_t'}{\gamma_t} = 0$) the loss function for the mean soil moisture coincides with $\rho_t(s)$, and this implies that the mean soil moisture decreases closely following a typical trajectory of the soil-drying process. When $\frac{\lambda_t'}{\gamma_t}$ increases, $f(\langle s \rangle_t)$ assumes forms more and more different from the original loss function, with a tendency towards a linearization of $\rho_t(s)$. This tendency is more evident for larger $\frac{\lambda_t'}{\gamma_t}$ values since in this case the range of s spanned by the jumps increases with rainfall intensity and decreases with the storage capacity. As a result, the effective loss function for $\langle s \rangle$, $f(\langle s \rangle)$, resembles a smoothing of $\rho_t(s)$.

The proposed correction function was tested by Laio et al. (2002) on several combinations of environmental conditions. As expected, the accuracy of the approximation declines when the noise is very intense, i.e., when the active soil depth Z_r is lower than 20–25 cm and the rainfall intensity $\theta_t = \lambda_t \alpha_t$ exceeds 0.5–0.6 cm day^{-1}. Such conditions, however, are seldom encountered, especially in water-controlled ecosystems.

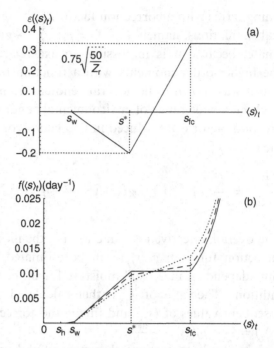

Figure 8.1 The correction function $\epsilon(\langle s \rangle_t)$ (see text for details) as a function of the mean soil moisture value (a) and the overall loss function for the mean soil moisture process (b) with zero (continuous line), low (dashed line), and high (dotted line) $\frac{\lambda_t'}{\gamma_t}$ values. After Laio et al. (2002).

Analytical expressions for the evolution of mean soil moisture

Within the previous framework, the differential Eq. (8.10) for the mean soil moisture assumes the form

$$\frac{d\langle s \rangle_t}{dt} = \frac{\lambda_t'}{\gamma_t} - \rho_t(\langle s \rangle_t) - \frac{\lambda_t'}{\gamma_t}\epsilon(\langle s \rangle_t) \tag{8.12}$$

or, by introducing in Eq. (8.12) the expressions for $\rho_t(\langle s \rangle_t)$ and $\epsilon(\langle s \rangle_t)$,

$$\frac{d\langle s \rangle_t}{dt} = \begin{cases} \frac{\lambda_t'}{\gamma_t} - \left[\eta_t - 0.2\frac{\lambda_t'}{\gamma_t}\right]\frac{\langle s \rangle_t - s_w}{s^* - s_w} & s_w < \langle s \rangle_t \leq s^* \\ 1.2\frac{\lambda_t'}{\gamma_t} - \eta_t - \frac{\lambda_t'}{\gamma_t}0.75\sqrt{\frac{50}{Z_r}}\frac{\langle s \rangle_t - s^*}{s_{fc} - s^*} & s^* < \langle s \rangle_t \leq s_{fc} \quad (8.13) \\ 1.2\frac{\lambda_t'}{\gamma_t} - \eta_t - \frac{\lambda_t'}{\gamma_t}0.75\sqrt{\frac{50}{Z_r}} - \frac{K_s(e^{\beta(\langle s \rangle_t - s_{fc})} - 1)}{nZ_r(e^{\beta(1 - s_{fc})} - 1)} & s_{fc} < \langle s \rangle_t \leq 1, \end{cases}$$

where $\eta_t = \frac{E_{max,t}}{nZ_r}$.

The first two equations above are first-order linear differential equations whose solution is

$$\langle s \rangle_t = \langle s \rangle_{t_i} e^{-\int_{t_i}^t A(t')dt'} + e^{-\int_{t_i}^t A(t')dt'} \int_{t_i}^t B(t')e^{\int_{t_i}^{t'} A(t'')dt''} dt', \qquad (8.14)$$

where $\langle s \rangle_{t_i}$ represents the value of the mean soil moisture at an initial time t_i,

$$A(t) = \begin{cases} \dfrac{\eta_t - 0.2\frac{\lambda'_t}{\gamma_t}}{s^* - s_w} & s_w < \langle s \rangle_t \leq s^* \\ \dfrac{\lambda'_t}{\gamma_t} 0.75 \sqrt{\dfrac{50}{Z_r}} \dfrac{1}{s_{fc} - s^*} & s^* < \langle s \rangle_t \leq s_{fc} \end{cases} \qquad (8.15)$$

and

$$B(t) = \begin{cases} \dfrac{\lambda'_t}{\gamma_t} + \dfrac{s_w}{s^* - s_w}(\eta - 0.2\frac{\lambda'_t}{\gamma_t}) & s_w < \langle s \rangle_t \leq s^* \\ 1.2\dfrac{\lambda'_t}{\gamma_t} - \eta_t + \dfrac{\lambda'_t}{\gamma_t} 0.75 \sqrt{\dfrac{50}{Z_r}} \dfrac{s^*}{s_{fc} - s^*} & s^* < \langle s \rangle_t \leq s_{fc}. \end{cases} \qquad (8.16)$$

The change of variable $y = e^{\beta \langle s \rangle_t}$ allows us to write the third differential equation in Eq. (8.13) as a canonical Bernoulli equation, whose solution is

$$\langle s \rangle_t = \langle s \rangle_{t_i} - \frac{1}{\beta} \int_{t_i}^t C(t')dt' - \frac{1}{\beta} \ln\left[1 - De^{\beta \langle s \rangle_{t_i}} \int_{t_i}^t e^{-\int_{t_i}^{t'} C(t'')dt''} dt' \right], \qquad (8.17)$$

where

$$C(t) = \beta\left\{ \eta_t - \frac{K_s}{nZ_r[e^{\beta(1-s_{fc})} - 1]} - 1.2\frac{\lambda'_t}{\gamma_t} + \frac{\lambda'_t}{\gamma_t} 0.75 \sqrt{\frac{50}{Z_r}} \right\} \qquad (8.18)$$

and

$$D = -\beta \frac{K_s e^{-\beta s_{fc}}}{nZ_r[e^{\beta(1-s_{fc})} - 1]}. \qquad (8.19)$$

The complete solution is obtained by linking together the results of Eqs. (8.14) and (8.17). The time dependence of $A(t)$, $B(t)$, and $C(t)$ derives from the possible variability with time of the parameters λ_t, γ_t, and η_t, that in turn reflects the influence of climatic fluctuations at different time scales. Some characteristic features of the coefficients $A(t)$, $B(t)$, $C(t)$, and D are worth noting: (i) all the coefficients are inversely proportional to the value of the active soil depth through the parameters $\gamma_t = \frac{nZ_r}{\alpha_t}$ and $\eta_t = \frac{E_{max,t}}{nZ_r}$; (ii) through the parameters K_s and β, the soil characteristics influence the values of $C(t)$ and D and not those of $A(t)$ and $B(t)$, as expected from the assumption that no leakage occurs below s_{fc}; (iii) the rainfall frequency λ_t and the average rainfall depth α_t always appear coupled together by the term $\frac{\lambda_t}{\gamma_t}$.

8.1.2 Examples of transient mean soil moisture dynamics

The approximate solution of the Eq. (8.4) was tested by means of numerical simulations of the full Eq. (8.1) in the hypothesis that λ_t, α_t, and $E_{max,t}$ do not change with time. Figure 8.2 shows some examples of the evolution of the average soil moisture for cases going from wet (upper panels) to drier (lower panels) initial conditions. The black dots are the daily values obtained by averaging 5000 numerically simulated realizations of Eq. (8.1), the continuous lines represent the solutions of Eqs. (8.14) and (8.17), and the dashed line the solution that one would obtain with the Taylor expansion Eq. (8.9) truncated to the second order. The lower part of each graph shows the relative errors of

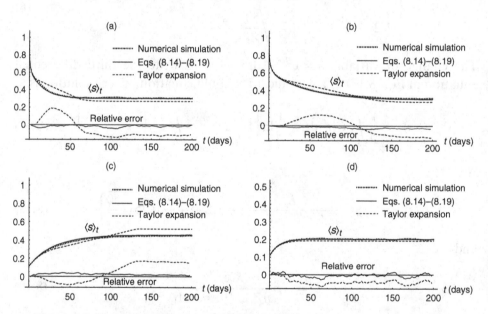

Figure 8.2 Comparison between calculated and simulated mean soil moisture trajectories under constant climatic conditions. Common parameters to all the graphs are $s_h = 0.08$, $s_w = 0.11$, $s^* = 0.31$, $s_{fc} = 0.52$, $K_s = 100\,\mathrm{cm\,day^{-1}}$, $n = 0.43$, $b = 4.38$ (typical parameters for a loamy sand (see Table 2.1)), $E_{max,t} = 0.45\,\mathrm{cm\,day^{-1}}$, and $\Delta = 0.2\,\mathrm{cm}$. In (a) and (b) the initial condition is soil saturation ($s = 1$), the daily rainfall rate is $0.4\,\mathrm{cm\,day^{-1}}$ ($\lambda_t = 0.2\,\mathrm{day^{-1}}$ and $\alpha_t = 2\,\mathrm{cm}$) and the active soil depth is 50 cm (a) and 100 cm (b). In (c) $\langle s \rangle_{t_i} = s_h$, $Z_r = 100\,\mathrm{cm}$, $\lambda_t = 0.3\,\mathrm{day^{-1}}$ and $\alpha_t = 2\,\mathrm{cm}$. In (d) $\langle s \rangle_{t_i} = s_h$, $Z_r = 50\,\mathrm{cm}$, $\lambda_t = 0.15\,\mathrm{day^{-1}}$ and $\alpha_t = 1.5\,\mathrm{cm}$. After Laio et al. (2002).

the two methods with respect to the numerical simulations. In all cases, representing different values of the average rooting depth Z_r and different climatic conditions, Eqs. (8.14) and (8.17) yield good results, with the relative error always in the $\pm 5\%$ range. In contrast, the use of the Taylor expansion gives rise to relevant errors, both in the reproduction of the transient phase and of the steady-state condition. An increase in the order of the expansion to orders higher than two causes a worsening of the results, due to the already-mentioned problems related to the convergence of the Taylor expansion. The fact that the results also worsen when the expansion is truncated to the first order evidences the incorrectness of the deterministic approximation $\langle \rho_t(s) \rangle = \rho_t(\langle s \rangle_t)$, that completely neglects the stochastic fluctuations around the mean.

Figure 8.2 also allows one to discriminate, at least qualitatively, between a transient and a stationary phase for the mean soil moisture dynamics, the latter being the part of the trajectory that is nearly horizontal. The time intervals to reach the steady-state condition range from ~ 30 days in Figure 8.2d to more than 100 days in Figure 8.2b, depending on the values of Z_r, α_t, and λ_t. The estimate of the duration of the transient phase is important; for example, to understand which environmental conditions allow a partitioning of the year into a dormant and a growing season and provide steady-state conditions for the soil moisture process during the two separate periods. A qualitative analysis of the trends of the mean soil moisture under different environmental conditions suggests that the main determinants for the duration of the transient are the active soil depth Z_r and the relative position along the $\langle s \rangle_t$ axis of the initial and the steady-state values.

Seasonal fluctuations in evapotranspiration

The effect of variable climatic conditions on mean soil moisture was analyzed first for the case when the rainfall rate is constant throughout the year, while the potential evapotranspiration varies periodically. Following Milly (1994a), a simple sinusoidal representation is used as a first approximation,

$$\eta_t = \frac{E_{\max,t}}{nZ_r} = \frac{E_{\max,0}}{nZ_r}[1 + \delta_{et}\sin(\omega_{et}t + \phi_{et})], \tag{8.20}$$

where $E_{\max,0}(\text{cm day}^{-1})$ is the average value around which the potential evapotranspiration rate fluctuates, t is the time of year in days (with $t = 0$ at the start of the hydrologic year, e.g., September 1st), δ_{et} (nondimensional) is the ratio of the amplitude of the harmonic to $E_{\max,0}$, $\omega_{et}(\text{day}^{-1})$ is the frequency of the sinusoid, and ϕ_{et} is the phase shift. In the example of Figure 8.3, $E_{\max,0} = 0.35\,\text{cm day}^{-1}$, $\delta_{et} = 0.5$, $\omega_{et} = \frac{2\pi}{365}\,\text{day}^{-1}$ (period of one year) and $\phi_{et} = \frac{5}{12}2\pi$ (maximum of the potential evapotranspiration in July). The resulting

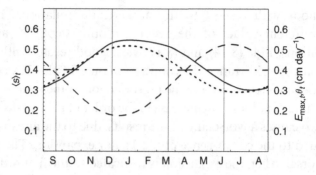

Figure 8.3 Mean soil moisture dynamics forced by seasonal variations in the potential evapotranspiration. The parameters are the same as in Figure 8.2a, b, except $E_{\max,t}$ which has the form of Eq. (8.20) with $E_{\max,0} = 0.35\,\text{cm day}^{-1}$, $\delta_{\text{et}} = 0.5$, $\omega_{\text{et}} = \frac{2\pi}{365}\,\text{day}^{-1}$, and $\phi_{\text{et}} = \frac{5}{12}2\pi$. Dashed line: $E_{\max,t}$; chain line: constant rainfall rate θ_t; continuous line: $\langle s \rangle_t (Z_r = 100\,\text{cm})$; dotted line: $\langle s \rangle_t (Z_r = 50\,\text{cm})$. After Laio et al. (2002).

average soil moisture trajectories are approximately in counter-phase with respect to the forcing, as one would expect considering that an increase in the maximum evapotranspiration rate corresponds to an increased water loss from the soil. However, a non-negligible delay occurs between the two curves. This delay is greater for the case of a deep soil (continuous line) compared to that of a shallow one (dotted line).

After investigating several other cases, the root depth was found to be the key variable in determining these delays. With the aim of highlighting this aspect, a quantitative analysis of the role of Z_r can be carried out using the estimates (deduced in the Appendix at the end of this chapter) of the delay time, t_d, and the ratio, r, of the amplitude of the seasonal oscillation of $\langle s \rangle_t$ to that of the forcing.

The delay time t_d (see Appendix) is plotted in Figure 8.4a. As expected, it increases with the active soil depth Z_r, arriving at values of 40–50 days for soil depths of 150 cm. An increase in the rainfall rate θ_t (and a simultaneous increase in the potential evapotranspiration rate, $E_{\max,0} = \theta_t$) reduces the delay times, because the response to changes in the forcing becomes faster when the noise intensity increases. A similarity between the delay times and the transient durations is evident, and it can be explained considering that both are consequences of the presence of the dampening effect of nZ_r in Eq. (8.1) and, implicitly, in Eq. (8.12). When $nZ_r = 0$ the response to the forcing is instantaneous, and no delays or transient conditions occur, while an increase in nZ_r reduces the response times in a complicated manner involving the soil and climate parameters (e.g., Figure 8.4a).

Using the ratio r (see Appendix C to this chapter), it can be shown (Figure 8.4b) that under typical rainfall conditions there is a decrease of the range of

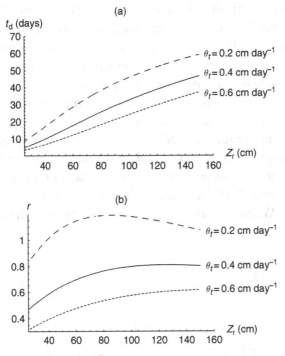

Figure 8.4 Delay times (a) and ratio of the range of variability of $\langle s \rangle_t$ to that of the forcing (b) as a function of the active soil depth for different daily rainfall rates. The parameters are the same as in Figure 8.2a, b, with a maximum evapotranspiration that varies as in Eq. (8.20) with $E_{\text{max},0} = \theta_t$, $\delta_{\text{et}} = 0.5$, $\omega_{\text{et}} = \frac{2\pi}{365}$ day^{-1}, and $\phi_{\text{et}} = 0$. After Laio et al. (2002).

variations of average soil moisture with respect to that of the maximum evapotranspiration ($r < 1$). In contrast, in very dry conditions r tends to become greater than 1 and an amplification occurs. This situation, however, is limited to dormant seasons with $E_{\text{max},0}$ as low as 0.2 cm day^{-1}. The values of r are always lower for shallower soils (see also Figure 8.3). As a consequence, under the same climatic conditions, $\langle s \rangle_t$ tends to be higher for deeper soils (Figure 8.3), with clear implications for plant water stress.

Seasonal fluctuations in rainfall and evapotranspiration regimes

Similarly to the case of the maximum evapotranspiration, the seasonal variations of daily rainfall can be represented in the form

$$\frac{\lambda_t}{\gamma_t} = \frac{\theta_t}{nZ_r} = \frac{\theta_0}{nZ_r}\left[1 + \delta_p \sin\left(\omega_p t + \phi_p\right)\right], \tag{8.21}$$

where θ_0 is the annual average of the rainfall rate, and δ_p, ω_p, and ϕ_p are the amplitude, frequency, and phase shift of the sinusoid, respectively. When the maximum evapotranspiration is nearly constant throughout the year, and θ_t assumes the form of Eq. (8.21), the mean soil moisture trajectories (not shown) are approximately in phase with respect to the rainfall forcing, with similar features (delay times, role of Z_r, etc.) to those found for the case with constant rainfall and variable maximum evapotranspiration.

The effects of a simultaneous variation of the daily rainfall and maximum evapotranspiration rates deserve more attention. In Figure 8.5a the phase shift of the sinusoid representing rainfall, ϕ_p, is chosen to have rainfall with opposite phase to evapotranspiration (e.g., Mediterranean climate), while in Figure 8.5b the phases are the same, $\phi_p = \phi_{et}$, as is typical of a continental climate (e.g., Nylsvley, Section 5.1). The mean soil moisture dynamics are quite different in the two cases. When rainfall and $E_{max,t}$ are of opposite phase there are very

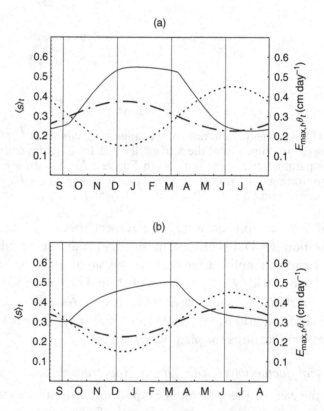

Figure 8.5 Seasonal fluctuations of the mean soil moisture (continuous line) under varying potential evapotranspiration (dotted line, Eq. (8.20) with $E_{max,0} = 0.3\,\text{cm day}^{-1}$, $\delta_{et} = 0.5$, $\omega_{et} = \frac{2\pi}{365}\,\text{day}^{-1}$) and rainfall (chain line, Eq. (8.21) with $\theta_0 = 0.3\,\text{cm day}^{-1}$, $\delta_p = 0.25$, and $\omega_p = \frac{2\pi}{365}\,\text{day}^{-1}$). The phase shift of rainfall is $\phi_p = \frac{11}{12}2\pi$ in (a) and $\phi_p = \frac{5}{12}2\pi$ in (b). The active soil depth is $Z_r = 100\,\text{cm}$, the other parameters are the same as in Figure 8.2. After Laio et al. (2002).

persistent wet- and dry-soil conditions, the first lasting from December to April and the latter from June to September. When rainfall and maximum evapotranspiration are in phase, the range of variation of $\langle s \rangle_t$ is smaller and the mean soil moisture slowly increases from October to March and then decreases slowly from April to September. More frequent runoff events are present in the case of Figure 8.5a in which many months have $\langle s \rangle_t$ close to field capacity. Following Budyko's (1974) representation of the annual water balance (see also Section 2.6.2), the two above situations have the same dryness index $\frac{E_{\max,0}}{\theta_0}$, but the ratio of annual evapotranspiration to rainfall is lower in the first case, as shown by the increased probability of runoff (see also Milly, 1994a).

The case of Figure 8.5b appears to be more favorable to plants, due to the higher mean soil moisture values during the growing season; in contrast, in the case of Figure 8.5a it becomes very important for plants to concentrate the growing effort in the early growing season or to increase the water storage capacity with deeper roots, so that the water stored in the dormant season lasts longer (e.g., Sala et al., 1982a, 1982b; Weltzin and McPherson, 1997; Rodríguez-Iturbe et al., 2001b).

A case study with real data

The solution for the mean soil moisture process, Eqs. (8.14) and (8.17), may also be used with real data of rainfall and potential evapotranspiration. As an example, $\langle s \rangle_t$ was computed in Figure 8.6 using monthly rainfall and maximum evapotranspiration data from Torino (northern Italy). The potential evapotranspiration is calculated from temperature data using the adjusted Thornthwaite equation (e.g., Chow, 1964). Both the rainfall and temperature

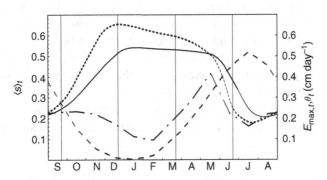

Figure 8.6 Mean soil moisture dynamics in Torino (Italy). The chain line represents the rainfall rate, θ_t, and the dashed line the maximum evapotranspiration rate, $E_{\max,t}$. The parameters are the same as in Figure 8.2, with $Z_r = 100$ cm and $K_s = 100$ cm day^{-1} for the continuous line and $Z_r = 30$ cm, $K_s = 10$ cm day^{-1} for the dotted line. After Laio et al. (2002).

monthly data used in Figure 8.6 are 30-year averages (1930–60), collected by the River Po Hydrographic Office. The resulting maximum evapotranspiration rate (dashed line in Figure 8.6) presents a clear maximum in the summer months, while rainfall (chain line) has a sharp peak in May, two minima (during summer and late winter), and a relatively flat plateau from August to November. The resulting mean soil moisture dynamics shows very persistent wet conditions from December to May, while the dry period, in terms of $\langle s \rangle_t$, only lasts a couple of months (July–August). Again, the mean soil moisture is considerably influenced by the depth of the soil and by its permeability. When a shallower and less permeable soil is considered (dotted line), the delays with respect to the forcings decrease and $\langle s \rangle_t$ assumes higher values during the wet months and lower values during the dry ones, thus favoring runoff production and worsening, at least on average, the plant water status during the growing season.

8.1.3 Discussion

The results of the analyses of the time-varying mean soil moisture, $\langle s \rangle_t$, for a wide range of climatic, pedologic, and plant conditions suggest the following conclusions (Laio et al., 2002):

(1) What determines the mean soil moisture evolution is mostly the interplay between the rainfall and evapotranspiration rates.
(2) The response of the mean soil moisture to the rainfall and evapotranspiration forcings occurs with a relevant time delay, which is mainly determined by the active soil depth, Z_r.
(3) The duration of these delays, for a given value of Z_r, is controlled by the rainfall rate θ_t; the lower θ_t the longer the delay in the mean soil moisture response.
(4) The combination of these factors allows an evaluation of the periods when the runoff events are more probable, as well as of the soil characteristics that are more favorable to plant growth under unsteady climatic conditions.
(5) Finally, depending on the conditions existing in each particular case, the presence of the mentioned delays in the $\langle s \rangle_t$ response may limit the validity of the steady-state solution for the soil moisture dynamics. The approximation is always good in the last part of the growing (or dormant) season, but its extension to the whole season requires special attention, especially when the soil is shallow and the rainfall rates are very high.

8.2 Interannual rainfall fluctuations and soil moisture dynamics

Interannual rainfall fluctuations represent the second form of temporal variability analyzed in this chapter. They are of crucial importance for the long-term dynamics of soil moisture and plant water stress. The results presented here are summarized after D'Odorico et al. (2000) and Ridolfi et al. (2000b). Chapter 11 will build upon the present analysis to study the impact of interannual hydrologic fluctuations on the evolution and spatial structure of vegetation.

8.2.1 The impact of interannual rainfall fluctuations

As a first step towards the characterization of the interannual soil moisture variability, D'Odorico et al. (2000) analyzed the characteristics of the rainfall regime during the growing season of the last 60 or 70 years in southwest Texas, where the impacts on vegetation of the pronounced interannual rainfall fluctuations have been well documented (e.g., Archer et al., 1988; Rodríguez-Iturbe et al., 1999b, 1999c; see also Section 5.2). Following the stochastic representation of rainfall as a marked Poisson process (see Chapter 2), the interannual variability of the rainfall regime is characterized through the sequence of the values of the average storm depth, α, and the frequency of the storm arrivals, λ, for each growing season.

The parameters λ and α were evaluated for a set of locations during a growing season extending from the beginning of May to the end of September. Table 8.1 shows the period of observation for each station which oscillates between 49 and 95 years. The interannual average and variance of λ and α are then estimated from the series of their individual yearly values. The variability of λ appears to be somewhat weaker than that of α, as suggested by the coefficients of variation of these two parameters (Table 8.1). This is not a general result, however, since D'Odorico et al. (2000) found contrasting behavior for many stations in Texas. Figure 8.7 shows a typical example of the time series of annual precipitation, along with the time series of α and λ. For purposes of illustration, the different years were also divided into two populations with rain above or below the average. The histograms of relative frequencies of λ and α (Figure 8.8) are always unimodal and, in all of the cases, two-parameter gamma distributions provide a good fit of the data (Figure 8.9).

An analysis of the autocorrelation of λ and α was also performed for the same locations and an example is given in Figure 8.10. In general, the autocorrelations are very weak for both parameters. In most of the stations, the autocorrelation is zero at lag one year, although in some of them there is a small positive correlation for λ at lag one year. The interannual autocorrelation of total rainfall during the growing season is generally insignificant. Examples of the cross-correlation functions between λ and α, and between these two parameters and total seasonal rainfall, are also given in Figure 8.10. The cross-correlation between λ and α appears to be very weak or nonexistent, so that these parameters may be modeled as independent of each other. As

Table 8.1 *Mean, variance, and coefficient of variation of* λ *and* α *during the growing season* (*May–September*) *for different locations in southern Texas. After D'Odorico et al.* (2000).

Station	Period	$\langle\lambda\rangle$ (day^{-1})	Var[λ] (day^{-2})	CV[λ]	$\langle\alpha\rangle$ (mm)	Var[α] (mm^2)	CV[α]	Θ (mm)
Austin	1948–97	0.21	0.0021	0.21	12.1	16.4	0.33	392
Beeville	1945–81	0.22	0.0026	0.22	12.0	27.1	0.40	451
Corpus Christi	1948–97	0.21	0.0026	0.21	12.9	21.0	0.35	427
Del Rio	1948–97	0.17	0.0021	0.27	9.7	7.0	0.27	254
Galveston	1946–94	0.26	0.0028	0.20	13.3	24.1	0.37	529
Houston	1921–97	0.28	0.0032	0.20	12.2	10.2	0.26	532
Houston*	1948–97	0.28	0.0028	0.19	14.5	22.2	0.32	618
Kingsville	1945–81	0.21	0.0023	0.23	12.4	31.7	0.45	404
Luling	1903–97	0.22	0.0022	0.21	12.1	13.4	0.30	399
San Angelo	1948–97	0.18	0.0016	0.22	10.0	9.5	0.31	280
San Antonio	1948–97	0.21	0.0021	0.22	12.4	13.4	0.29	393

* Hobby airport

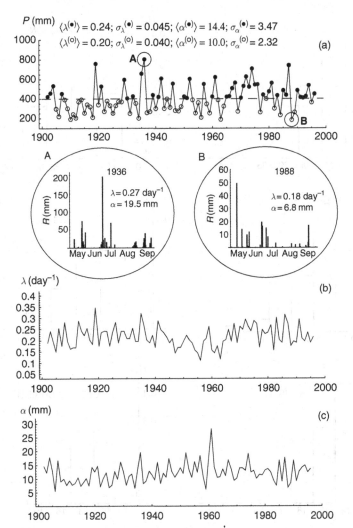

Figure 8.7 Analysis of the rainfall regime during the growing season at Luling (Texas) based on daily precipitation data: (a) time series of total seasonal rain (1 May to 30 September). (b) Time series of the estimated rate of storm arrivals, λ. (c) Time series of the average storm depth, α. After D'Odorico et al. (2000).

expected, the total seasonal rainfall is highly correlated with the values of λ and α in the same year.

8.2.2 *Probabilistic behavior of mean soil moisture*

The mean value of soil moisture during a growing season, $\langle s \rangle$, was used by D'Odorico et al. (2000) as the key parameter characterizing the impact of

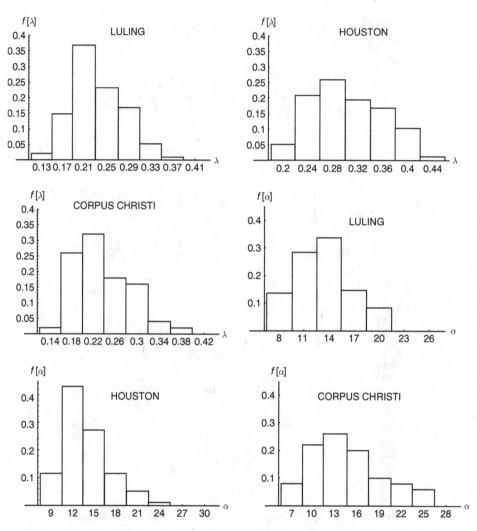

Figure 8.8 Histograms of the rate of arrival of storms, λ, and of average storm depth, α, estimated from daily data of precipitation for some locations in Texas during the growing season. After D'Odorico et al. (2000).

climate on the water balance and vegetation in arid and semi-arid regions. The interannual variability in the rainfall regime leads to fluctuations in $\langle s \rangle$ through all the mechanisms embedded in the dynamics of the water balance; $\langle s \rangle$ thus depends in a nonlinear way on climate, soil, and vegetation characteristics. D'Odorico et al. (2000) used the simpler model of Section 2.6.1 (e.g., Eq. (2.56)), but similar results are to be expected also for the more general model of Chapter 2.

A Monte Carlo procedure was implemented to numerically estimate the probability distribution of $\langle s \rangle$ resulting from the random interannual

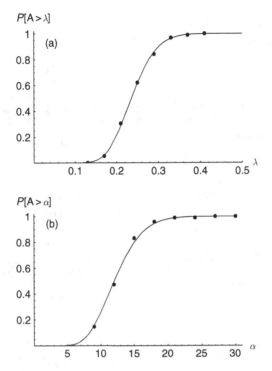

Figure 8.9 Probability distributions of (a) λ and (b) α estimated by fitting a gamma distribution to the data from Luling (Texas). After D'Odorico et al. (2000).

fluctuations of λ and α. Pairs of values, $\{\lambda, \alpha\}$, were repeatedly sampled from their gamma distributions and the corresponding values of $\langle s \rangle$ were estimated from Eq. (2.56).

Figure 8.11 shows an example of the distribution of $\langle s \rangle$ for given climate, soil, and vegetation characteristics, with different hypotheses for the coefficients of variation of α and λ. For high values of the coefficient of variation of any or both of these parameters, one observes the emergence of a bimodal behavior driven by the variability of the climatic parameters. This behavior suggests that the system tends to switch between two preferential states, one characterized by high average soil moisture and the other by low average soil moisture conditions. This feature has important implications for the dynamics of vegetal ecosystems because it implies that vegetation may frequently remain in states far from average conditions. The bimodal behavior disappears when the fluctuations become weaker, clearly showing the importance of the intensity of the fluctuations.

In systems forced by fluctuations of the same intensity, different combinations of climate, soil, and vegetation lead to different probability distributions

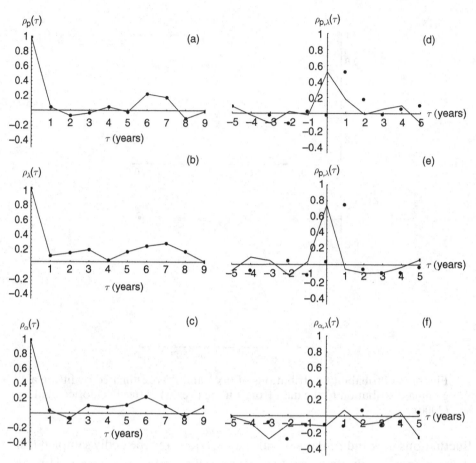

Figure 8.10 Autocorrelations and cross-correlations characteristic of the rainfall regime at Luling (Texas) during the growing season. (a) Autocorrelation of the total seasonal rain. (b) Autocorrelation of the rate of storm arrivals, λ. (c) Autocorrelation of the average storm depth, α. (d) Cross-correlation of the total seasonal rain and the rate of storm arrivals, λ. (e) Cross-correlation of the total seasonal rain and the average storm depth, α. (f) Cross-correlation of the rate of storm arrivals, λ, and the average storm depth, α. After D'Odorico et al. (2000).

of $\langle s \rangle$. As an example, the role of climate can be investigated by estimating the probability density function of $\langle s \rangle$ for different mean values of α and λ while keeping unchanged their coefficients of variation as well as all the other parameters representative of soil and vegetation. Figure 8.12a shows that the larger the mean seasonal rainfall, the higher the peak in the mode corresponding to the wet regime. In a similar way, drier climates lead to an increasingly higher dry mode until the other mode disappears. For the climate of Kingsville, Texas (solid line of Figure 8.12a), we observe the emergence of

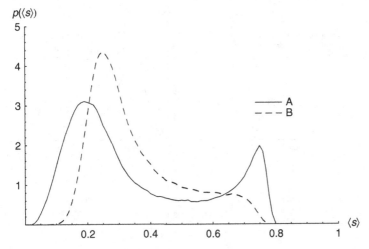

Figure 8.11 Probability density function of the average soil moisture during the growing season. The parameters for soil and vegetation are as follows: $n = 0.431$; $Z_r = 1.40 \, \text{m}$; $K_s = 9.5 \cdot 10^{-6} \, \text{m s}^{-1}$; $s_1 = 0.8$; $s^* = 0.36$; $E_{max} = 3.2 \, \text{mm day}^{-1}$. The rainfall is characterized by $\langle \alpha \rangle = 12.4 \, \text{mm}$ per storm and $\langle \lambda \rangle = 0.21$ storms day^{-1}, with coefficients of variation: (A) $CV[\alpha] = 0.45$; $CV[\lambda] = 0.23$. (B) $CV[\alpha] = 0.22$; $CV[\lambda] = 0.11$. After D'Odorico et al. (2000).

two preferential states with the mode corresponding to the dry regime more pronounced than the one for the humid regime.

The probabilistic structure of $\langle s \rangle$ is also very sensitive to the soil and vegetation characteristics. Figure 8.12b shows the probability distribution of $\langle s \rangle$ computed for different values of E_{max}. As expected, we observe that lower rates of evapotranspiration enhance the wet mode, while higher values of E_{max} correspond to larger values in the dry mode until the wet mode finally disappears. The role of the active soil depth is shown in Figure 8.12c. Notice how the double mode behavior tends to disappear for shallow soils. The probability distributions of $\langle s \rangle$ for different types of soil are shown in Figure 8.12d. The occurrence of a second mode on the wet regime seems to be more important for finer-grained soils because of their ability to retain the water for longer. The dry mode is much more pronounced in coarser-grained soils.

As D'Odorico et al. (2000) notice, the bimodal character of the probability distribution of the average soil moisture under fluctuating climatic conditions results primarily because of nonlinearity of the losses and the upper bound on the amount of rainfall that can infiltrate depending on the soil moisture state (see Chapter 2). The emergence of a double mode is the signature of a dynamics switching between two preferential states, characterized by either dry or wet average soil moisture conditions. This fact has important climatic and ecologic

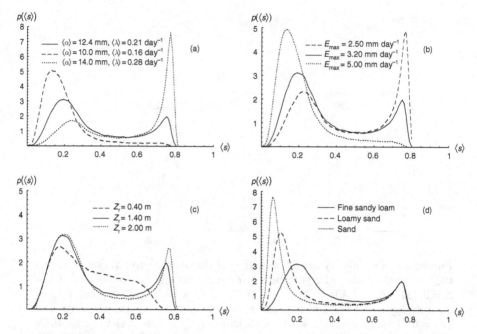

Figure 8.12 Probability density function of the average soil moisture during the growing season. $CV[\alpha] = 0.45$, $CV[\lambda] = 0.23$, and the parameters that are not indicated are the same as in Figure 8.11. After D'Odorico et al. (2000).

implications. In particular the occurrence of bimodal characteristics in the distribution of $\langle s \rangle$ enhances the impacts of climatic fluctuations on ecosystems in arid and semi-arid environments, especially when the dry mode corresponds to values of soil moisture close to the permanent wilting point of some species. It also suggests that the description of the soil moisture and ecosystem dynamics by an equilibrium average state may be misleading, since such conditions may correspond to a state which takes place with relatively low probability. The equilibrium concept itself is not the most appropriate one for a system subject to switching between two preferential states that favor or restrict the growth of vegetation, with important consequences for the ecosystem structure.

The possible occurrence of double modes in the probability distribution of soil moisture was previously reported by Rodríguez-Iturbe et al. (1991) through another stochastic model of soil water balance in a completely different framework. Their results showed how preferential states may arise in the soil moisture dynamics at the continental scale as a consequence of the coupling between soil surface and atmosphere, while the present analysis focuses on a much smaller scale where there is no recycling of moisture between the land and the atmosphere. Since here the mean soil moisture during the growing season is statically linked to the rainfall parameters for that same period, the rate of switching of $\langle s \rangle$ between the two possible states is determined only by the interannual rainfall fluctuations.

8.2.3 Interannual fluctuations and plant water stress

Ridolfi et al. (2000b) analyzed the role of interannual rainfall fluctuations in the mean crossing properties of soil moisture. Like D'Odorico et al. (2000), they also employed the simpler model of Section 2.6.1, and extended the analysis of the rainfall variability to eight stations located in different regions of the United States.

They focused on the impact of the interannual rainfall variability on the mean duration and frequency of periods of water stress. Analogously to the case of $\langle s \rangle$, the statistical nature of λ and α, assumed to be random variables that change from year to year according to their (gamma) distributions, confers a probabilistic structure to $\overline{T}_\xi$ and $\overline{n}_\xi$ that depends on climate, soil, and vegetation properties.

Analytical considerations

The case when only one of the two parameters α or λ presents random interannual fluctuations may be approached analytically. Ridolfi et al. (2000b) considered the mean duration of the periods below ξ constrained to the length of the season, i.e., $\overline{T}'_\xi = \min\left[T_{\text{seas}}, \overline{T}_\xi\right]$, when only α fluctuates from year to year while λ remains fixed. The mean of this random variable is given by

$$\mu_{\overline{T}'_\xi} = \int_0^{T_{\text{seas}}} f_{\overline{T}'_\xi}\left(\overline{T}'_\xi\right) \overline{T}'_\xi \mathrm{d}\overline{T}'_\xi + p_{\overline{T}'_\xi}(T_{\text{seas}}) T_{\text{seas}}. \tag{8.22}$$

The first term of the r.h.s. represents the contribution of the continuous part of the pdf of $\overline{T}'_\xi$ to $\mu_{\overline{T}'_\xi}$, while the second term is the contribution from the discrete probability corresponding to $\overline{T}'_\xi = T_{\text{seas}}$. In what follows $\overline{T}'_\xi = g(\alpha)$ indicates the functional relationship of $\overline{T}'_\xi$ and α. As shown in Figure 8.13, for very low values of α, $\overline{T}'_\xi = T_{\text{seas}}$, while for $\alpha \geq \alpha_{\text{seas}}$, the relationship decays monotonically, reflecting the fact that $\overline{T}'_\xi$ becomes smaller when the mean rain depth α becomes larger. The density component $f_{\overline{T}'_\xi}\left(\overline{T}'_\xi\right)$ may then be written as

$$f_{\overline{T}'_\xi}(\overline{T}'_\xi) = -\frac{f_\alpha(\alpha)}{\frac{\mathrm{d}g(\alpha)}{\mathrm{d}\alpha}}, \tag{8.23}$$

where the numerator is the pdf of α and the negative sign accounts for the fact that $g(\alpha)$ is a monotonically decreasing function. The derivation of $f_{\overline{T}'_\xi}(\overline{T}'_\xi)$

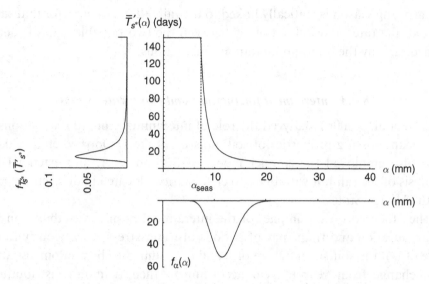

Figure 8.13 Derivation of the probability distribution of the seasonal mean duration of periods with water stress $\overline{T}'_\xi$ based on its functional dependence on α. α is assumed to follow a two-parameter gamma distribution; the values used in this figure are $\overline{\alpha} = 12\,\mathrm{mm\ day^{-1}}$, $\mathrm{CV}[\alpha] = 0.2$, $\overline{\lambda} = 0.3\ \mathrm{storm\ day^{-1}}$, $\mathrm{CV}[\lambda] = 0$, $nZ_r = 300\,\mathrm{mm}$, $E_{\max} = 4\,\mathrm{mm\ day^{-1}}$, $\xi = s^* = 0.35$ ($s_1 = 0.8$, $K_s = 90\,\mathrm{cm\ day^{-1}}$, $T_{\mathrm{seas}} = 150$ days). After Ridolfi et al. (2000b).

using the functional dependence of $\overline{T}'_\xi$ on α (see Chapter 3), and a two-parameter gamma distribution for α is schematically shown in Figure 8.13 for some particular parameter values. One observes that the part of $f_\alpha(\alpha)$ corresponding to the interval $[0, \alpha_{\mathrm{seas}}]$ yields an atom of probability at $\overline{T}'_\xi = T_{\mathrm{seas}}$. This atom may be more or less pronounced depending on the characteristics of climate, soil, and vegetation. In many cases (e.g., wet climates or low soil moisture thresholds ξ) the atom of probability at T_{seas} is negligible, but in arid or semi-arid climates, it may be important for the characterization of the periods when plants are under stress. Thus all values of $\overline{T}'_\xi \geq T_{\mathrm{seas}}$ are concentrated at T_{seas}.

Substituting Eq. (8.23) into Eq. (8.22), one obtains

$$\mu_{\overline{T}_\xi} = -\int_\infty^{\alpha_{\mathrm{seas}}} \frac{f_\alpha(\alpha)}{\frac{dg}{d\alpha}} g(\alpha) \frac{dg}{d\alpha}\, d\alpha + p_{\overline{T}'_\xi}(T_{\mathrm{seas}}) T_{\mathrm{seas}}$$

$$= \int_{\alpha_{\mathrm{seas}}}^\infty f_\alpha(\alpha) g(\alpha)\, d\alpha + p_{\overline{T}'_\xi}(T_{\mathrm{seas}}) T_{\mathrm{seas}}, \qquad (8.24)$$

where the limits correspond to the cases $\overline{T}'_\xi = 0, \alpha = \infty$, and $\overline{T}'_\xi = T_{\text{seas}}$, $\alpha = \alpha_{\text{seas}}$. Similarly, the variance is obtained as

$$\sigma^2_{\overline{T}'_\xi} = \int_{\alpha_{\text{seas}}}^\infty f_\alpha(\alpha) g^2(\alpha) d\alpha + \mu^2_{\overline{T}'_\xi}\left[1 - p_{\overline{T}'_\xi}(T_{\text{seas}})\right]$$

$$- 2\mu_{\overline{T}'_\xi} \int_{\alpha_{\text{seas}}}^\infty f_\alpha(\alpha) g(\alpha) d\alpha \qquad (8.25)$$

$$+ p_{\overline{T}'_\xi}(T_{\text{seas}})(T_{\text{seas}} - \mu_{\overline{T}'_\xi})^2.$$

The atom of probability, $p_{\overline{T}'_\xi}(T_{\text{seas}})$, in Eqs. (8.22) and (8.25) may be estimated as

$$p_{\overline{T}'_\xi}(T_{\text{seas}}) = \int_0^{\alpha_{\text{seas}}} f_\alpha(\alpha) d\alpha. \qquad (8.26)$$

Equations (8.24) and (8.25) allow the numerical evaluation of the coefficient of variation of $\overline{T}'_\xi$ for any particular parameter values. Identical relationships to Eqs. (8.24) and (8.25) are valid for the case when λ is considered a random variable and α is assumed to be constant in time. Both equations are still valid with the substitution of $f_\lambda(\lambda)$ for $f_\alpha(\alpha)$ and $\overline{T}'_\xi = r(\lambda)$ for $\overline{T}'_\xi = g(\alpha)$.

Figure 8.14 shows the influence of variability in λ on the atom of probability $p_{\overline{T}'_\xi}(T_{\text{seas}})$. The dependence is highly nonlinear with the strongest sensitivity of $p_{\overline{T}'_\xi}(T_{\text{seas}})$ with respect to $\overline{\lambda}$ in the range of values where those parameters are most commonly found in nature. It is also apparent that the sensitivity to $\overline{\lambda}$ is larger than that to σ^2_λ. The impact on $p_{\overline{T}'_\xi}(T_{\text{seas}})$ when varying $\overline{\alpha}$ (not shown) is quite similar to the one obtained when varying $\overline{\lambda}$.

Figure 8.14b shows the impact on $p_{\overline{T}'_\xi}(T_{\text{seas}})$ of the active soil depth and the evapotranspiration rate. Both variables strongly affect the atom of probability. Since especially in arid and semi-arid climates $p_{\overline{T}'_\xi}(T_{\text{seas}})$ affects all the statistical characteristics of $\overline{T}'_\xi$, the analyses above make evident the crucial role of evapotranspiration and soil depth in modulating the impact of climate fluctuations.

Figure 8.15 summarizes the dependence of the coefficient of variation of the seasonal mean duration of period of soil moisture below ξ on climate, soil, and vegetation. $\text{CV}[\overline{T}'_\xi]$ is normalized using $\text{CV}[\lambda]$, to show how much the non-linear dynamics amplifies the interannual variability of climate. Since the ratios $\text{CV}[\overline{T}'_\xi]/\text{CV}[\lambda]$ are practically always above 1 and reach values around 2–4, interannual climate fluctuations always induce a strong amplification of the fluctuations of the seasonal mean duration of periods with water stress. The crucial role of soil depth and evapotranspiration emerges again. Figure 8.15c shows the influence of the threshold ξ on the relative fluctuations of $\overline{T}'_\xi$ as

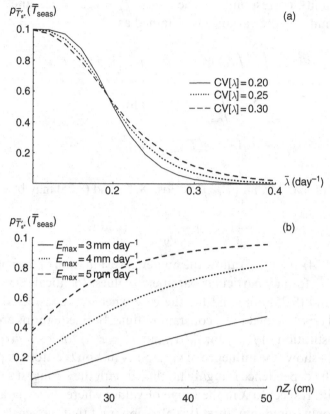

Figure 8.14 Impact of climate, soil, and vegetation characteristics on the atom of probability, $p_{\overline{T}'_\xi}(\overline{T}_{seas})$, when only λ is randomly changing from year to year ($\xi = s^* = 0.35, s_1^\xi = 0.8, K_s = 90\,\mathrm{cm\ day}^{-1}, T_{seas} = 150$ days). (a) $\overline{\alpha} = 10\,\mathrm{mm}$, $\mathrm{CV}[\alpha] = 0, nZ_r = 300\,\mathrm{mm}, E_{max} = 4\,\mathrm{mm\ day}^{-1}$; (b) $\overline{\lambda} = 0.2, \mathrm{CV}[\lambda] = 0.25$, $\overline{\alpha} = 0.2, \mathrm{CV}[\alpha] = 0$. After Ridolfi et al. (2000b).

a function of $\overline{\lambda}$. It is interesting to note the pronounced amplification of the climate fluctuations on the fluctuations of $\overline{T}'_\xi$ for a particular range of values of $\overline{\lambda}$ and how these fluctuations depend on soil and vegetation characteristics. The impact of the interannual climatic fluctuations is highly sensitive to the threshold considered (i.e., s^* or s_w). Analogous conclusions can be drawn when the variability in α is taken into account (Ridolfi et al., 2000b).

Monte Carlo simulations

Monte Carlo simulations were performed by Ridolfi et al. (2000b) to investigate the effect of the combined variability of λ and α. Fifty thousand pairs of random samples of α and λ from their respective two-parameter gamma distributions yield stable and smooth probability distributions for the seasonal

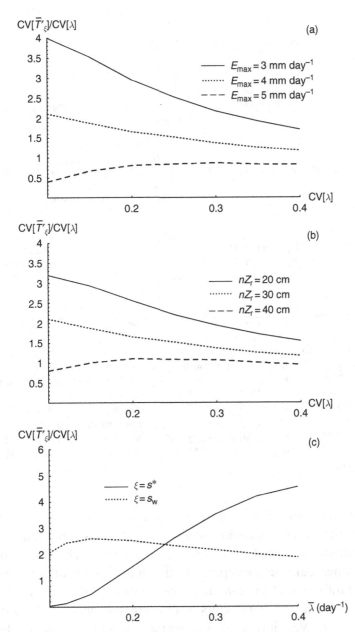

Figure 8.15 Impact of climate, soil, and vegetation characteristics on the coefficient of variation of the seasonal mean duration of periods with water stress. Only λ is randomly varied from year to year ($T_{\text{seas}} = 150$ days). (a) $\bar{\lambda} = 0.2$ storm day^{-1}, $\alpha = 10$ mm storm^{-1}, $\text{CV}[\alpha] = 0$, $nZ_r = 300$ mm, $\xi = s^* = 0.35$; (b) $\bar{\lambda} = 0.2$ storm day^{-1}, $\alpha = 10$ mm day^{-1}, $\text{CV}[\alpha] = 0$, $E_{\text{max}} = 4$ mm day^{-1}, $\xi = s^* = 0.35$ ($s_1 = 0.8$, $K_s = 90$ cm day^{-1}); (c) $\text{CV}[\lambda] = 0.25$, $\alpha = 10$ mm storm^{-1}, $\text{CV}[\alpha] = 0$, $nZ_r = 300$ mm, $E_{\text{max}} = 4$ mm day^{-1}, $s^* = 0.35$, $s_w = 0.15$. After Ridolfi et al. (2000b).

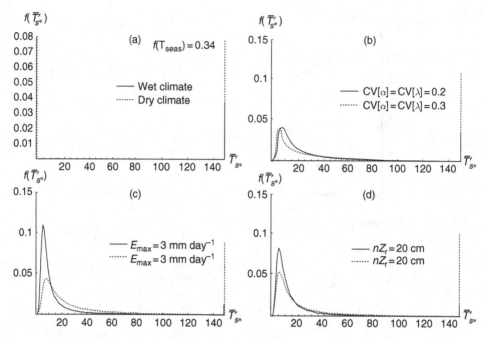

Figure 8.16 Impact of climate, soil, and vegetation characteristics on the probability distribution of the seasonal mean duration of water stress for $\xi = s^* = 0.35 (s_1 = 0.8, K_s = 90 \, \text{cm day}^{-1}, T_{\text{seas}} = 150 \, \text{days})$. (a) Wet climate: $\overline{\alpha} = 12 \, \text{mm}, \text{CV}[\alpha] = 0.25, \overline{\lambda} = 0.35, \text{CV}[\lambda] = 0.25$; dry climate: $\overline{\alpha} = 10 \, \text{mm}$, $\text{CV}[\alpha] = 0.25, \overline{\lambda} = 0.25, \text{CV}[\lambda] = 0.25 (nZ_r = 300 \, \text{mm}, E_{\text{max}} = 4 \, \text{mm day}^{-1})$; (b) $\overline{\alpha} = 12 \, \text{mm}, \overline{\lambda} = 0.3, nZ_r = 300 \, \text{mm}, E_{\text{max}} = 4 \, \text{mm day}^{-1}$; (c) $\overline{\alpha} = 12 \, \text{mm}$, $\text{CV}[\alpha] = 0.25, \overline{\lambda} = 0.35, \text{CV}[\lambda] = 0.25, nZ_r = 300 \, \text{mm}$;   (d) $\overline{\alpha} = 12 \, \text{mm}$, $\text{CV}[\alpha] = 0.25, \overline{\lambda} = 0.35, \text{CV}[\lambda] = 0.25, E_{\text{max}} = 4 \, \text{mm day}^{-1}$.   After   Ridolfi et al. (2000b).

mean duration of periods with water stress. Figure 8.16 shows typical probability distributions corresponding to an incipient stress condition, $\xi = s^*$, for different climate, soil, and vegetation characteristics. The results obtained previously when considering separately the impacts of α and λ are all confirmed. The influence of the climate is always strong: both the mean and the variance of the climatic parameters have a great impact on the probability distribution of $\overline{T}_{\xi}'$. Very important also are the roles of the active soil depth and the evapotranspiration rate. When the threshold level is the permanent wilting point (not shown), the atom of probability at T_{seas} is strongly reduced and in many conditions it is practically absent. Aside from this, the distributions confirm what is obtained for $\xi = s^*$.

The pdf of $\overline{n}_{\xi}$ is computed directly from Eq. (3.26). Differently from $\overline{T}_{\xi}'$, the results for $\overline{n}_{\xi}$ depend on the existence of a steady-state condition (see Chapter 3).

Thus the value of the initial soil moisture content may be a very important factor for the value of $\bar{n}_\xi$, especially in regions with relatively cold and humid winters, where transient conditions may last for a considerable part of the growing season (see Sections 3.3 and 7.1).

Figure 8.17 shows the influence of climatic variability and the role of soil and vegetation on $f(\bar{n}_\xi)$. Three qualitatively different cases may be distinguished: (i) the temporal evolution of soil moisture often remains above (e.g., wet climate) or below (e.g., dry climate) the threshold, and thus the crossings during the growing season are rare. In this case the distribution has a strong mode at $\bar{n}_\xi = 0$ with a sharp decrease for $\bar{n}_\xi \geq 0$; (ii) the temporal evolution of soil moisture involves frequent crossings of the threshold so that the mode shifts toward values of $\bar{n}_\xi$ different from zero and a number of stress periods are expected during the season; (iii) most commonly, the pdf is bimodal with one mode at zero (e.g., lack of crossing during the season) and another at a value that

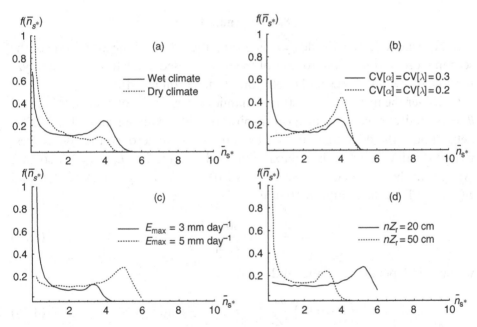

Figure 8.17 Impact of climate, soil, and vegetation characteristics on the probability distribution of the seasonal frequency of water stress for $\xi = s^* = 0.35 (s_1 = 0.8, K_s = 90 \text{ cm day}^{-1}, T_{\text{seas}} = 150 \text{ days})$. (a) Wet climate: $\bar{\alpha} = 12 \text{ mm}, \text{CV}[\alpha] = 0.25, \bar{\lambda} = 0.35, \text{CV}[\lambda] = 0.25$; dry climate: $\bar{\alpha} = 10 \text{ mm}, \text{CV}[\alpha] = 0.25, \bar{\lambda} = 0.20, \text{CV}[\lambda] = 0.25 (nZ_r = 300 \text{ mm}, E_{\text{max}} = 4 \text{ mm day}^{-1})$; (b) $\bar{\alpha} = 12 \text{ mm}, \bar{\lambda} = 0.3, nZ_r = 300 \text{ mm}, E_{\text{max}} = 4 \text{ mm day}^{-1}$; (c) $\bar{\alpha} = 12 \text{ mm}, \text{CV}[\alpha] = 0.25, \bar{\lambda} = 0.35, \text{CV}[\lambda] = 0.25, nZ_r = 300 \text{ mm}$; (d) $\bar{\alpha} = 12 \text{ mm}, \text{CV}[\alpha] = 0.25, \bar{\lambda} = 0.35, \text{CV}[\lambda] = 0.25, E_{\text{max}} = 4 \text{ mm day}^{-1}$. After Ridolfi et al. (2000b).

depends on the climate, soil, and vegetation characteristics. Two qualitatively different preferential states coexist during the evolution of soil moisture and the mean of the distribution is not very representative of the impact of interannual climatic fluctuations. The coefficient of variation for both the seasonal mean duration of periods with water stress, $\overline{T}'_\xi$, and the mean seasonal frequency of these periods, $\overline{n}_\xi$, is almost always greater than that of the climate fluctuations. The soil moisture dynamics amplifies the variability of $\overline{T}'_\xi$ and $\overline{n}_\xi$ compared to that existing in the interannual fluctuations of the rainfall regime.

Both the active soil depth and the evapotranspiration rate have fundamental roles in modulating the impact of climatic variability on the soil moisture evolution. Relatively small changes in soil, plant, and vegetation parameters may lead to significant differences in $\overline{T}'_\xi$ and $\overline{n}_\xi$. Vegetation also has an additional impact resulting from its control of the permanent wilting point and the moisture content at which water stress starts.

8.3 Appendix C

This Appendix derives the delay time t_d and the ratio r of the amplitude of the seasonal oscillation of $\langle s \rangle_t$ to that of the forcing used in Section 8.1 to analyze the role of Z_r in the seasonal evolution of mean soil moisture.

Consider the mean soil moisture dynamics driven by a constant rainfall rate $\theta_t = \theta_0$ and a maximum evapotranspiration rate evolving as in Eq. (8.20). Setting both the derivative of $\langle s \rangle_t$ at $t = 0$ and ϕ_{et} to zero, so that the corresponding value of $\langle s \rangle_{t=0}$ is a maximum, and considering $E_{max,0} = \theta_0$ so that $\langle s \rangle_{t=0}$ is in the range (s^*, s_{fc}) (see Eq. (8.13)), the resulting values of $A(t)$ and $B(t)$ from Eqs. (8.15) and (8.16) are

$$A(t) = \frac{\lambda'_t}{\gamma_t} 0.75 \sqrt{\frac{50}{Z_r}} \frac{1}{s_{fc} - s^*} = A, \tag{8.27}$$

which is independent of time, and

$$B(t) = 0.2 \frac{\theta_0}{nZ_r} + As^* - \frac{\delta_{et}\theta_0}{nZ_r} \sin(\omega_{et}t). \tag{8.28}$$

The average soil moisture is therefore, from Eq. (8.14),

$$\langle s \rangle_t = \langle s \rangle_0 e^{-At} + \left(0.2 \frac{\theta_0}{AnZ_r} + s^*\right)(1 - e^{-At}) - \frac{\delta_{et}\theta_0\omega_{et}}{nZ_r(A^2 + \omega_{et}^2)} e^{-At}$$
$$+ \frac{\delta_{et}\theta_0[\omega_{et}\cos(\omega_{et}t) - A\sin(\omega_{et}t)]}{nZ_r(A^2 + \omega_{et}^2)}, \tag{8.29}$$

and the condition of zero derivative in Eq. (8.13) reads

$$B(t_m) - A\langle s \rangle_{t_m} = 0.2\frac{\theta_0}{nZ_r} + A(s^* - A\langle s \rangle_{t_m}) - \frac{\delta_{et}\theta_0}{nZ_r}\sin(\omega_{et}t_m) = 0, \quad (8.30)$$

where t_m is the abscissa of the minima (maxima) of the mean soil moisture. The value of $\langle s \rangle_{t_m}$ from Eq. (8.29) can be substituted in Eq. (8.30), to obtain, after some mathematics ($\langle s \rangle_0 = s^* + \frac{0.2\theta_0}{AnZ_r}$ from Eq. (8.30)),

$$A\cos(\omega_{et}t_m) + \omega_{et}\sin(\omega_{et}t_m) - Ae^{-At_m} = 0. \quad (8.31)$$

The delay time (in days) is obtained as the first solution of Eq. (8.31) minus the value of the abscissa of the maximum of $E_{max,t}$, which is $\frac{365}{4}$, namely $t_d = t_m - \frac{365}{4}$.

Note that Eq. (8.31) is independent of the amplitude factor δ_{et}. This allows one to extend the analysis to any value of the parameters Z_r, λ_t and α_t, because there will always exist a δ_{et} value small enough to restrict the trajectory in the range (s^*, s_{fc}), which is the essential condition of validity of Eq. (8.31).

Consider now the ratio r of the amplitude of the seasonal oscillation of $\langle s \rangle_t$ to that of the forcing: r equals the difference between $\langle s \rangle_0$ and the ordinate of the minimum of the average soil moisture trajectory. One may find the latter from Eq. (8.30) and thus

$$\langle s \rangle_{t_m} = \langle s \rangle_0 - \frac{\delta_{et}\theta_0 \sin(\omega_{et}t_m)}{nZ_r A}. \quad (8.32)$$

The range of variations of the potential evapotranspiration is $\frac{\delta_{et}\theta_0}{nZ_r}$, and therefore

$$r = \frac{\sin(\omega_{et}t_m)}{A}. \quad (8.33)$$

Notice that the term $\frac{\delta_{et}\theta_0}{nZ_r}$ cancels out, i.e., r does not depend on the range of variations of the maximum evapotranspiration rate. This allows the comparison of different r values under different climatic conditions.

The values of r are always smaller for shallower soils, but this effect is probably limited to the analyzed range (s^*, s_{fc}), due to the presence of the term $-\frac{\lambda_t}{\gamma_t}0.75\sqrt{\frac{50}{Z_r}\frac{\langle s \rangle_t - s^*}{s_{fc} - s^*}}$, which tends to lower the value of the positive derivatives when Z_r is low.

9

Spatial scale issues in soil moisture dynamics

This chapter focuses on the role of the spatial dimension in the dynamics of water balance, with special emphasis on vegetation response and the resulting feedbacks. In fact, notwithstanding the importance of the results obtained in the preceding chapters for the spatially lumped description of soil moisture, there are important cases in which the lateral and vertical fluxes of soil moisture may become essential in the water balance of a site.

The first part of the chapter will discuss the role of the vertical dimension in soil moisture dynamics. Historically, this issue has been the focus of much attention in the ecohydrology context, especially in connection with the vertical partitioning of the water available to vegetation. The well-known Walter hypothesis for tree–grass coexistence in savannas (Walter, 1971; Eagleson and Segarra, 1985) has perhaps contributed to overemphasize the role played by the vertical dimension in soil moisture dynamics on vegetation patterns (see Scholes and Walker, 1993; and also Chapter 11). Although some climate and soil combinations may indeed enhance the importance of the vertical dimension (e.g., as in the case of the Patagonia steppe or the shrublands of Arizona; Burgess, 1995; Paruelo and Sala, 1995; Golluscio et al., 1998), its relevance at the daily and seasonal time scale is usually less dramatic than at the hourly time scale, so that vertically integrated models of soil moisture dynamics can often provide an accurate enough description in what concerns plant stress and ecosystem function. Likewise, the interaction of the active soil layer with the water table is of secondary importance in most water-controlled ecosystems, unless strong topographic gradients or plants with very deep tap roots (phreatotypes) are present.

The assessment of the role of the vertical fluxes of soil moisture involves the use of Richards' equation with time-varying boundary conditions plus suitable plant uptake and evaporation models. The mathematical difficulty is such that the analyses need to be carried out via numerical experimentation. The investigation is nevertheless very important for testing the adequacy of the simple spatially lumped model and investigating its applicability to specific cases of interest. The analysis presented here closely follows the recent paper by Guswa et al. (2002).

Regarding the impact of the lateral soil moisture fluxes, in areas that are mostly flat this tends to be very local, but wherever relevant topographic features exist, such as a river basin with a drainage network and its accompanying hillslope system, the lateral fluxes may be an important factor for the spatial distribution of soil moisture and its temporal evolution. Later in this chapter we will focus on the role of subsurface, unsaturated, lateral water flow and its links to climate, soil, and hillslope characteristics at the daily time scale, following the numerical analysis by Ridolfi et al. (2003a), who explored the role of lateral fluxes in the probabilistic behavior of soil moisture along a hillslope. Besides suggesting acceptable approximations for the full spatial problem, this line of research may provide, in the future, the means to link the spatial structure of the soil moisture field and its inherent temporal fluctuations with the organization and scaling that has been found in the interlocked systems of hillslopes and channels which make up the river basin (Rodríguez-Iturbe and Rinaldo, 1997).

9.1 An assessment of the role of the vertical distribution of soil moisture

This section describes a critical assessment of the role of the vertical dimension in the local water balance, through a comparison of the simple soil moisture model at a point introduced in Chapter 2 with a more complete model which explicitly accounts for the vertical variability of soil moisture within the root zone. The first model will be referred to as the zero-dimensional or bucket model and the second as the vertical or Richards model. The section is taken from Guswa et al. (2002).

9.1.1 The zero-dimensional and Richards models

The vertically lumped model introduced in Chapter 2 represents soil moisture dynamics at the daily time scale with a volume-balance equation applied over the root zone

$$nZ_r \frac{d\bar{s}}{dt} = I(\bar{s}, t) - L(\bar{s}) - E(\bar{s}) - EV(\bar{s}), \qquad (9.1)$$

where the average saturation over the root zone Z_r is now indicated by $\bar{s}$ to avoid confusion with the local soil moisture value along the vertical direction,

$s = s(z)$, which will be used in the Richards model. For the sake of comparison with the vertically explicit model, in Eq. (9.1) transpiration and evaporation rates, $E(\bar{s})$ and $EV(\bar{s})$, are accounted for separately.

In the Richards model, the soil column is resolved in the vertical dimension, and the soil moisture dynamics is described by Richards equation

$$\frac{\partial(ns)}{\partial t} - \frac{\partial}{\partial z}\left(K\frac{\partial h}{\partial z}\right) + \frac{\partial K}{\partial z} = -e' - \sigma' \qquad (9.2)$$

where s is the local saturation, K is the unsaturated hydraulic conductivity (with units of length per time), h is the pressure head in the water (with units of length), and z is positive downward. The functions e' and σ' are the local rates of evaporation and plant uptake in units of depth of water per depth of soil per unit time. Evaporation is assumed to take place over a depth, Z_e, and uptake of water by the plant occurs over the depth of the root zone, Z_r.

For a soil column that is discretized uniformly in space, the volume balance for layer i becomes

$$\Delta z \frac{\partial(ns)_i}{\partial t} - K_{i+\frac{1}{2}}\left(\frac{h_{i+1} - h_i}{\Delta z} - 1\right) - K_{i-\frac{1}{2}}\left(\frac{h_i - h_{i-1}}{\Delta z} - 1\right) = -e_i - \sigma_i \qquad (9.3)$$

where Δz is the spatial discretization, and e_i and σ_i are the rates of evaporation and plant uptake from layer i in units of depth per time. Figure 9.1 gives a schematic representation of the Richards model.

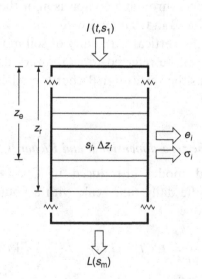

Figure 9.1 Schematic representation of the one-dimensional Richards model. After Guswa et al. (2002).

The retention curve relating the pressure head and saturation employed by Guswa et al. (2002) is given by

$$h(s) = h_e \cdot \left(\frac{s - s_h}{1 - s_h} \right)^{-b},$$

           (9.4)

where h_e is the entry pressure head, s_h is the hygroscopic saturation, and the exponent, b, describes the shape of the curve. The unsaturated hydraulic conductivity is given by

$$K(s) = K_s \cdot \left(\frac{s - s_h}{1 - s_h} \right)^{2b+3},$$

           (9.5)

where K_s is the hydraulic conductivity when the soil is fully saturated. Notice that Guswa et al. (2002) use water potential, h, in units of length. The relation with Ψ_s (units of pressure) is simply given by $h = \Psi_s/(g\rho_w)$, where g is the gravitational acceleration and ρ_w is the water density.

Infiltration and drainage

In the Richards model, precipitation is introduced as a boundary condition at the surface of the soil column. Each rainfall event is characterized by an intensity and duration. The duration for a given storm is taken from a uniform distribution, and the intensity is calculated so that the depth of water is equal to the depth of precipitation minus interception specified in the bucket model. This rainfall rate is set as a flux boundary condition at the top of the soil column, provided that the intensity is less than the potential infiltration rate for the soil (the infiltration rate if the water pressure at the surface were atmospheric). If the rainfall rate exceeds the potential infiltration rate, then the boundary flux is set equal to the potential rate, and the remaining precipitation runs off. This formulation allows both Hortonian and Dunne mechanisms of runoff generation.

Leakage of soil water to depths greater than the root zone is governed by Richards equation. The bottom boundary of the soil column is made deep enough so as to have minimal impact on the soil moisture dynamics within the root zone.

Plant uptake and evaporation

The vertical resolution of the Richards model allows the use of a more sophisticated uptake function that depends not only on the total water within the root zone but also on its spatial distribution. The water uptake by the plant is based on the models of Gardner (1960), Cowan (1965), Federer (1979, 1982),

Lhomme (1998), and others. Thus plant water uptake is taken as proportional to the difference in water potential between the soil and the plant. This type of plant uptake model is sometimes referred to as a type I model (e.g., Cardon and Letey, 1992; Shani and Dudley, 1996). The uptake is limited by two resistances: one associated with water movement through the root tissue and the other associated with movement of the water through the soil to the roots. The local uptake function is described mathematically by

$$\sigma_i = \Delta z_i \cdot \frac{(h_i - H_p)}{r_{s,i} + r_{r,i}}, \tag{9.6}$$

with the restriction that σ_i cannot be negative. In Eq. (9.6), $r_{s,i}$ and $r_{r,i}$ are the local soil and root resistances, respectively, and Δz_i is the thickness of the soil layer. The driving force for the uptake is a difference between the water potential in the soil, h_i, and the water potential in the plant, H_p (units of length). Variations of H_p within the plant are ignored.

As in Lhomme (1998), the local soil resistance is taken as inversely proportional to the unsaturated hydraulic conductivity and the local root density (length of roots per volume of soil)

$$r_{s,i} = \frac{C_s}{K(s_i) \cdot RW_i}, \tag{9.7}$$

where C_s is a dimensionless constant that accounts for root diameter, geometry, and arrangement. The local root resistance is inversely proportional to the root density

$$r_{r,i} = \frac{C_r}{RW_i}, \tag{9.8}$$

where C_r is a constant parameter of the plant with units of time per length. Re-writing the local root density as the product of the average root density over the entire root zone times a local, relative, root density gives the following expression for the local uptake

$$\sigma_i = \Delta z_i \cdot rw_i \cdot \frac{(h_i - H_p)}{\frac{C_s}{K(s_i)RW_0} + \frac{C_r}{RW_0}}, \tag{9.9}$$

where RW_0 is the average root density over the entire root zone, and rw_i is the relative root density as a function of depth. The local uptake function, Eq. (9.9), contains three unknowns: H_p, C_s, and C_r. Guswa et al. (2002) chose the values for these unknowns so that the plant uptake from the bucket and Richards models is the same. Notice that here the model of soil–root resistance is similar but not equal to that of Eq. (6.4). However, the differences between the two models are negligible if one considers constant the root area index (R_{AI}).

The value of the plant water potential, H_p, is subject to two constraints. The first is that H_p must be greater than or equal to the plant wilting potential, h_w (i.e., the pressure head corresponding to s_w from the bucket model). This ensures that the plant cannot extract water from regions in which the water potential is less than h_w. Additionally, the plant cannot transpire more over a day than the atmospheric demand, T_{max}. Therefore, the potential in the plant is assumed to be one of two values: the plant wilting potential or the potential for which the extraction over the entire root zone equals the transpiration demand, whichever is larger. Thus, either the plant extracts enough water from the entire root zone to meet demand, or the plant extracts as much as it can without its potential dropping below h_w.

The bucket model incorporates a reduction in the plant uptake when the root zone saturation drops below s^*. The uptake in the Richards model, however, depends not only on the total soil moisture but also on its spatial distribution. At one extreme, an average saturation of s^* can be achieved by a uniform saturation of s^* over the entire root zone. In this case, the total transpiration over the root zone is given by

$$E = \sum_{z_i \leq Z_r} \sigma_i = Z_r \cdot \frac{(h^* - H_p)}{\frac{C_s}{K(s^*)RW_0} + \frac{C_r}{RW_0}}. \tag{9.10}$$

For comparison with the bucket model, Guswa et al. (2002) stipulate that for this saturation profile the plant uptake over the entire root zone is equal to the transpiration demand when the plant potential is at its lowest value, h_w,

$$T_{max} = Z_r \cdot \frac{(h^* - h_w)}{\frac{C_s}{K(s^*)RW_0} + \frac{C_r}{RW_0}}. \tag{9.11}$$

If the potential in the soil were to drop below h^*, i.e., if the average saturation were to drop below s^*, then the extraction would not meet the demand because the plant potential cannot go below h_w (Guswa et al., 2002). This constraint corresponds to the s^* point used in the bucket model.

If the soil moisture is distributed nonuniformly over the root zone, the plant may compensate for some of its roots being dry by extracting additional water from roots in a wet region. The ability to do so is a function of the magnitude of the root resistance, r_r, since the soil resistance will be negligible at high saturations. If the root resistance is small, the plant can extract water at high rates from wet regions to compensate for portions of the root zone that are dry. This behavior is expressed by Guswa et al. (2002) using a factor, f, defined as the minimum fraction of the roots that must be at full saturation ($h = 0$) in order for the plant to withdraw enough water to meet the transpiration demand if extraction from elsewhere in the soil column is zero, i.e.,

$$T_{max} = f \cdot Z_r \cdot \frac{-h_w}{\frac{C_s}{K_s \, RW_0} + \frac{C_r}{RW_0}}. \tag{9.12}$$

If f is close to one, the plant has little ability to compensate; if f is close to zero, the plant can readily compensate for dry regions within the root zone. Specification of f enables the simultaneous solution of Eqs. (9.11) and (9.12) for C_r and C_s (Guswa et al., 2002).

Unlike the plant uptake, evaporation from the upper soil layers is presumed to be a function of the local saturation only. The extraction from a layer is described by

$$e_i = \Delta z_i \cdot ew_i \cdot \frac{EV(s_i)}{\sum_{i=1}^{m_e} \Delta z_i \cdot ew_i}, \tag{9.13}$$

where $EV(s_i)$ is the same function used in the bucket model with the local saturation as its argument, m_e is the number of soil layers over which evaporation is nonzero, and ew_i is a depth-dependent weight. If the soil layers from the ground surface to Z_e were uniformly saturated, the total evaporation as a function of saturation would be equal to that for the bucket model.

Since the two models are intended to be interpreted at the daily time scale, only a rudimentary diurnal variation of σ and e was included by Guswa et al. (2002). The extraction function is set to zero for 12 hours each day, and for the remaining 12 hours the values of T_{max} and E^* used in Eqs. (9.9)–(9.13) are set to twice those used in the bucket model.

Simulation and solution

Because of the highly nonlinear nature of the governing equations for the Richards model, an analytical solution is infeasible in this case. To simulate soil moisture dynamics and plant uptake, Guswa et al. (2002) used a one-step, fully implicit, temporal scheme with finite-difference approximations in space. The mass-conserving modified Picard iterations, described in Celia et al. (1990), are employed for solution of the nonlinear equations.

9.1.2 Key differences between the models

The two model formulations described before have some key differences that fall into three categories (Guswa et al., 2002): (i) infiltration dynamics,

(ii) runoff generation, and (iii) extraction functions for evaporation and plant uptake.

The first difference is related to the temporal and spatial resolution of the two models. The bucket model is a point model, and therefore cannot represent the migration of a wetting front through the soil column. Consequently, only an average saturation over the root zone can be resolved, giving no information about the spatial distribution of that water within the soil. The Richards model represents the vertical movement explicitly and resolves the spatial distribution of soil moisture.

Even if one were to use the Richards model with a single layer only, however, it would not be equivalent to the bucket model. In the bucket model, the intensity of rain events is ignored, and all water reaching the ground surface is presumed to enter the soil column until the root zone is fully saturated. In the Richards model, the precipitation rate must be specified, and runoff can occur via either the Dunne or Hortonian mechanism.

The last difference pertains to how the water losses due to evaporation and plant uptake are implemented. In the bucket model, these are simple functions of the average saturation over the entire root zone. In the Richards model, evapotranspiration is a function of the local saturation, the distribution of soil moisture over the entire root zone, and the weighting functions, $rw(z)$ and $ew(z)$. The driving force for the plant uptake is a potential difference, and for comparison purposes the function is constrained to match the important end points of the bucket model (Guswa et al., 2002).

Non-dimensional groups

The impact of the differences described in the previous section can be characterized by nondimensional groups of parameters (Guswa et al., 2002). The simplified representation of infiltration dynamics in the bucket model ignores both the temporal and spatial distributions of infiltration. The impact of the temporal simplification can be quantified with a temporal infiltration index

$$I_{I,t} = \frac{\min(\alpha/t_\mathrm{p}, K_\mathrm{s})}{E_{\max}}, \qquad (9.14)$$

where t_p is the characteristic duration of a rain event. $I_{I,t}$ is the ratio of the characteristic infiltration rate to the maximum rate of evapotranspiration, $E_{\max} = T_{\max} + E^*$. If $I_{I,t}$ is much greater than one, then infiltration occurs much faster than evapotranspiration, and representation of the process as instantaneous will not impact the results significantly.

Similarly, the impact of the spatial distribution of infiltration can be characterized by a spatial infiltration index

$$I_{I,x} = \frac{\overline{Z}_i}{Z_r},$$ (9.15)

where

$$\overline{Z}_i = \frac{\alpha}{n(s_{fc} - s_h)}.$$ (9.16)

$\overline{Z}_i$ is a measure of the depth of infiltration from a characteristic storm. Thus, $I_{I,x}$ is the ratio of the infiltration depth to the total depth of the bucket model. A small value of this ratio indicates that the spatial distribution of soil moisture after a rain event can be far from uniform, and this may lead to differences in the predictions of the two models. If both $I_{I,t}$ and $I_{I,x}$ are large, then differences in the model results due to their different representations of infiltration dynamics are likely to be small.

Under some climate and soil conditions, the omission of Hortonian overland flow from the bucket model may lead to differences in results when compared to the Richards model. The impact of this omission can be characterized by a runoff index,

$$R_I = \frac{K_s}{\alpha/t_p}.$$ (9.17)

If R_I is large, Hortonian runoff is unlikely to occur, and differences between the models due to runoff generation will be insignificant.

The impact of the different representations of evaporation and transpiration between the bucket and Richards models depends on whether the climate is characteristically wet or dry. This can be quantified using a modified dryness index (see Eq. (2.60)), i.e., the ratio of the maximum daily evapotranspiration rate to the average rainfall rate minus interception (Guswa et al., 2002),

$$D'_I = \frac{E_{max}}{\lambda'\alpha},$$ (9.18)

where λ' is, as usual, the rate of storm arrivals after interception has been subtracted. A dryness index greater than one indicates that evapotranspiration will be limited by the amount of water supplied through precipitation, not the atmospheric demand. In this case, leakage is likely to be small. If runoff is also negligible, then at steady-state, or over a long enough time, the average evapotranspiration rate will equal the rainfall rate. Therefore, predictions of average evapotranspiration from all models that are mass-conserving and have a plateau of E_{max} will be the same. Differences may arise in the rate of approach to steady-state, the partitioning between evaporation and transpiration, the time history of the average root zone saturation, and the timing and intensity of evapotranspiration.

As a way to quantify the differences in timing and intensity of evaporation and transpiration, Guswa et al. (2002) compared the characteristic rates of these processes immediately following a precipitation event, i.e.,

$$\Gamma_E = \frac{E_b^*}{E_R^*} \tag{9.19}$$

$$\Gamma_T = \frac{T_b^*}{T_R^*}, \tag{9.20}$$

where E_b^* and E_R^* are the evaporation rates just after a rain event for the bucket and Richards models, respectively, and T_b^* and T_R^* are the transpiration rates.

In a dry climate with infrequent rain events, between storms the soil moisture will decay to a value close to the wilting point saturation, below which the loss rate is very small. Consequently, for the bucket model applied to water-controlled ecosystems, the evapotranspiration rate, $E_b^* + T_b^*$, can be approximated as the rate at an average saturation equal to the saturation at the wilting point plus the change in saturation due to a characteristic rain event

$$E_b^* + T_b^* = E(s_w + \Delta \bar{s}) + T(s_w + \Delta \bar{s}), \tag{9.21}$$

where $\Delta \bar{s} = \alpha/nZ_r = 1/\gamma$ is the average jump in relative saturation due to a rainfall event.

For the Richards model, the evapotranspiration rate immediately following a storm can be characterized by the evaporation and transpiration rates calculated from Eqs. (9.13) and (9.9), presuming that the soil column is saturated to field capacity to a depth, $\overline{Z}_i$, and that there is no extraction from elsewhere in the soil column. This rate can be approximated by

$$E_R^* + T_R^* = f_E \cdot E(s_{fc}) + T_{max} \cdot \min\left(1, \frac{f_T}{f}\right), \tag{9.22}$$

with

$$f_E = \frac{\int_0^{\overline{Z}_i} ew(z)dz}{Z_e}, \tag{9.23}$$

$$f_T = \frac{\int_0^{\overline{Z}_i} rw(z)dz}{Z_r}. \tag{9.24}$$

In Eq. (9.22), f is the minimum fraction of roots under fully saturated conditions needed to achieve a total uptake of T_{max}; f_T and f_E are the fraction of the roots and the fraction of the zone of evaporation, respectively, that are wetted by a characteristic precipitation event.

If Γ_E and Γ_T are close to one, the losses from the bucket and Richards models are similar. Values of Γ larger than one indicate that the bucket model predicts a higher loss rate than the Richards model, and vice versa for a value less than one. Note that if $I_{I,x}$ is greater than one, a typical precipitation event will cover the entire soil column with a saturation greater than s_{fc}. Since s_{fc} is generally larger than s^*, this implies that, following a storm, the Richards and bucket models will both predict an evapotranspiration rate equal to E_{max}. Therefore, if $I_{I,x}$ is greater than one, there is no need to compute Γ_E and Γ_T.

Parameters of comparison

Guswa et al. (2002) limited the comparison of the two models to a specific plant, soil, and climate condition for which extensive data exist, and focused on the ability of the two models to represent the soil moisture dynamics and plant uptake for *Burkea africana,* a woody species from an African savanna (e.g., Scholes and Walker, 1993; see also Section 5.1).

Table 9.1 presents the parameter values used to describe the climate, including the rate of storm arrivals and mean storm depth, as well as the soil, using a combination of two soil types presented in Scholes and Walker (1993). Because of the slight modification of the functional forms for the unsaturated conductivity and retention curves, the parameter values differ slightly from those used in Section 5.1.

Table 9.2 presents the values used to describe the plant, *Burkea africana.* Guswa et al. (2002) used beta distributions to describe the weighting functions, ew(z) and rw(z)

$$\mathrm{rw}(z), \mathrm{ew}(z) = \frac{\Gamma(A+B)}{\Gamma(A) \cdot \Gamma(B)} \cdot z^{(A-1)} \cdot (1-z)^{(B-1)} \tag{9.25}$$

where A and B are non-negative shape parameters. Data on the root density distribution for *Burkea africana* are from Scholes and Walker (1993). Figure 9.2 presents the functions rw(z) and ew(z) for the simulations conducted by Guswa et al. (2002).

Because f is an unknown parameter, they chose two values in the comparison for *Burkea africana*: 0.1 and 0.8. To demonstrate how this parameter affects transpiration, Figure 9.3 presents the local uptake per unit of roots ($\sigma_i / \Delta z_i \cdot \mathrm{rw}_i \cdot \mathrm{RW}_0$) as a function of the local saturation, under the condition

Table 9.1 *Values of climate and soil parameters for the Nylsvley savanna used in the comparison by Guswa et al. (2002); see also Section 5.1.*

Parameter	Symbol	Value	Units
Storm arrival rate	λ	1/6	day^{-1}
Mean rainfall depth	α	1.5	cm
Min storm duration (Richards' model)	–	0.05	days
Max storm duration (Richards model)	–	0.15	days
Porosity	n	0.42	–
Saturated conductivity	K_s	109.8	cm day^{-1}
Hygroscopic saturation	s_h	0.02	–
Field capacity	s_{fc}	0.29	–
Entry pressure head	h_e	3.0	cm
Retention curve parameter	b	2.25	–
Drainage curve parameter	β	9.0	–

that $H_p = h_w$, for three values of f. The ordinate is normalized by $T_{max}/Z_r \cdot RW_0$, the average uptake per unit of roots when $T = T_{max}$. This figure shows that, locally, the plant can withdraw up to ten times the amount needed per unit of roots to meet the demand, T_{max}, when f is 0.1. As f moves closer to one, however, the plant loses its ability to withdraw water at high rates to compensate for the spatial variability of soil moisture.

As another interpretation of the plant uptake, Figure 9.4 displays the rate of uptake over the entire root zone, normalized by T_{max}, as a function of the depth, Z_i, to which the soil column is saturated to field capacity. The remaining portion of the soil column is taken to be at the wilting saturation, and, therefore, not contributing to the plant uptake. When $f = 0.1$, only the top 20% of the soil column needs to be wetted in order for the plant to uptake T_{max} (the fraction is not 10% due to the nonuniformity of the root distribution). When $f = 0.8$, more than 70% of the soil column needs to be saturated. Under the presumed soil-moisture profile, the average saturation over the root zone is a function of Z_i; also plotted in Figure 9.4 is the corresponding uptake predicted by the bucket model. When $Z_i = 0.27 \cdot Z_r$, the average root zone saturation equals s^*, and the transpiration for the bucket model equals T_{max}.

Table 9.2 *Values of vegetation parameters for Burkea africana, a woody species of the Nylsvley savanna. After Guswa et al. (2002).*

Parameter	Symbol	Value	Units
Depth of interception	Δ	0.2	cm
Maximum daily evaporation rate	E^*	0.15	cm day^{-1}
Maximum daily transpiration rate	T_{max}	0.325	cm day^{-1}
Potential at the point of stomatal closure	h^*	730	cm
Saturation at the point of stomatal closure	s^*	0.105	–
Minimum plant potential	h_w	31 600	cm
Saturation at the wilting point	s_w	0.036	–
Evaporation depth	Z_e	30	cm
Root zone depth	Z_r	100	cm
Parameter 1 for ew distribution	A	0.9	–
Parameter 2 for ew distribution	B	5.0	–
Parameter 1 for rw distribution	A	2.0	–
Parameter 2 for rw distribution	B	2.0	–
Mean root density	RW_0	0.02	cm cm^{-3}
Fraction of roots needed to supply T_{max}	f	0.1, 0.8	–

Three realizations of climate for a 200-day growing season were simulated by Guswa et al. (2002). The three realizations comprise a wet, average, and dry season, all generated from the same basic stochastic process. The differences among the realizations are due to variations in storm depth; all three have similar storm arrival rates. In none of the three cases did the rainfall rate exceed the saturated hydraulic conductivity; therefore, Hortonian runoff does not occur. For all simulations, the initial saturation is uniform over the root zone with a value of 0.1. The vertical resolution of the Richards model is 1 cm, and the saturation at twice Z_r is held fixed at s_{fc}.

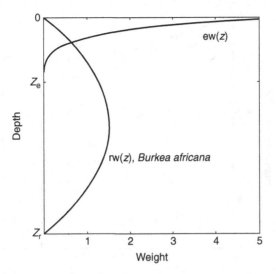

Figure 9.2 Weights for evaporation and transpiration as functions of depth. After Guswa et al. (2002).

Table 9.3 presents the values of the nondimensional groups, discussed above, for the particular climate, soil, and vegetation conditions under consideration. Inspection of such values indicates some anticipated behavior. First, the large value of the runoff index indicates that differences in runoff generation mechanisms are not important for this comparison. Second, the

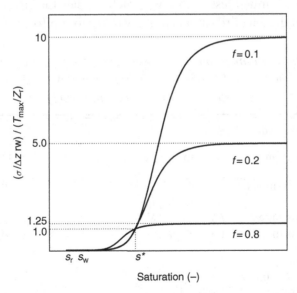

Figure 9.3 Normalized transpiration per unit of roots versus saturation for the Richards model when $H_p = h_w$. After Guswa et al. (2002).

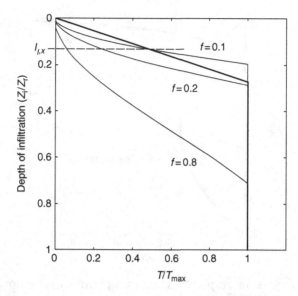

Figure 9.4 Transpiration rate as a function of depth of infiltration, Z_i, presuming a uniform saturation of s_{fc} from the ground surface to Z_i, and a uniform saturation of s_w below Z_i. The heavy line represents the bucket model; the light solid lines represent the Richards model for various values of f. After Guswa et al. (2002).

temporal infiltration index is much greater than one, indicating that representation of rain events as instantaneous is reasonable. The spatial infiltration index, however, is much less than one, which implies that the distribution of soil moisture in the Richards model is likely to be nonuniform. This may affect

Table 9.3 *Values of dimensionless numbers quantifying the comparison of the bucket and Richards models of soil moisture dynamics for* Burkea africana *at Nylsvley. After Guswa et al. (2002).*

Nondimensional group	Symbol	Value
Temporal infiltration index	$I_{I,t} = \min(K_s, \alpha/t_p)/E_{\max}$	32
Spatial infiltration index	$I_{I,x} = \overline{Z}_i/Z_r$	0.13
Runoff index	$R_I = K_s/(\alpha/t_p)$	7.3
Modified dryness index	$D_I' = E_{\max}/(\lambda'\alpha)$	2.1
Ratio of post-storm evaporation	$\Gamma_E = E_b^*/E_R^*$	0.64
Ratio of post-storm transpiration, $f = 0.8$	$\Gamma_T = T_b^*/T_R^*$	8.4
Ratio of post-storm transpiration, $f = 0.1$	Γ_T	1.0

the match between the models, since the uptake in the Richards model depends on the local saturation. The value of Γ_E less than one indicates that evaporation immediately following a storm will be higher for the Richards model than the bucket model. The value of Γ_T close to one for $f = 0.1$ indicates that the bucket model may be a good approximation to this Richards model; the value of Γ_T larger than one for the Richards model with $f = 0.8$ indicates that, in terms of transpiration losses, the bucket model predicts higher rates of uptake immediately following a rain event. The difference in Γ_E and Γ_T indicates that the partitioning of infiltration between evaporation and transpiration may be different for the bucket and Richards models.

9.1.3 Results

Table 9.4 presents a summary of the results obtained by Guswa et al. (2002) from the simulations conducted with the bucket and Richards models. The results include the values of λ' and α estimated from the climate realizations, the depth of total infiltration over the growing season, the cumulative transpiration and evaporation over the growing season – as depths and as percentages of total infiltration – and the temporally averaged root zone saturation. For the simulations with the Richards model, capillary rise into the root zone contributed 1.8 cm of moisture over the growing season; this accounts for the difference in infiltration values between the bucket and Richards models. Though leakage is negligible, in each simulation the sum of evaporation and transpiration does not equal the infiltration because of changes in storage in the root zone over the growing season.

For all climate realizations, the bucket model predicts the highest values of transpiration over the growing season (Guswa et al., 2002). The bucket model also predicts the lowest values of average saturation over the root zone. Across the three seasons (dry, average, and wet), the fraction of infiltration that ends up as transpiration is nearly constant for the bucket model. For the Richards model, however, the fraction increases with increasing wetness of the climate. For example, the transpiration goes from 54% to 57% to 66% of infiltration for the case with $f = 0.1$. Correspondingly, the fraction of infiltration that goes as evaporation decreases from the dry to the wet climate. For the dry climate, losses are split evenly between evaporation and transpiration, while for the wet climate the ratio is nearly one to two. The match between the bucket and Richards models improves as the climate gets wetter and as the value of f gets smaller, i.e., as the plant's ability to compensate for spatial variability in the soil-moisture distribution improves.

Table 9.4 *Cumulative infiltration, transpiration, evaporation, and average saturation over a growing season of 200 days as predicted by the bucket and Richards models. After Guswa et al. (2002).*

Season	λ' (day^{-1})	α (cm)	f	Model	Infiltr. (cm)	Transp. (cm (%))	Evap. (cm (%))	Sat. (%)
Dry	0.14	1.2	–	Bucket	35.2	23.5 (67%)	14.4 (41%)	7
			0.8	Richards	37.0	17.9 (48%)	19.4 (52%)	10
			0.1		37.0	20.1 (54%)	17.0 (46%)	9
Average	0.15	1.5	–	Bucket	45.4	30.8 (68%)	17.2 (38%)	8
			0.8	Richards	47.3	25.0 (53%)	22.6 (48%)	12
			0.1		47.3	27.1 (57%)	20.3 (43%)	10
Wet	0.14	2.2	–	Bucket	58.2	39.6 (68%)	20.4 (35%)	12
			0.8	Richards	60.1	36.5 (61%)	20.4 (34%)	17
			0.1		60.2	40.0 (66%)	19.2 (32%)	14

Average values over the entire growing season do not give the complete picture, however. Figure 9.5 shows a time history of the vertically integrated root zone saturation over the growing season under average climate conditions, as predicted by the bucket model and Richards model with $f = 0.8$ and 0.1. This figure is in general agreement with the results mentioned above: the trace for the bucket model is lower than the two others, and it is closer to the Richards model with $f = 0.1$ than the case with $f = 0.8$. The lowest saturations for the Richards model simulations level out near 7%, while those for the bucket model drop down to 4%. Except for the behavior at very low saturations, the results from the bucket model and the Richards model with $f = 0.1$ are in close agreement. The predictions of the Richards model with $f = 0.8$ follow the same trends, but the details of the trace are different (Guswa et al., 2002).

Figure 9.6 presents a time history of the daily transpiration rate for the same three cases. While the bucket model and the Richards model with $f = 0.1$ predict daily transpiration rates that reach T_{max}, the transpiration rate for the Richards model with $f = 0.8$ never does. The frequency of the transpiration fluctuations for the Richards model with $f = 0.8$ is also much lower than it is for the two other cases. In general, the transpiration predictions from the

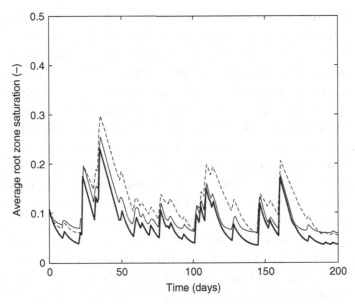

Figure 9.5 Traces of root zone saturation over the growing season for the average climate realization. The heavy line is the result from the bucket model; the light solid line is the result from the Richards model with $f = 0.1$; the dashed line is the result from the Richards model with $f = 0.8$. After Guswa et al. (2002).

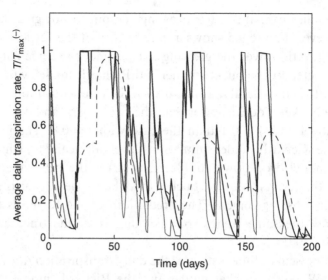

Figure 9.6 Traces of daily transpiration over the growing season for the average climate realization. The heavy line is the result from the bucket model; the light solid line is the result from the Richards model with $f = 0.1$; the dashed line is the result from the Richards model with $f = 0.8$. After Guswa et al. (2002).

bucket model and the Richards model with $f = 0.1$ are in good agreement. After a significant rain event, however, the decline of the transpiration rate is faster for the bucket model (see examples near days 120 and 170). This model shows a rapid decrease in total transpiration from $T_{\max}$ to nearly zero.

In addition to the temporal evolution of transpiration, Guswa et al. (2002) also investigated the dependence of total evapotranspiration on the average root zone saturation. Figure 9.7 presents this relationship for the Richards model with $f = 0.1$ for the average climate realization. Each discrete point corresponds to the total evapotranspiration and average saturation over one day. Plotted on the same figure is a solid line depicting the corresponding relationship for the bucket model. This figure shows that the relationship between total evapotranspiration and average saturation for the Richards model is not single-valued. For an average root zone saturation of 0.1, the evapotranspiration rate ranges from around 0.25 cm day^{-1} up to $E_{\max} = 0.475$ cm day^{-1}. Despite some differences at low saturations, the upscaled evapotranspiration functions for the two models are not too different (Guswa et al., 2002).

Figure 9.8 presents the relationship between saturation and evapotranspiration for the bucket model and the Richards model with $f = 0.8$. In this case, evapotranspiration as a function of average saturation exhibits even greater

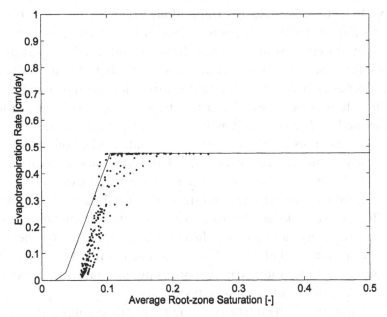

Figure 9.7 Relationship between total evapotranspiration and average root zone saturation for the average climate realization. Points represent upscaled results from the Richards model with $f = 0.1$, and the solid line represents the function used in the bucket model. After Guswa et al. (2002).

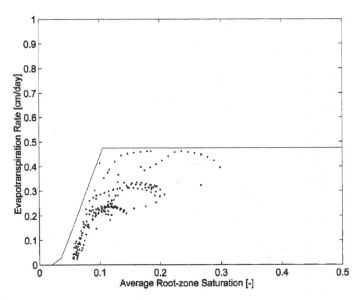

Figure 9.8 Relationship between total evapotranspiration and average root zone saturation for the average climate realization. Points represent upscaled results from the Richards model with $f = 0.8$, and the solid line represents the function used in the bucket model. After Guswa et al. (2002).

variation. In fact, one can identify traces along which the average saturation is decreasing but the total evapotranspiration is increasing. This results from propagation of wetting fronts through the soil column and the plateau of the extraction function – that is, the same amount of water spread over twice the depth can produce higher rates of evapotranspiration. As shown in this figure, the relationships between evapotranspiration and root zone saturation for the bucket and Richards models with $f = 0.8$ are quite different.

Generally, the plots for the wet and dry climate realizations are similar to those already presented. However, the relationship between evapotranspiration rate and average saturation for the Richards model with $f = 0.1$ coupled with the wet climate shows some interesting behavior. Figure 9.9 presents these results. The figure looks similar to that for the average climate conditions, except for a few traces along which the evapotranspiration drops below $E_{\max}$ even for saturations well above s^*. This phenomenon results from the limited spatial extent of evaporation. The evapotranspiration rate after a large storm remains at $E_{\max}$ for a number of days during which the upper soil layers quickly dry out due to the intensity of the evaporative uptake. At this point, the evaporation is reduced even though the average saturation of the soil column is still fairly high. The figure shows traces of the evapotranspiration rate approaching the value of $E_{\max} - E^* = T_{\max}$.

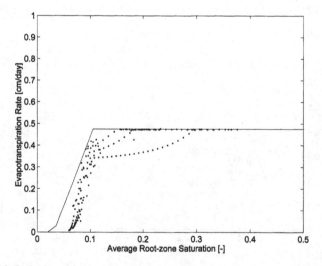

Figure 9.9 Relationship between total evapotranspiration and average root zone saturation for the wet climate realization. Points represent upscaled results from the Richards model with $f = 0.1$, and the solid line represents the function used in the bucket model. After Guswa et al. (2002).

9.1.4 Discussion

The two models of soil moisture dynamics discussed before present differences in the infiltration dynamics, runoff generation, and uptake for evapotranspiration (Guswa et al., 2002). Because of the climate and soil characteristics, differences in runoff generation are negligible for this comparison, as indicated by a large value of R_I. The temporal infiltration index is also much greater than one, so the representation of rainfall as instantaneous shots of water in the bucket model is reasonable. Therefore, the differences in the results can be attributed to the spatial variability in soil moisture and its effect on local losses due to evaporation and transpiration. The biggest differences in the predictions from the bucket and Richards models are in the relationship between evapotranspiration and average root zone saturation, the timing and intensity of transpiration, and the partitioning of uptake between evaporation and transpiration. Whether or not these differences are significant will depend on the objective of the modeling study. For example, if one is interested in the time history of soil moisture, the relationship between evapotranspiration and saturation is important, but the partitioning of the losses between evaporation and transpiration is less significant. Conversely, if one is specifically interested in the time course of transpiration, the above relationship may be a most significant quantity. Therefore, the appropriate use of the bucket or Richards model will depend on the goals of the study.

For the behavior of the bucket and Richards models to match, the upscaled relationships between evapotranspiration and average saturation must be similar. For all cases analyzed by Guswa et al. (2002), the evapotranspiration at low average saturations is higher for the bucket model than for the Richards model. Part of this difference is due to the difference in the steepness of the supply curve and part to the limited depth over which evaporation takes place. For a uniform saturation profile, Figure 9.3 gives the total, normalized, plant uptake when the saturation is less than s^*. As the saturation goes from s^* to s_w, the transpiration rate depicted by this curve decays more quickly than that for the bucket model. This difference in the steepness of the upscaled transpiration curves between the bucket and Richards models is consistent with the different rates of decline in the transpiration rate after a storm, depicted in Figure 9.6. For the Richards model, the steepness of the curve is directly related to the unsaturated hydraulic conductivity. The shape of the relative permeability curve influences both the rate of change of the transpiration rate as a function of soil moisture and also the saturation below which uptake effectively ceases. Even though transpiration does not go to zero until the wilting saturation, the steepness of the supply curve causes rates of uptake at saturations below 5% to be negligible. Since relative permeability is a difficult quantity to measure at low saturations, whether the losses from the Richards model at low saturations are a better representation than those from the bucket model is not clear. To obtain a better match between the two models, one could reassign s_w from $s(h_w)$, as determined from the retention curve, to a value closer to 7%, at which point the unsaturated hydraulic conductivity is a factor of 50 less than the value at s^*. Alternately, one could use a different representation of the relative permeability curve for the plant uptake function, or a different uptake function altogether, that would bring the supply function for the Richards model closer to that for the bucket model. Of course, the appropriate representation depends on available data and their reliability.

In addition to the differences arising from the formulation of the uptake function, the vertical variability of evaporation leads to differences in the relationship between total evapotranspiration and average saturation for the two models. Because evaporation occurs over just a fraction of the root zone, the average saturation over the entire root zone could be much higher than hygroscopic even when evaporation is nearly zero (Guswa et al., 2002).

Figures 9.7, 9.8, and 9.9 also indicate that the relationship between transpiration and average saturation is not unique. Federer (1982) shows similar scatter in the relationship between transpiration and average saturation. Because of the variation in root density and the nonlinearity of the retention and relative permeability curves with respect to saturation, the transpiration is strongly dependent on the distribution of soil moisture. The total uptake from a soil column with an average saturation of 15% can be very different if the water is distributed evenly over the soil column versus that if the top 15% of an otherwise dry root zone is fully saturated. As the resistance to flow in the root tissue decreases, the plant can compensate for the spatial variability of soil moisture, and the nonuniqueness of transpiration as a function of saturation is less severe for smaller values of f. For *Burkea africana* with $f = 0.1$, representation of the relationship between transpiration and average saturation with a single-valued function would be reasonable. For the case with $f = 0.8$, however, such a representation is not appropriate (Guswa et al., 2002).

Because of the shallow depth over which evaporation occurs, the spatial variability in soil moisture over this depth is likely to be regulated purely by the process of drying. Thus, in this shallow layer, there is likely to be a one-to-one relationship between the distribution of soil moisture and the average saturation. Consequently, one would anticipate that representing evaporation as a function of an average saturation would be appropriate, provided that the average saturation were calculated over the depth of evaporation only. Representation of evaporation as a function of the average saturation over the entire root zone is not appropriate when only a fraction of that depth is affected by the process of evaporation.

Figure 9.6 shows that the timing and intensity of transpiration for the Richards model is a strong function of the parameter f. When the root resistance is small, i.e., when f is small, the plant can extract water from wet regions at high rates to compensate for roots in dry parts of the soil column (Guswa et al., 2002). In such a case, the transpiration rate fluctuates rapidly. When f is close to one, however, the changes in transpiration rate over time are more subdued. For the cases under consideration here, predictions of the transpiration rate from the bucket model more closely match those from the Richards model when $f = 0.1$ than when $f = 0.8$. This is consistent with the values of Γ_T for these two cases: 1.0 and 8.4, respectively. This consistency is depicted graphically in Figure 9.4, which shows that, when $Z_i = \overline{Z}_i$ (i.e., $Z_i/Z_r = I_{l,x}$), the transpiration rates for the bucket model and the Richards model with $f = 0.1$ are close, but the rate for the bucket model is far greater than the rate for the Richards model with $f = 0.8$.

Though f is not known for *Burkea africana*, a sense of the appropriateness of the values can be obtained by comparing the associated values of C_r to measured values for other plant species. The values of C_r for the Richards model are $7.8 \cdot 10^4$ days cm^{-1} when $f = 0.8$ and $9.7 \cdot 10^3$ days cm^{-1} when $f = 0.1$. Steudle et al. (1987) subjected many samples of young maize roots (with a diameter of approximately 1 mm) to an applied pressure gradient and measured the resulting root conductivities. When converted to the same units as C_r, the values range from $1.3 \cdot 10^4$ to $3.1 \cdot 10^5$ days cm^{-1}. The values used in the Richards model for *Burkea africana* are comparable, but given the range of measured values and the difference in species, a definitive value of f for *Burkea africana* cannot be obtained (Guswa et al., 2002).

A final possible difference between the bucket and Richards models is in the partitioning of evapotranspiration between evaporation and transpiration. If the environment were not water limited, i.e., if the index of dryness, D'_I, were much less than one, indicating deep and frequent storms, the ratio of transpiration to evaporation would be T_{max}/E^*, or 2.167 in this case. The approach to this asymptote for the water-controlled ecosystem under consideration varies between the bucket and the Richards models. Since the soil column dries substantially between storms, both models predict that shallower storms will lead to a greater fraction of the uptake going as evaporation. In the bucket model, this is due to the shape of the loss function at low saturations. For the Richards model, the weighting functions depicted in Figure 9.2 indicate that evaporation is a large component of local uptake at shallow depths; for large storms, however, more water infiltrates to depths at which the plant can use it, as indicated in Table 9.4. Following a characteristic storm, Γ_E indicates that the evaporative losses for the Richards model are about 50% larger than those for the bucket model. Over a season, the Richards model predicts that the split between transpiration and evaporation is approximately 55% and 45% of total evapotranspiration, respectively, for the average climate realization. For the bucket model the split is 68% and 38%.

The comparison presented above dealt with a specific case in which the root depth was relatively large (e.g., 100 cm), so that the differences between the two models were particularly contrasted. In other simulations, not shown by Guswa et al. (2002), with more superficial root distributions the match between the two models is always good regardless of the value of the parameter f, as long as the value of the rooting depth Z_r in the bucket model is a good representation of the effective root zone.

9.2 Stochastic soil moisture dynamics along a hillslope

The study of the spatial dynamics of soil moisture along a hillslope was approached by Ridolfi et al. (2003a) using the same framework of the vertically lumped model of Chapter 2. This section is taken from their work. Before discussing the specific model they used, a quick glance at the various approaches pursued in the literature of hillslope hydrology is an instructive introduction to the complexity of the problem at hand and underlines the necessity for a simplified and scope-limited treatment.

9.2.1 Challenges and approaches to hillslope hydrology

In the last decades hillslope hydrology has attracted much attention (e.g., Kirkby, 1978, 1988; O'Loughlin, 1990; Western and Grayson, 1998; Bronstert, 1999), mostly focusing on runoff processes at the hillslope scale and the modeling of infiltration, subsurface storm flow, surface runoff (e.g., Freeze, 1980; Wigmosta et al., 1994; Shakya and Chander, 1998; Bronstert, 1999; Beldring et al., 2000), soil macroporosity (e.g., Beven and Germann, 1982; Mosley, 1982; Germann and Beven, 1986; Bronstert and Plate, 1997), saturated and unsaturated lateral fluxes (e.g., Stagnitti et al., 1982; Sloan and Moore, 1984; Salvucci and Entekhabi, 1995; Bronstert and Plate, 1997), and saturated overland flow (e.g., Fan and Bras, 1998; Ogden and Watts, 2000). Some investigations have studied greater spatial and temporal scales (e.g., Western and Grayson, 1998; Yeh and Eltahir, 1998; Bronstert, 1999; Western et al., 1999; Zhu and Mackay, 2001) and the simulation of continuous time water balances (e.g., Stagnitti et al., 1982; Bronstert, 1999), while others have related hillslope hydrology to specific environmental and engineering applications, such as landslide triggering, agricultural land use, interactions between riparian zones and hillslopes, and pollutant transport (e.g., Guzzetti, 1998; Bronstert, 1999). The interest in hillslope hydrology is also evidenced by the increasing number of research projects on experimental hillslopes, including measurements of soil moisture (e.g., Ritsema et al., 1996; Grayson et al., 1997; Famiglietti et al., 1998; Western and Grayson, 1998; Western et al., 1999) and studies of the links between soil texture and water fluxes (Tsuboyama et al., 1994; Noguchi et al., 1999).

Hillslope hydrology is especially challenging because a number of processes interact at different scales, significantly contributing to the complexity of the system. In the context of this section some of the most important are the following (Ridolfi et al., 2003a):

- Soil properties and their spatial heterogeneity (Crave and Gascuel-Odoux, 1997; Munro and Huang, 1997; Famiglietti et al., 1998; Western et al., 1999; Zhu and Mackay, 2001). This includes the nonlinear dependence of plant water potential and water uptake on soil texture (Chapters 2 and 4), microstructures of the soil surface (Beven and Germann, 1982), layers with different hydraulic properties and fissures, macroporosity due to roots, decayed root holes, animal burrows, worm holes and other structural channels (Beven and Germann, 1982; Mosley, 1982; Bronstert, 1999; Casanova et al., 2000), and soil erosion and production (Carson and Kirkby, 1972; Anderson, 1988; Noguchi et al., 1999).
- Different types of climate characteristics which, although spatially uniform over the hillslope, may trigger other mechanisms that generate spatial dynamics (e.g., Western et al., 1999).
- Lateral redistribution of water along the hillslope due to the formation of a saturated zone in the soil (Sloan and Moore, 1984; Anderson and Burt, 1990; Salvucci and Entekhabi, 1995; Bronstert, 1999).
- Lateral subsurface flow in the unsaturated zone (McCord and Stephens, 1987; Genereux and Hemond, 1990; Grayson et al., 1997; Yeh and Eltahir, 1998; Bronstert, 1999), especially during wet periods when the soil lateral hydraulic conductivity is non-negligible because of interconnected macroporosities (Beven and Germann, 1982; Niemann and Edgell, 1993; Bronstert and Plate, 1997).
- Type and spatial patterns of vegetation along a hillslope (Hawley et al., 1983; Wigmosta et al., 1994; Famiglietti et al., 1998; Western et al., 1999), which are both the cause and consequence of soil moisture dynamics.
- Infiltration of runoff generated in the uphill part of the slope (Bronstert and Plate, 1997; Corradini et al., 1998). This process occurs at the rainstorm time scale and depends on the properties of the surface soil layer (e.g., crusted and sealing soils; Luk et al., 1993; Zhu et al., 1997; Corradini et al., 2000) as well as on the type and spatial distribution of vegetation on the hillslope.

- Longitudinal hillslope profile and form (Moore et al., 1988; Nyberg, 1996; Grayson et al., 1997; Famiglietti et al., 1998; Western et al., 1999).
- Three-dimensional hillslope geometry and the presence of spurs and hollows (Beven, 1989; Band, 1991; Fan and Bras, 1998; Western et al., 1999).
- Geographic position of the hillslope and exposure to sun and winds, which may strongly affect the rate of evapotranspiration, the spatial distribution of vegetation, and soil properties (Hanna et al., 1982; Cerda, 1997; Famiglietti et al., 1998; Western et al., 1999; Casanova et al., 2000; Zhu and Mackay, 2001).
- Boundary conditions, especially at the bottom of the hillslope.

9.2.2 Hypotheses and model

From the previous discussion it is clear that the complexity of the phenomena involved and their difficult quantification hamper the possibility of a general theory. When studying hillslope hydrology, it is thus convenient to isolate some mechanisms and explore them separately making simplifying assumptions. With such a perspective, Ridolfi et al. (2003a) explored the effectiveness of the lateral water fluxes within the root zone in creating long-term spatial gradients of soil moisture on a hillslope, the conditions in which lateral fluxes are important at the seasonal time scale as well as how these are controlled in a nonlocal way by all the hillslope characteristics.

Figure 9.10 shows the scheme of the hillslope geometry together with the reference frame adopted. The hillslope profile is expressed by the following power law (Carson and Kirkby, 1972)

$$z = a\left[1 - \left(\frac{x}{L}\right)^{\frac{1-l}{m}+1}\right], \tag{9.26}$$

where z is the elevation, x is the horizontal distance from the divide, and a and L are the height and length of the vertical and horizontal projections of the hillslope, respectively. This equation, which represents the hillslope form occurring in transport-limited conditions (Carson and Kirkby, 1972), describes a wide variety of real hillslope geometries (Figure 9.11).

The soil is assumed to have uniform depth, Z_r, corresponding to the rooting depth, and be placed above a bedrock or basement characterized by very low hydraulic conductivity. Due to the shallowness of the soil layer and to the time scales of interest, the dynamics of soil moisture is integrated over the depth Z_r. Consequently, the development of a water table and the consequent partitioning between saturated and unsaturated flow are not considered (e.g., Fan and Bras, 1998).

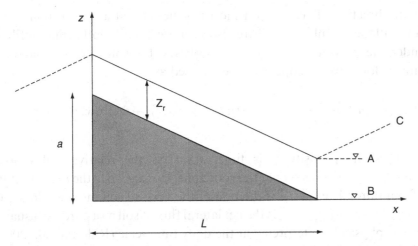

Figure 9.10 Scheme of the idealized hillslope considered in this section. The capital letters indicate the three boundary conditions at $x = L$, discussed in the text. After Ridolfi et al. (2003a).

The vegetation type is supposed to be uniform along the hillslope and constant in time. The soil is considered vertically and laterally homogeneous, while the rainfall regime is uniform over the hillslope. These hypotheses allow one to neglect the downhill infiltration of the runoff generated in the uphill portion of the hillslope. In fact, the homogeneity of soil and vegetation along the hillslope implies that soil moisture is monotonically increasing in the

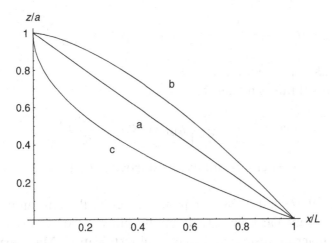

Figure 9.11 Hillslope profiles used in the analysis by Ridolfi et al. (2003a). The three shapes correspond to $\{l, m\}$ equal to $\{1, 1\}$ (case a), $\{0, 1\}$ (case b) and $\{2, 2\}$ (case c). After Ridolfi et al. (2003a).

downhill direction. Thus, when runoff is generated at a certain point of the hillslope all the downhill points are also saturated and runoff cannot infiltrate.

Under the previous simplifying hypotheses, the soil moisture balance at a point x along the hillslope can be expressed as

$$nZ_r \frac{\partial s(x,t)}{\partial t} = \varphi[s(x,t);t] - \chi[s(x,t)] + \Phi_{lat}[s(x,t)], \qquad (9.27)$$

where n is the porosity, t is time, $s(x,t)$ is the relative soil moisture ($0 \leq s(x,t) \leq 1$) averaged over the root zone, $\varphi[s(x,t);t]$ is the rate of infiltration from rainfall, $\chi[s(x,t)]$ is the rate of the vertical soil moisture losses (see Section 2.1), and $\Phi_{lat}[s(x,t)]$ is the net lateral flux of soil moisture. As usual, the process is physically interpreted at the daily time scale (Ridolfi et al., 2003a).

Due to the assumption of a shallow soil overlying a low-permeability basement (as, for example, in Fan and Bras, 1998, or Yeh and Eltahir, 1998), the losses due to leakage $L[s(x,t)]$ were neglected by Ridolfi et al. (2003a). However, when the leakage term is significant, it can be easily incorporated into the model as a vertical percolation, using for the hydraulic conductivity the same model as in Chapter 2.

Despite the possible existence of a certain degree of anisotropy in the soil properties, most of the studies in the literature refer to isotropic conditions. In Ridolfi et al. (2003a) no distinction is made between lateral and vertical hydraulic conductivity. The dependence of hydraulic conductivity on soil moisture is modeled as an exponential decay from the saturated value K_s at $s = 1$ to zero at field capacity s_{fc},

$$K(s) = \begin{cases} 0 & s \leq s_{fc} \\ \frac{K_s}{e^{\beta(1-s_{fc})}-1}[e^{\beta(1-s_{fc})} - 1] & s > s_{fc}, \end{cases}$$

where $\beta = (2b + 4)$ (see Section 2.1.6).

The net lateral flux is modeled as

$$\Phi_{lat}(s) = Z_r \frac{\partial}{\partial x}\left\{ K_{lat}(s)\left[S(x) + \frac{\partial \Psi_s}{\partial x} \right] \right\}, \qquad (9.28)$$

where $K_{lat}(s)$ is the lateral hydraulic conductivity, $K_{lat}(s) = K(s)$, and $S(x)$ is the local slope.

Equations (9.27) and (9.28) adequately describe the unsaturated Darcian flow in porous media; however, its applicability in soils with a connected structure of macroporosity requires careful justification. Many experimental results (e.g., Beven and Germann, 1982; Mosley, 1982; Bronstert, 1999; Noguchi et al., 1999) indicate that a significant portion of the total flow along the hillslope takes place in the macropores, where the rate of flow is about 100–400 times faster than in the soil matrix (Bronstert and Plate, 1997). Consequently a higher hydraulic conductivity is experimentally measured for the soil profile as a whole. Several approaches have been developed to account for these two different mechanisms (the matrix flow and the macropore flow) of subsurface flow in the hillslope models (Sloan and Moore, 1984; Germann and Beven, 1985, 1986; Wigmosta et al., 1994; Shakya and Chander, 1998; Bronstert, 1999; Beldring et al., 2000) and their reliability often depends on the scales of interest. If it is assumed that there is not a dense network of macropores and that the matrix and macropore flow are not independent but in continuous exchange of soil moisture (e.g., Beven and Germann, 1982; Noguchi et al., 1999), the long-term behavior of the hillslope can be studied using a diffusive model based on Richards' equation (i.e., without an explicit description of the macropore flow), provided the hydraulic conductivity is increased to indirectly account for the effect of macropores (Sloan and Moore, 1984; Bronstert and Plate, 1997; Stagnitti et al., 1992; Wigmosta et al., 1994; Yeh and Eltahir, 1998; Ogden and Watts, 2000; Zhu and Mackay, 2001).

9.2.3 Numerical simulations

Ridolfi et al. (2003a) performed a number of numerical simulations of stochastic Eq. (9.27) to investigate the role of the lateral infiltration under the hypotheses specified before when driven by different conditions of climate, soil, and vegetation. The influence of the geometry of the hillslope is investigated by employing the three hillslope shapes (Eq. (9.26)) shown in Figure 9.11, with hillslope length, L, equal to 250 m and height, a, equal to 50 m. Other cases with different L and a, but similar mean slopes, were also considered, qualitatively confirming the results reported by Ridolfi et al. (2003a).

Two typical rainfall regimes, a humid climate ($\lambda = 0.4 \, \text{day}^{-1}$ and $\alpha = 13 \, \text{mm}$) and a dry climate ($\lambda = 0.1 \, \text{day}^{-1}$ and $\alpha = 13 \, \text{mm}$), as well as two different types of soil, loamy sand and loam, were considered (Table 9.5) with values of saturated hydraulic conductivity amplified by one order of magnitude (e.g., Sloan and Moore, 1984) to partially incorporate the effect of macroporosity. As explained before, this increment of hydraulic conductivity is important to simulate the real hydraulic behavior of the hillslope, especially concerning macropore flow (Beven and Germann, 1982; Mosley, 1982). Such an artifact also accounts for the underestimation of the duration of saturated conditions due to the assumption of instantaneous storms. It should be noted that the values of saturated hydraulic conductivity in Table 9.5 are similar (or lower) to those quoted by other studies of soil moisture

Table 9.5 *Parameters describing the soil characteristics used by Ridolfi et al.*
(2003a).

Type of soil	$\overline{\Psi}_s$ (Log) (MPa)	b	$K_s^\dagger$ (cm day^{-1})	n	β	s_h	s_{fc}
Loamy sand	$-0.17 \cdot 10^{-3}$	4.38	1000	0.42	12.7	0.08	0.52
Loam	$-1.43 \cdot 10^{-3}$	5.39	200	0.45	14.8	0.19	0.65

† The K_s values account for macroporosity and anisotropy effects along the hillslope.

dynamics along a hillslope (e.g., Stagnitti et al., 1992; Shakya and Chander, 1998; Fan and Bras, 1998; Noguchi et al., 1999; Beldring et al., 2000; Ogden and Watts, 2000; Zhu and Mackay, 2001). A complete coverage of the hillslope by either trees or grass was used for the analysis. Several key parameters of the model depend on vegetation (Table 9.6), allowing one to account for the manifold action of plants on soil moisture dynamics.

The balance equation (Eq. (9.27)) was integrated using an explicit finite difference method according to the following scheme (Ridolfi et al. 2003a)

$$nZ_r \frac{s_i^{j+1} - s_i^j}{\Delta t} = \varphi(s_i^j; t_j) - \chi(s_i^j) + Z_r\Phi_{\text{lat}}(s_{i-1}^j, s_i^j, s_{i+1}^j), \qquad (9.29)$$

where the subscript i indicates the spatial discretization while the superscript j indicates the time step. The flux term in Eq. (9.29) can be expressed as a discrete representation of Eq. (9.28)

$$\Phi_{\text{lat}}\left[s_{i-1}^j, s_i^j, s_{i+1}^j\right] = K_{\text{lat}} \frac{s_{i-1}^j + s_i^j}{2} \frac{z_{i-1} - z_i}{\Delta x} + \frac{\Psi_{s,i-1}^j - \Psi_{s,i}^j}{\Delta x}$$
$$- K_{\text{lat}} \frac{s_i^j + s_{i+1}^j}{2} \frac{z_i - z_{i+1}}{\Delta x} + \frac{\Psi_{s,i}^j - \Psi_{s,i+1}^j}{\Delta x}. \qquad (9.30)$$

After an extensive numerical experimentation, the values $\Delta t = 1/24$ days and $\Delta x = L/100$ were chosen as adequate temporal and spatial discretization for all the simulations. Sometimes, particularly in the wet cases, some spurious oscillations of the average profile of soil moisture were observed near the bottom of the hillslope. To avoid this effect when soil moisture at any point i, s_i^{j+1}, was found to be smaller than s_{i-1}^{j+1}, its value was set to s_{i-1}^{j+1}. This is motivated by the fact that the profile of soil moisture is expected to be monotonically increasing with x.

Table 9.6 *Parameters of the two typical vegetation coverages considered in the simulations by Ridolfi et al. (2003a).*

| | Z_r | Δ | E_{max} | Ψ_{s,s_w} | Ψ_{s,s^*} | Loam | | Loamy sand | |
	(mm)	(mm)	(mm day^{-1})	(MPa)	(MPa)	s_w	s^*	s_w	s^*
Grass	300	1	4.7	−3.5	−0.09	0.23	0.46	0.10	0.24
Trees	700	2	4.3	−2.5	−0.12	0.25	0.44	0.11	0.22

Since the balance equation is a second-order differential equation in space, two boundary conditions are necessary. At the uphill side ($x = 0$) the boundary condition is of zero lateral flux due to symmetry (Figure 9.10).

The downhill boundary condition is less obvious. Although the real cases may be numerous, three conditions (see Figure 9.10) are generally used in the numerical simulations (e.g., Sloan and Moore, 1984). A first condition (indicated in Figure 9.10 with capital letter A) corresponds to the presence of a river or a lake; in this case the soil is saturated at $x = L$ and the boundary condition is $s(x = L, t) = 1$. A second condition is when the water table in the river is lower than the soil at the bottom of the hillslope (Figure 9.10, case B) and a humid front forms at $x = L$, from which water evaporates and leaks out in relation to the local soil moisture content. In such a case, the boundary condition is represented by the value of soil moisture in $x = L$. The third condition of zero flow at the bottom of the slope (Figure 9.10, case C) corresponds to the assumption that the hillslope is half of a symmetrical valley.

Since the goal of the investigation was the study of the stochastic dynamics of soil moisture on the hillslope in the absence of an external lateral input, the first condition was not considered by Ridolfi et al. (2003a). The second condition was simulated by putting in the last element of the spatial discretization an outgoing flux equal to a portion (varied from 5% to 20%) of the incoming one. No significant difference in the soil moisture dynamics along to the hillslope was observed between the last two conditions. Only for the last two or three elements of the spatial discretization at the downhill side were some differences observed, which however are not important for the purposes of this analysis. All the results were obtained using the following boundary conditions

$$\Phi_{\text{lat}}(x = 0, t) = 0; \quad \Phi(x = L, t) = 0 \quad \forall t. \tag{9.31}$$

Since the study deals with statistically steady-state conditions, the initial condition of soil moisture is not important. All the simulations were thus performed assigning

$$s(x, 0) = 0.5 \qquad \forall x, \tag{9.32}$$

as the initial condition and then discarding the first 200 days of the simulations (Ridolfi et al., 2003a). This time interval was always found to be longer than the duration of the initial transient. A number of 10 000 days after this first interval was sufficient to ensure stable average statistics, while 100 000 days were used for the soil moisture pdf's.

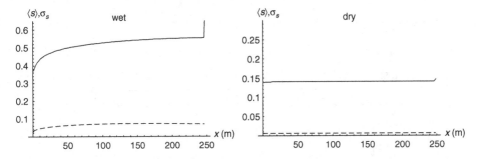

Figure 9.12 Spatial behavior of the mean (continuous line) and standard deviation (dashed line) of soil moisture along the planar hillslope for grass coverage and loamy sand under wet and dry climate conditions. After Ridolfi et al. (2003a).

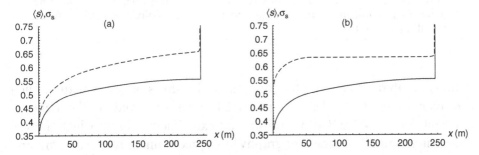

Figure 9.13 Mean soil moisture along the planar hillslope for wet climate. (a) Comparison between trees (dashed line) and grasses (continuous line) in the case of loamy sand; (b) comparison between different soil properties in the case of grass: loamy sand (continuous line) and loam (dashed line). After Ridolfi et al. (2003a).

9.2.4 Results

Figures 9.12–9.14 show the spatial behaviors of the temporal mean and standard deviation of soil moisture along the hillslope; these results correspond to different conditions of climate, soil, and vegetation for each of the three hillslope shapes considered before.

Two distinct regimes emerge from the analysis of Figure 9.12: a humid regime, in which soil water content is often sufficiently high to allow significant subsurface water flow along the hillslope and the development of gradients of soil moisture, and a dry regime with negligible lateral movement of water. In this second case, the slope does not induce any spatial variability in the distribution of soil moisture along the hillslope; soil moisture still reaches peak values close to saturation but their duration is too short to allow a

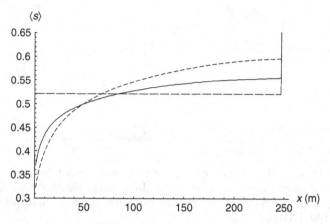

Figure 9.14 Mean soil moisture for the three hillslope shapes shown in Figure 9.11 under wet climate conditions for grass coverage and loamy sand. The dotted line corresponds to the concave hillslope, continuous line to the planar hillslope, and the dashed line to the convex hillslope. After Ridolfi et al. (2003a).

significant lateral flux and affect the spatial dynamics of soil moisture. Such two regimes resemble the two preferred states observed in the field by Grayson et al. (1997) and also confirm some experimental results which showed that in moist conditions the topography has a predominant role in determining spatial patterns of soil moisture (Crave and Gascuel-Odoux, 1997; Famiglietti et al., 1998; Western and Grayson, 1998; Western et al., 1999).

In humid conditions, the standard deviation always increases along the hillslope, in agreement with a number of pieces of experimental evidence (e.g., Robinson and Dean, 1993; Famiglietti et al., 1998). At the same time, standard deviation values are not always higher where mean soil moisture is higher. For example, the cases where the soil is more permeable always show a higher standard deviation, even if the mean soil moisture is lower compared to that of less permeable soils (not shown). This indicates that the statistical structure of the local temporal dynamics is not linked in a simple way to the mean soil moisture value, but other aspects of the dynamics need to be considered to fully characterize it. Such statistical properties are essential, for example, when studying vegetation water stress (Chapter 4).

The influence of vegetation is twofold and can be observed from the example reported in Figure 9.13a for the humid climate: First, it strongly interacts with climate and soil properties to determine the statistical characteristics of soil moisture. Coherently with what is observed in the analysis at a point

(Chapter 2), the mean value of soil moisture on the hillslope is greater in the case of trees. Second, vegetation also influences the lateral flux of soil moisture, thus amplifying or reducing the spatial gradients of soil moisture on the hillslope. When trees are present, the spatial gradients tend to be more pronounced for a significant portion of the hillslope. The rooting depth, in particular, contributes in an important manner to the differentiation of the spatial distribution of average soil moisture between trees and grasses.

The role of soil properties is also relevant (Figure 9.13b) and should be considered jointly with that of vegetation. Different soils are linked not only to different values of soil water content representative of the direct influence of soil type on the dynamics of soil moisture, namely s_h and s_{fc}, but also to different values of soil moisture characterizing the link between soil and vegetation, namely s_w and s^*. When the soil is more permeable all these values are lower. As a consequence, there is a lower probability of transpiration in conditions of water stress as well as a greater saturated lateral conductivity. These two factors lead to lower values of soil moisture and to stronger gradients of soil water content.

Regarding the influence of the hillslope shape, Figure 9.14 shows that increasing concavity amplifies the spatial gradient of soil moisture. In the case of humid climate and loam, the soil moisture increases more than 50% from the top to the bottom of the hillslope for both plane and concave shapes (dotted and continuous lines in Figure 9.14). The strongest spatial differences in the soil moisture are found in the top part of the hillslope.

For convex hillslopes (dashed line in Figure 9.14), the curvature balances the effect of the topographic gradient, a/L, and the mean gradient of soil moisture disappears. The small topographic gradient in the upper portion of the hillslope strongly reduces the lateral fluxes of water in this zone so that there is no significant water movement from the higher to the lower part of the hillslope. This confirms that the spatial dynamics of soil moisture does not depend only on local characteristics but also on the global dynamics operating on the entire hillslope.

In all the humid cases soil moisture suddenly increases at the bottom of the hillslope, coherently with the boundary condition of zero flux assigned at $x = L$. As a consequence, the downward water flow along the hillslope generates runoff in the final portion of the hillslope where soil is saturated and the excess of water coming from uphill strongly increases the soil moisture in the last element.

In the cases of significant spatial gradients of soil moisture, the variability of the soil moisture balance along the hillslope was also investigated by Ridolfi et al. (2003a). The different terms of the water balance are estimated as long-term averages of the corresponding components of the soil moisture dynamics (see also Chapter 2),

$$\langle \Phi \rangle = \langle R \rangle - \langle Q \rangle - \langle E_s \rangle - \langle E_{ns} \rangle, \tag{9.33}$$

where $\langle \Phi \rangle$ is the mean rate of the net lateral flux, $\langle R \rangle$ is the mean rate of the rainfall, $\langle Q \rangle$ represents the mean rate of runoff, and $\langle E_s \rangle$ and $\langle E_{ns} \rangle$ are the mean rates of evapotranspiration under stressed and unstressed conditions, respectively. A typical example of the water balance in humid conditions is shown in Figure 9.15, where the various components of the water balance are normalized by the mean rainfall rate. Consequently a normalized value of $\langle E_s \rangle + \langle E_{ns} \rangle + \langle Q \rangle$ lower than one indicates a net positive lateral downhill flux, while when this sum is larger than one, the soil receives from uphill more water than that which is discharged downhill.

A point marking the separation between the upper part of the slope that provides water downhill and the lower part that receives water by means of lateral flow is often present. Its position depends on soil properties and vegetation more than hillslope shape and moves downhill when there are more permeable soils and in the case of tree coverage. In the case of tree coverage and permeable soils, a lower production of runoff in the higher part of the hillslope is found. While runoff production significantly varies along the hillslope, the mean rates of stressed and unstressed evapotranspiration are spatially more uniform, except for a small portion of soil near the top of the hillslope.

Some examples of probability density functions of soil moisture are shown in Figure 9.16a–d. Because of the underlying impermeable layer which prevents deep infiltration, the soil moisture levels tend to be relatively high for the precipitation regimes considered. The net contribution of the lateral flux to

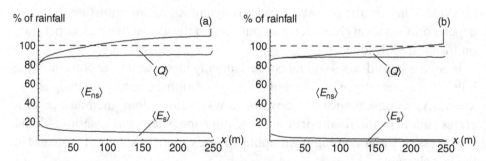

Figure 9.15 Soil moisture balance along the planar hillslope for grass coverage and loamy sand under wet climate conditions: (a) planar hillslope, (b) concave hillslope. The components of the balance are normalized by the total rainfall. After Ridolfi et al. (2003a).

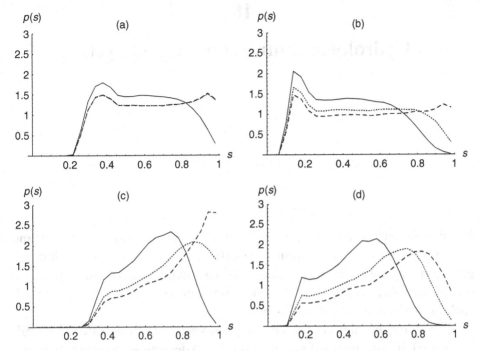

Figure 9.16 Examples of pdf's along the planar hillslope for different conditions of soil and vegetation under the wet climate conditions (continuous line, $x = 25$ m; dotted line, $x = 100$ m; dashed line, $x = 225$ m). (a) grass, loam; (b) grass, loamy sand; (c) trees, loam; (d) trees, loamy sand. After Ridolfi et al. (2003a).

the local water balance depends on the position of the point along the hillslope. For points placed uphill, where the net lateral flux is negative (see Figure 9.15a–b), the lateral flux is equivalent to a further water loss added to the point water balance; on the other hand, when a point is placed downhill, where the net lateral flux is positive, higher values of soil moisture become more probable. The threshold effect of s^* on the soil moisture pdf's often results in an increase of probability for low values of soil moisture, whose magnitude depends on the characteristics of the hillslope. Vegetation also strongly influences these probability distributions: the probability density function is practically flat for $s > s^*$ with grass coverage while in the case of trees there is a clearly defined maximum at higher soil moisture values. The reason for this is likely linked to the higher transpiration rates of grasses at high soil moisture values.

10

Hydrologic controls on nutrient cycles

In this chapter we turn to the study of the cycles of organic matter and nutrients in soils. Such dynamics are extremely important for the life and growth of vegetation in water-controlled ecosystems and, as will be seen, the role of hydrology is paramount with an intricate series of linkages and feedbacks at various levels within the system.

Since plants in water-controlled ecosystems are often both water and nutrient limited, it is difficult – if not impossible – to determine the extent to which net primary production is controlled by water or nutrient availability. Water scarcity heavily contributes to nutrient deficit through a series of different mechanisms, so that an understanding of plant stress requires a joint analysis of water and nutrient cycles. As noticed by Pastor et al. (1984), the nitrogen cycle needs to be explained through the water balance. The latter in turn depends on the existing type of plant, which determines the litter composition and so on.

Although the importance of the interaction between nutrient cycles and hydrology has been recognized for a long time, its consequences are far from being completely understood. Soil moisture, which is itself the result of the joint action of climate, soil, and vegetation, controls both directly and indirectly the soil carbon and nitrogen cycles. Directly, it impacts some of the most important phases of these cycles, such as decomposition, leaching, and plant uptake; indirectly, through its influence on vegetation growth, soil moisture affects the amount and composition of plant residues. By enhancing some processes and quenching others, the temporal patterns of soil moisture regulate the sequence of fluxes between different pools and determine the temporal dynamics of the other state variables of the system at different time scales.

The presence of different interacting processes with many nonlinearities and feedbacks makes the whole picture quite complex (Figure 10.1). In this chapter, we will mostly concentrate on the soil system, focusing only on the

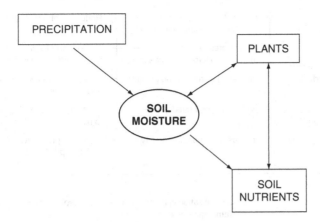

Figure 10.1 Central role of soil moisture in the soil–plant system

direct influence of soil moisture on the soil carbon and nitrogen cycle without directly considering vegetation growth. The plant characteristics necessary for the soil carbon and nitrogen dynamics (e.g., flux of added litter, rooting depth, water and nitrogen uptake, etc.) will be assigned using field data. This simplification does not represent a real limitation of the model because vegetation parameters may be expressed as dependent on time; the assumption of fixed vegetation characteristics is only a convenient approach to initially avoid further complications and uncertainties arising from the use of vegetation-growth models.

Before going into the specific topic of this chapter, it is worth mentioning that the interactions between hydrology, soil organic matter (SOM), and nutrient cycles are very broad (Figure 10.2). Despite their role in hydrology and ecology, many of these connections have been little investigated from a modelistic point of view. It is our conviction that their understanding and modeling will provide important insights for both disciplines.

Historically, the investigation of the role of soil moisture dynamics on the carbon and nitrogen cycles originated mainly from agricultural and ecological studies. Field and laboratory experiments were followed by quantitative models (e.g., Stevenson, 1986; Tate, 1987; Larcher, 1995; Brady and Weil, 1996; Schlesinger, 1997; Sala et al., 2000; Aber and Melillo, 2001). Pastor and Post (1986; see also Post and Pastor, 1996) investigated the influence of climate, soil moisture, and vegetation on carbon and nitrogen cycles in natural ecosystems. They found that soil moisture limitation increased with the lower water-holding capacity along a soil texture gradient and that low water availability altered the composition of the tree species and in turn nitrogen availability. Parton and coworkers (Parton et al., 1987, 1988, 1993; Schimel et al., 1996,

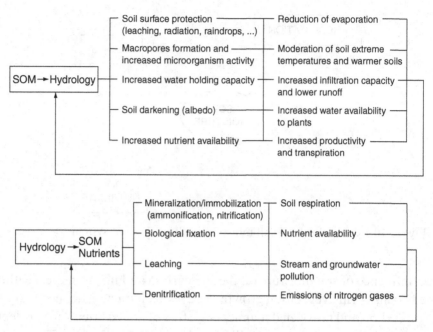

Figure 10.2 Interactions between soil organic matter (SOM) and hydrology.

1997) noticed a correlation between water and nitrogen limitations due to the water budget control of carbon and nitrogen fluxes. They further showed that, whereas biophysical models may have a memory of one or two years through the soil moisture storage and plant dynamics, the linkage of water, carbon, and nitrogen cycles can induce lagged effects over decades through the decomposition of soil organic matter. Other studies investigated the effect of climate and hydrology on SOM and nutrients (e.g., Jenkinson, 1990; Post et al., 1996; Aber and Driscoll, 1997; Gusman and Marino, 1999; Moorhead et al., 1999; Birkinshaw and Ewen, 2000), pointing out the importance of quantitative modeling in the analysis of the physical processes in the soil-nutrient system. More specifically, Moorhead et al. (1999) and Bolker et al. (1998) have emphasized the need for greater temporal resolution in the modeling of climate and litter-quality controls (most models work at the monthly time scale), as well as the need for a finer-scale description of the relationships between carbon and nitrogen dynamics during decomposition.

In the first part of this chapter, the soil moisture probabilistic model for soil moisture dynamics of Chapter 2 is coupled to a system of equations describing the temporal dynamics of carbon and nitrogen over the depth of active soil. The analytical formulation by Porporato et al. (2003a) provides the basis for the investigation of the hydrologic control of the various components of the soil nitrogen cycle. Following D'Odorico et al. (2003) an application of the model to the Nylsvley savanna in South Africa is presented, where the interesting connections between the hydrologic cycle and the soil nitrogen cycle are analyzed in detail. The chapter proceeds by briefly outlining some topics that may be interesting for future research, such as the problem of biogenic emission of N-oxides and the development of minimalistic models for soil carbon and nitrogen dynamics.

10.1 The carbon and nitrogen cycles in soils

10.1.1 Soil organic matter and the carbon cycle

Soil organic matter (SOM) is a complex and varied mixture of organic substances, with three main components: plant residues, microorganisms' biomass, and humus (Figure 10.3). Figure 10.4 shows the carbon cycle within the soil–plant–atmosphere system. Although it is only a small portion of the global carbon cycle, this portion of the carbon and nitrogen cycles has high circulation rates, which make it almost a closed cycle. Globally, the soil contains more carbon than the vegetation and atmospheric pools combined. Table 10.1 gives an indication of the amount and distribution of carbon in various types of natural soils of interest here.[1] The amount of organic matter in soils varies widely and usually increases with humidity and, to a lesser extent, with temperature. Aridisols (dry soils) are generally lowest in organic matter (mineral surface soils contain mere traces), while some humid vegetated A horizons (histosols) arrive at carbon levels as high as 20–30%. This is because, in poorly drained soils with high productivity, the little aeration inhibits organic matter decomposition and promotes the accumulation of SOM. For this reason, bogs and marshes are very important carbon pools in the terrestrial carbon cycle.

Plants take carbon from the atmosphere through photosynthesis; part of it is used by plants as a source of energy and then directly released by respiration, while the other part is assimilated by vegetation and later transfered as plant litter to the soil, where it becomes part of SOM. Soil moisture has an important long-term influence on the amount and quality of litter, especially on its carbon-to-nitrogen (C/N) ratio, which in turn affects the rates of SOM decomposition.

---

[1] Histosols are formed from materials high in organic matter and are typical of bogs and marshes; aridisols are soils of dry climates; mollisols are soils quite rich in organic matter formed under grasslands; alfisols form under forests or savannas in climates with slight-to-pronounced moisture deficit; spodosols are soils with alluvial accumulation of organic matter formed under forests in humid, temperate climates.

SOM decomposition involves an enzymatic oxidation that produces mineral compounds (e.g., ammonium) and carbon dioxide (CO_2), which is then returned to the atmosphere (soil respiration). While part of the carbon is lost as soil respiration and the simpler compounds are metabolized by soil microbes, the most complex ones are not metabolized but, along with other compounds polymerized by soil microbes, are combined to form the so-called humic substances, or resistant humus (Figure 10.3). Other less-resistant substances are also formed (nonhumic substances). Humic substances are very

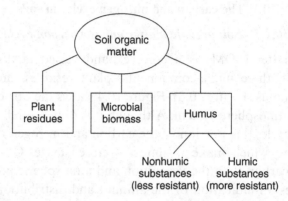

Figure 10.3 Scheme of the three main components of SOM, redrawn after Brady and Weil (1996).

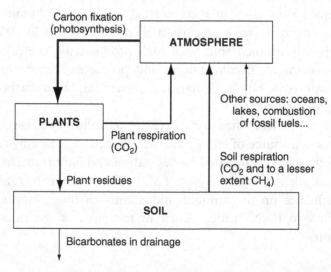

Figure 10.4 A simplified representation of the soil–plant–atmosphere carbon cycle (the thickness of the arrows and the dimension of the rectangles indicate the relative importance of the relevant fluxes or pools). After Porporato et al. (2003a).

Table 10.1 *Typical content of SOM in the upper meter of different soils, modified from Brady and Weil (1996).*

Soil order	Organic carbon ($kg\ m^{-2}$)	SOM ($kg\ m^{-2}$)	Organic nitrogen ($kg\ m^{-2}$)
Histosols	205	350	10.2
Aridisols	<3	<5	<0.3
Mollisols	13	22	1.1
Spodosols	15	26	1.2
Alfisols	7	12	0.6

stable and contribute to maintain high organic levels in soils, protecting the associated essential nutrients against mineralization and loss from the soil. Such a protection may be further helped by some clay and other inorganic components that combine with humic materials. The so-called half-time (i.e., the time required to destroy half the amount of a substance) of humic substances varies from decades to centuries; because of their stability, humic

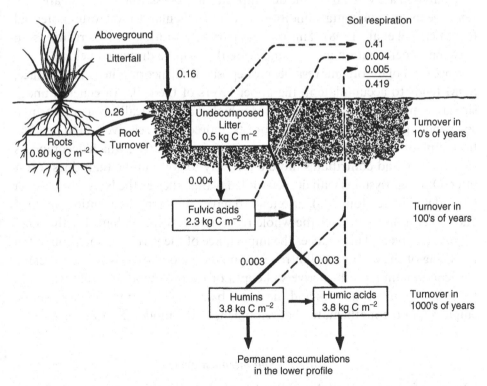

Figure 10.5 Turnover of litter and soil organic fractions in a grassland soil. Flux estimates are in $kg\ C\ m^{-2}\ year^{-1}$. After Schlesinger (1997).

substances comprise up to 60–80% of the soil organic matter. An example of the turnover of litter and soil organic fractions in a grassland soil is shown in Figure 10.5.

In soils of mature natural ecosystems the release of carbon as CO_2 is generally balanced in the long term by the input of plant residues, while at shorter time scales (e.g., seasonal-to-interannual) the carbon content is subject to fluctuations induced by climatic and hydrologic variability, so that the entire soil carbon cycle is quite sensitive to external disturbances (see Section 10.3).

The decomposition process is related to immobilization in a complex manner and is regulated by the SOM carbon-to-nitrogen (C/N) ratio and by the environmental conditions. It is usually modeled as first-order kinetics (i.e., proportional to the amount of substance to be decomposed and to the amount of existing decomposing bacteria). When plant residues are added to the soil and the conditions are favorable, the bacterial colonies grow fast, but as soon as the decomposable SOM is reduced, they starve and die easily. The decomposition of dead microbial cells is associated with the release of simple products, such as nitrates and sulfates. Different bacterial colonies exist, each specialized in a given part of the decomposition process. Such bacteria are very sensitive to environmental conditions and, in particular, to soil water potential (e.g., Fenchel et al., 1998). This in great part determines the close relationship between mineralization and soil moisture that will be discussed below.

Since most of the organic residues are deposited and incorporated at the surface, SOM tends to accumulate in the upper layers of the soils. In general, under similar climatic conditions, total SOM is higher and vertically more uniform in soils under grasslands than under forests; moreover, a relatively high proportion of plant residues in grasslands consists of root matter, which decomposes more slowly and contributes more effectively to soil humus than does forest litter. Other ecosystems with deeper-rooted plants, such as the Nylsvley savanna (considered in Section 10.3), also tend to have a uniform distribution of SOM and mineral nitrogen over the whole root layer (except perhaps for the most superficial layer). This reduces the importance of the vertical dimension in the modeling of the carbon and nitrogen dynamics and allows us to use the same vertical domain (i.e., the active soil depth or rooting depth Z_r) that was used for soil moisture dynamics and the water balance. This simplification will be employed in the mathematical development of the model (Section 10.2).

10.1.2 Soil nitrogen cycle

Although it is an essential nutrient for plants, soil nitrogen is mostly in the form of organic compounds that protect it from loss but leave it largely unavailable to vegetation. Plants almost only uptake mineral nitrogen in the form of ammonium (NH_4^+) and nitrate (NO_3^-), which are made available through SOM decomposition. For this reason, the nitrogen cycle is intimately linked to that of carbon. The greatest amount of nitrogen in terrestrial ecosystems is in the soil, which contains 10–20 times as much nitrogen as does the living vegetation. Soil organic matter typically contains about 5% nitrogen, while inorganic (i.e., mineral) nitrogen is usually not more than 1–2% of the total nitrogen in the soil. The atmosphere, which is composed of 78% nitrogen in the form of dinitrogen (N_2 being quite inert is not usable by most plants and animals), is practically a limitless reservoir of this element. Figure 10.6 provides a schematic representation of the main components of the nitrogen cycle in soils. The existence of an internal cycle is clear, which involves only soil and plants through nitrogen uptake and the production of SOM and dominates the nitrogen turnover at the daily-to-seasonal time scales. The other external fluxes, such as wet and dry deposition and the biological fixation, become important only in the long-term balance and will be neglected in this analysis.

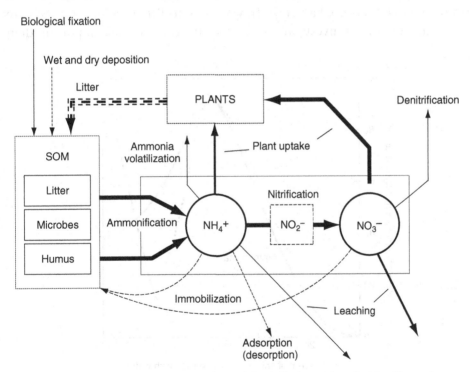

Figure 10.6 Schematic representation of the soil nitrogen cycle (the dimension of the arrows indicates the relative importance of the various fluxes in the cycle; the continuous lines refer to processes wherein the impact of soil moisture is relevant). After Porporato et al. (2003a).

For the nitrogen cycle, decomposition of SOM gives rise to ammonia and nitrate (ammonification and nitrification); approximately 1.5–3.5% of the organic nitrogen of a soil mineralizes annually, depending on environmental factors, such as pH, temperature, and soil moisture. The influence of soil moisture on mineralization is due mainly to the balance between aeration, which diminishes with soil moisture, and favorable humid conditions for microbial biomass (Figure 10.7). At high soil moisture levels anoxic conditions prevent bacteria from performing the aerobic oxidation necessary for decomposition. The reason for the reduction of the decomposition rate at low soil moisture levels is twofold (Stark and Firestone, 1995; Fenchel et al., 1998). First, a lack of substrate supply due to the reduced water content reduces microbial activity; as the pores within solid matrices dry and the water film coating the surfaces becomes thinner, diffusion paths become more tortuous and the rate of substrate diffusion to microbial cells declines. Second, low water contents induce low soil water potentials which lower the intracellular water potential and in turn reduce the hydration and activity of enzymes. Interestingly, the water stress of soil microbes has similar characteristics to plant water stress (see Chapter 4). In water-controlled ecosystems, which are the main focus of our investigation, temperature is usually less important than

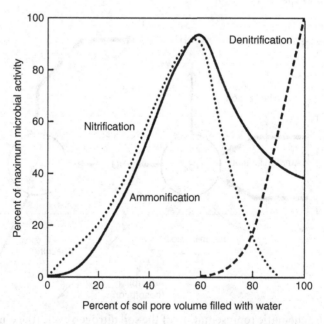

Figure 10.7 Rate of microbial activity related to the various phases of nitrogen transformation as a function of soil water content; after Brady and Weil (1996).

soil moisture, especially at the daily time scale. The most favorable conditions for mineralization are found at around 20–30 °C; mineralization practically ceases outside the temperature range 5–50 °C; and also tends to decline in acid soils.

When the conditions are favorable nitrification is quite rapid. In hot and dry environments, sudden water availability can cause a flush of soil nitrate production, which may greatly influence the growth patterns of natural vegetation (see Figure 10.20). For this reason, under warm and hot conditions, nitrate is the predominant form of nitrogen in most soils. Release of ammonia gas (NH_3) may be of certain importance (ammonia volatilization), especially in drying and hot sandy soils with ammonium accumulation in the soil top layers. The presence of nitrite (NO_2^-), instead, is always negligible, as its transformation to NO_3^- is practically immediate. This is important, since NO_2^- is quite toxic.

As already mentioned, the mineralization rate also depends on the composition of plant residues and, in particular, on their C/N ratio. The growth of microbial colonies takes place with fixed proportions of carbon and nitrogen so that their C/N ratio remains practically constant: thus, assuming the biomass C/N ratio is 8, for each part of nitrogen metabolized by microbes 24 parts of carbon are needed, of which 8 are metabolized and 16 are respired as CO_2 (Brady and Weil, 1996). As a consequence, if the nitrogen content of the organic matter being decomposed is high (i.e., C/N < 24), mineralization proceeds unrestricted and mineral components in excess are released into the soil. On the contrary, when the litter is nitrogen poor (i.e., C/N > 24), microbes can use some of the mineral nitrogen through the process of immobilization. If mineral nitrogen is not available, then mineralization may be halted. The modeling of this delicate balance is discussed later in Section 10.2.4. The overall dependence of mineralization rate on environmental conditions (soil moisture, temperature, pH, etc.) and litter composition (e.g., C/N ratio) is summarized by the data in Figure 10.8.

Nitrate is easily soluble in water and, although this facilitates its uptake by plants, it also makes it prone to losses by leaching at high soil moisture levels. Differently, the positive charge of ammonium ions attracts them to the negatively charged surfaces of clays and humus, thus partially protecting them from leaching. Although held in an exchangeable form, this may be a problem for plant uptake, since the rate of release of the fixed ammonium is often too slow to fulfill plant needs. At high soil moisture levels the process of denitrification may take place (see Figure 10.7), releasing greenhouse nitrogen gases (see Section 10.4).

Plant nitrogen requirements are met by two different mechanisms of root uptake: either passively through the soil solution during the transpiration

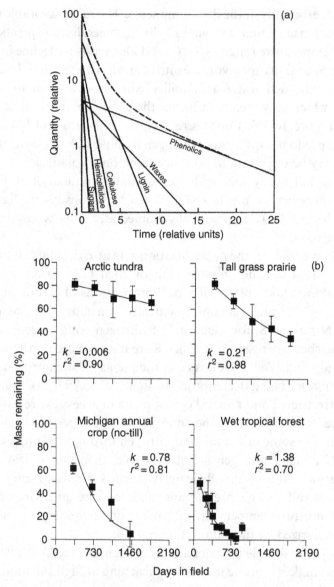

Figure 10.8 Time courses of decay of plant litter. (a) Breakdown of the various organic components of litter; after Chapman (1976) and Larcher (1995). (b) Mass loss from litter bags containing the same wheat litter placed in four different ecosystems; after Sala et al. (2000).

process or actively by a diffusive flux driven by concentration gradients produced by the plant itself (e.g., Wild, 1988; Engels and Marschner, 1995; Larcher, 1995). As this second mechanism is energy absorbing, the active uptake seems to take place only when the nitrogen demand by the plant is higher than the passive supply by transpiration. If both mechanisms are insufficient to meet such a demand, the plant is under conditions of nitrogen deficit. On the other hand, if the concentration in the soil solution is high, the passive uptake may even overtake the actual plant demand. Plants appear to have little control on passive uptake and nitrogen excess may even result in toxicity (Brady and Weil, 1996).

10.2 Modeling the carbon and nitrogen cycle in water-controlled ecosystems

Porporato et al. (2003a) proposed an analytical model for the analysis of soil carbon and nitrogen cycles and their relationship with stochastic soil moisture dynamics. The relevance of this type of mechanistic model of nutrient dynamics has been recently stressed by Entekhabi et al. (1999, page 2048): "possible areas of improvement [of our understanding of the basic hydrologic processes] include better representation of the limitations imposed by nutrient deficits. This will require the incorporation of nutrient cycles, especially the nitrogen cycle, in models that account for the coupling between hydrologic and ecological processes."

Under the same framework used for the description of the soil moisture dynamics (Chapter 2), Porporato et al. (2003a) considered vertically averaged values of carbon and nitrogen concentrations over the active soil depth, Z_r. As in Section 9.1, transpiration is separated from evaporation, since the transpiration term is used also in the model of plant nitrogen uptake.

The highly intertwined carbon and nitrogen cycles, described before, are modeled employing five separate pools for each main component of the system (Figure 10.9). In particular, the SOM is divided into three pools, representing litter, humus, and microbial biomass, respectively, while the inorganic nitrogen in the soil is divided in ammonium (NH_4^+) and nitrate (NO_3^-). For the sake of model simplicity, no distinction among the different bacterial colonies is made and all of them are included in a single biomass pool; similarly, because of the high rate of nitrification, the presence of nitrite (NO_2^-) is usually very low and is neglected here. The structure of the model is in agreement with the recommendations of Jenkinson (1990) and Bolker et al. (1998), who studied the influence of changing the number of compartments in the description of SOM and suggested using more than one and fewer than five pools for SOM (we use three). The use of a single pool for microbial biomass is justified by the scarce quantitative information available on soil microbial colonies.

Only the input of the added litter and the losses due to soil respiration, leaching, and plant uptake are considered among the external fluxes to the soil

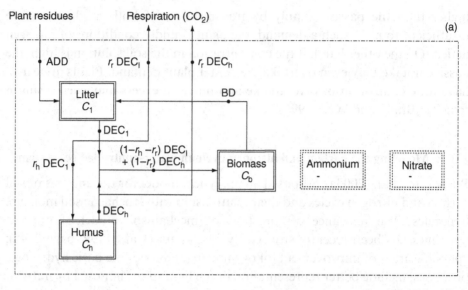

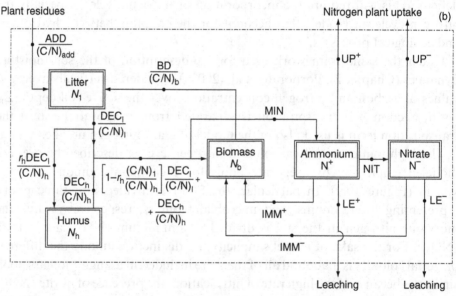

Figure 10.9 Schematic representation of the main components of the model. (a) Soil carbon cycle; (b) soil nitrogen cycle. See text for details. After Porporato et al. (2003a).

system, neglecting in this first analysis other fluxes that are less important at the daily-to-seasonal time scale, such as ammonium adsorption and desorption by clay colloids, volatilization or absorption of ammonium, nitrogen input due to wet and dry deposition, biological fixation, and denitrification.

Since differences of some orders of magnitude are to be expected between the decomposition rate of the faster (proteins) and that of the slower (lignin) organic components (Brady and Weil, 1996), two or more pools of organic matter are considered (e.g., Jenkinson, 1990; Hansen et al., 1995). The separate consideration of litter and humus addresses this aspect since, in general, litter compounds have a faster decomposition rate than humic ones. Although a range of components is present in the SOM, the litter and humus pools are assumed to be characterized by unique values of the C/N ratio and of resistance to microbial decomposition, representing weighted averages of the various components (Porporato et al., 2003a).

As explained in Section 10.1, the C/N ratios of the pools containing organic matter are very important in the dynamics of the entire carbon and nitrogen cycles. Both the humus and the biomass C/N ratio remain approximately constant in time, while the litter C/N ratio may be highly variable. Typically, the humus C/N ratio, $(C/N)_h$, is in the order of 10–12, that of biomass, $(C/N)_b$, is usually close to 8–12 (depending on the type of microbial community), and the one of the litter, $(C/N)_l$, ranges from 20 to over 50. As will be seen, the constancy of $(C/N)_b$ and the variability of $(C/N)_l$ play an essential role in controlling the rates of decomposition, mineralization, and immobilization (e.g., Hansen et al., 1995; Brady and Weil, 1996).

All the components of the soil carbon and nitrogen cycles considered in the model of Porporato et al. (2003a) are represented in Figure 10.9. Eight state variables in terms of mass per unit volume of soil (e.g., grams of carbon or nitrogen per m^3 of soil) are needed to characterize the system, namely:

C_l carbon concentration in the litter pool;
C_h carbon concentration in the humus pool;
C_b carbon concentration in the biomass pool;
N_l organic nitrogen concentration in the litter pool;
N_h organic nitrogen concentration in the humus pool;
N_b organic nitrogen concentration in the biomass pool;
N^+ ammonium concentration in the soil;
N^- nitrate concentration in the soil.

The temporal dynamics of such variables is controlled by a system of as many coupled differential equations that describe the balance of carbon and nitrogen in the various pools. All the equations represent balances of fluxes of nitrogen or carbon in terms of mass per unit volume per unit time (e.g., g m^{-3} day^{-1}). Since many of the fluxes are heavily dependent on soil moisture content, the system is coupled to the soil moisture evolution, Eq. (2.3). In this way, the main hydrologic control of the carbon and nitrogen cycles is explicitly considered.

10.2.1 The litter pool

According to the fluxes depicted in Figure 10.9, the carbon balance equation
for the litter pool can be written as

$$\frac{dC_l}{dt} = \text{ADD} + \text{BD} - \text{DEC}_l. \tag{10.1}$$

The term ADD is the external input into the system, representing the rate at
which carbon in plant residues is added to the soil and made available to the
attack of the microbial colonies. Its temporal variability is dependent on the
evolution of vegetation biomass. The term BD represents the rate at which
carbon returns to the litter pool due to the death of microbial biomass.
Porporato et al. (2003a) simply used a linear dependence on the amount of
microbial biomass, i.e., $\text{BD} = k_d C_b$, without explicitly considering the influence of environmental factors, such as temperature and soil moisture.

DEC$_l$ represents the carbon output due to microbial decomposition, modeled using first-order kinetics (e.g., Hansen et al., 1995; Gusman and Marino,
1999; Birkinshaw and Ewen, 2000),

$$\text{DEC}_l = [\varphi f_d(s) k_l C_b] C_l, \tag{10.2}$$

where the first-order rate in square brackets is rather complicated owing to its
dependence on many different factors. The coefficient φ is a nondimensional
factor accounting for a possible reduction of the decomposition rate when the
litter is very poor in nitrogen and immobilization is not sufficient to integrate
the nitrogen required by bacteria. We will return later to this point. The factor
$f_d(s)$ is another nondimensional factor, describing soil moisture effects on
decomposition. Following Cabon et al. (1991) and Gusman and Marino
(1999), the soil moisture control of aerobic microbial activity and decomposition is modeled via a linear increase up to field capacity and a hyperbolic
decrease up to soil saturation

$$f_d(s) = \begin{cases} \frac{s}{s_{fc}} & s \leq s_{fc}, \\ \frac{s_{fc}}{s} & s > s_{fc}. \end{cases} \tag{10.3}$$

As shown in Figure 10.10, Eq. (10.3) expresses well the qualitative behavior
shown in Figure 10.7. The value of the constant k_l defines the rate of decomposition for the litter pool as a weighted average of the decomposition rates of
the different organic compounds in the plant residues. Its average value is
usually much higher than the corresponding value for humus. The rate of
decomposition depends also on the microbial biomass concentration, C_b

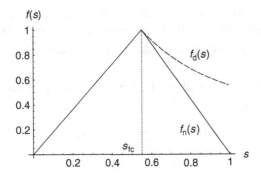

·Figure 10.10 Modeled dependence of decomposition and nitrification as a function of the relative soil moisture. After Porporato et al. (2003a).

(e.g., Hansen et al., 1995), whose evolution is described later. The relationship between the concentration of microbial biomass and its activity is assumed to be linear.

The nitrogen balance in the litter pool is similar to Eq. (10.1), with each term divided by the C/N ratio of its respective pool, i.e.,

$$\frac{dN_l}{dt} = \frac{ADD}{(C/N)_{add}} + \frac{BD}{(C/N)_b} - \frac{DEC_l}{(C/N)_l}. \tag{10.4}$$

$(C/N)_{add}$, the C/N ratio of added plant residues, ranges from 10/1 in legumes and young green leaves to more than 200/1 in sawdust (Brady and Weil, 1996). This large variability may produce pronounced changes in the C/N ratio of the litter pool, which has a very important role in regulating decomposition, immobilization, and mineralization.

10.2.2 The humus pool

The balance equation for carbon in the humus pool is

$$\frac{dC_h}{dt} = r_h\, DEC_l - DEC_h, \tag{10.5}$$

where the only input is represented by the fraction r_h of the decomposed litter undergoing humification (see Figure 10.9). The coefficient r_h, which is sometimes referred to as the "isohumic coefficient" (Wild, 1988), is in the range 0.15–0.35 (Brady and Weil, 1996), depending on the composition of plant residues.

The output due to humus decomposition is modeled in the same way as litter decomposition,

$$DEC_h = [\varphi f_d(s) k_h C_b] C_h, \tag{10.6}$$

where the value of the constant k_h, encompassing the various components of the humus pool (see Figure 10.3), is much smaller than the corresponding value for the litter pool, because of the greater resistance to microbial attack of the humic substances.

The nitrogen balance equation may be simply obtained by dividing Eq. (10.5) by $(C/N)_h$, i.e.,

$$\frac{dN_h}{dt} = r_h \frac{DEC_l}{(C/N)_h} - \frac{DEC_h}{(C/N)_h}.$$

(10.7)

As noticed by Porporato et al. (2003a), this fact implies the assumption that the products of the humification process from litter have the same characteristics, and thus also the same C/N ratio, as the soil humus. As a consequence, the value of $(C/N)_h$ remains constant in time, making Eq. (10.7) redundant. Moreover, the fraction r_h cannot exceed $\frac{(C/N)_h}{(C/N)_l}$, as follows from the obvious condition that the fraction of nitrogen entering the humus pool from litter decomposition cannot exceed the total nitrogen flux decomposed from the litter, i.e., $r_h \frac{DEC_l}{(C/N)_h} \leq \frac{DEC_l}{(C/N)_l}$.

10.2.3 The biomass pool

The carbon balance in the biomass pool is given by

$$\frac{dC_b}{dt} = (1 - r_h - r_r)DEC_l + (1 - r_r)DEC_h - BD.$$

(10.8)

The input is represented by the fraction of organic matter incorporated by the microorganisms from litter and humus decomposition (see Figure 10.9). The constant r_r ($0 \leq r_r \leq 1 - r_h$) defines the fraction of decomposed organic carbon that goes into respiration (CO_2 production), usually estimated in the interval 0.6–0.8 (Brady and Weil, 1996). The only output is BD, already defined when discussing Eq. (10.1).

The balance of the nitrogen component in the biomass may be expressed as

$$\frac{dN_b}{dt} = \left[1 - r_h \frac{(C/N)_l}{(C/N)_h}\right]\frac{DEC_l}{(C/N)_l} + \frac{DEC_h}{(C/N)_h} - \frac{BD}{(C/N)_b} - \Phi.$$

(10.9)

The first two terms on the r.h.s. represent the incoming nitrogen from decomposition and do not contain r_r because the respiration process only involves the carbon component. The coefficient in brackets preceding the first term on the r.h.s. accounts for the nitrogen fraction that goes into the humified litter. As already discussed regarding the first term on the r.h.s. of Eq. (10.7), the fraction $\frac{(C/N)_l}{(C/N)_h}$ is usually greater than one, because the humified litter is richer in nitrogen than the litter itself. The third term of Eq. (10.9) is the output of nitrogen due to microbial death, while the fourth term, Φ, takes into account the contribution due to either the net mineralization or the immobilization. Such a term is now discussed in detail, since it is relevant to the entire dynamics and furthermore synthesizes the governing role of the SOM C/N ratio.

10.2.4   Mineralization and immobilization rates

The term Φ in Eq. (10.9) attains positive or negative values in relation to the difference between the rate of gross mineralization and the total rate of immobilization of NH_4^+, IMM^+, and NO_3^-, IMM^-, i.e.,

$$\Phi = MIN - IMM, \qquad (10.10)$$

where

$$IMM = IMM^+ + IMM^-. \qquad (10.11)$$

Since the net fluxes among the various pools are what is important to the nitrogen balance, only the net amounts of mineralization and immobilization need to be modeled. This can be done as if they were mutually exclusive processes. Thus, when $\Phi > 0$, we will assume that

$$\begin{cases} MIN = \Phi \\ IMM = 0, \end{cases} \qquad (10.12)$$

while, when $\Phi < 0$,

$$\begin{cases} MIN = 0 \\ IMM = -\Phi. \end{cases} \qquad (10.13)$$

The switch between the two states is determined by the condition that $(C/N)_b$ is constant in time. Accordingly, when the average C/N ratio of the biomass is lower than the value required by the microbial biomass, the decomposition results in a surplus of nitrogen, which is not incorporated by the bacteria, and net mineralization takes place. If, instead, the decomposing organic matter is nitrogen-poor, bacteria try to meet their nitrogen requirement by increasing the immobilization rate from ammonium and nitrate, thus impoverishing the mineral nitrogen pools. In the case when the nitrogen supply from immobilization is not enough to ensure a constant C/N ratio for the biomass, the rates of decompositions are reduced below their potential values by means of the parameter φ (e.g., see Eqs. (10.2) and (10.6)) as follows.

The condition of constant C/N ratio for the biomass pool is implemented analytically as

$$\frac{d(C/N)_b}{dt} = \frac{dC_b}{dt} - \frac{dN_b}{dt}(C/N)_b = 0, \tag{10.14}$$

which, using Eqs. (10.8) and (10.9), yields (Porporato et al., 2003a)

$$
\begin{aligned}
\Phi &= DEC_h\left[\frac{1}{(C/N)_h} - \frac{1-r_r}{(C/N)_b}\right] \\
&\quad + DEC_l\left[\frac{1}{(C/N)_l} - \frac{r_h}{(C/N)_h} - \frac{1-r_h-r_r}{(C/N)_b}\right] \\
&= \varphi f_d(s)C_b\left\{k_h C_h\left[\frac{1}{(C/N)_h} - \frac{1-r_r}{(C/N)_b}\right] \right. \\
&\quad \left. + k_l C_l\left[\frac{1}{(C/N)_l} - \frac{r_h}{(C/N)_h} - \frac{1-r_h-r_r}{(C/N)_b}\right]\right\}.
\end{aligned}
\tag{10.15}
$$

Notice that Eq. (10.15), and not Eq. (10.9), is the real dynamic equation to be associated with Eq. (10.8) for the biomass evolution. The previous condition, in fact, makes Eq. (10.9) redundant, as the biomass C/N ratio is set constant.

When the term in curly brackets of Eq. (10.15) is positive, net mineralization takes place, while no net immobilization occurs, as indicated by Eq. (10.12). In such conditions, humus and litter decomposition proceed unrestricted and the parameter φ is equal to 1.

In the opposite case, when the term in curly brackets of Eq. (10.15) is negative, net mineralization is halted and immobilization sets in (Eq. (10.13)). The latter is partitioned proportionally between ammonium and nitrate on the basis of their concentrations and according to two suitable coefficients, k_i^+ and k_i^-, i.e.,

$$
\begin{cases}
IMM^+ = \dfrac{k_i^+ N^+}{k_i^+ N^+ + k_i^- N^-}\, IMM \\[2ex]
IMM^- = \dfrac{k_i^- N^-}{k_i^+ N^+ + k_i^- N^-}\, IMM.
\end{cases}
\tag{10.16}
$$

The rate of immobilization may be limited by environmental factors, biomass concentration, and – especially – by insufficient mineral nitrogen. For this reason, we assume the existence of an upper bound for the rate of mineralization, i.e.,

$$\text{IMM} \leq \text{IMM}_{\max}, \tag{10.17}$$

which after Eq. (10.15) becomes

$$
\begin{aligned}
-\varphi f_{\mathrm{d}}(s) C_{\mathrm{b}} &\left\{ k_{\mathrm{h}} C_{\mathrm{h}} \left[\frac{1}{(C/N)_{\mathrm{h}}} - \frac{1 - r_{\mathrm{r}}}{(C/N)_{\mathrm{b}}} \right] \right. \\
&\left. + k_{\mathrm{l}} C_{\mathrm{l}} \left[\frac{1}{(C/N)_{\mathrm{l}}} - \frac{r_{\mathrm{h}}}{(C/N)_{\mathrm{h}}} - \frac{1 - r_{\mathrm{h}} - r_{\mathrm{r}}}{(C/N)_{\mathrm{b}}} \right] \right\} \leq \text{IMM}_{\max}.
\end{aligned}
\tag{10.18}
$$

Since immobilization is a conversion of inorganic nitrogen ions into organic form, operated by microorganisms that incorporate mineral ions to synthesize cellular components (Brady and Weil, 1996, page 405), it is assumed that it depends on the concentration of the microbial biomass and soil moisture similarly to the decomposition process,

$$\text{IMM}_{\max} = (k_i^+ N^+ + k_i^- N^-) f_{\mathrm{d}}(s) C_{\mathrm{b}}. \tag{10.19}$$

Two possible regimes thus exist for immobilization. In the first one, immobilization is unrestricted, as the immobilization rate is lower than this maximum value, and the coefficient φ is equal to 1. This means that the bacteria can meet their nitrogen requirement and decompose the organic matter at a potential rate. The second regime occurs when the requirement of mineral nitrogen to be immobilized becomes higher than the maximum possible rate and thus the mineralization rate needs to be reduced. In the model this is accomplished by reducing φ to a value lower than one, so that the immobilization rate is equal to $\text{IMM}_{\max}$. By imposing the equality in Eq. (10.19), i.e., $\text{IMM} = \text{IMM}_{\max}$, one obtains the value of φ by which the decomposition rates must be reduced (Eq. (10.2) and (10.6), respectively),

$$
\varphi = -\frac{k_i^+ N^+ + k_i^- N^-}{k_{\mathrm{h}} C_{\mathrm{h}} \left[\frac{1}{(C/N)_{\mathrm{h}}} - \frac{1 - r_{\mathrm{r}}}{(C/N)_{\mathrm{b}}} \right] + k_{\mathrm{l}} C_{\mathrm{l}} \left[\frac{1}{(C/N)_{\mathrm{l}}} - \frac{r_{\mathrm{h}}}{(C/N)_{\mathrm{h}}} - \frac{1 - r_{\mathrm{h}} - r_{\mathrm{r}}}{(C/N)_{\mathrm{b}}} \right]}. \tag{10.20}
$$

Equations (10.15) and (10.20) regulate the entire dynamics of decomposition, mineralization, and immobilization. Their functioning is illustrated by the numerical examples in Figure 10.11. In the first example (upper rows of numbers in the figure) the litter C/N ratio is rather low (20) so that the decomposed organic matter from litter and humus (node A in Figure 10.11) is nitrogen-rich, i.e., with an overall C/N ratio lower than 8. As a consequence, a fraction of nitrogen must go into net mineralization ($\Phi > 0$) and the inequality in Eq. (10.19) is satisfied with $\varphi = 1$. In the second example $(C/N)_{\mathrm{l}} = 30$ so that the C/N ratio in A is higher than 8. Immobilization of some mineral

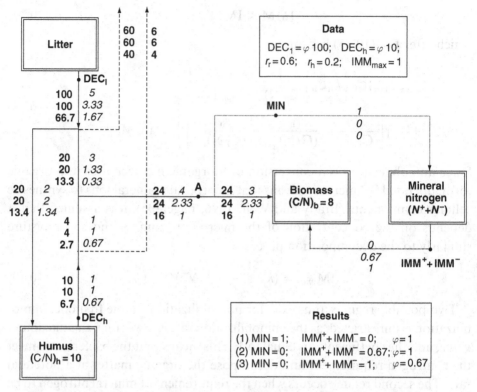

Figure 10.11 Numerical examples of decomposition, mineralization, and immobilization rates. The three examples have the common features reported in the upper right corner of the figure, while they are different in the value of the C/N ratio of the litter pool, which is equal to 20 in the first example (upper rows of numbers), to 30 in the second (central rows), and to 40 in the third (lower rows). The boldface numbers refer to carbon fluxes, the italic numbers to nitrogen fluxes. The resulting mineralization and immobilization rates, along with the corresponding values of φ, are reported in the lower right corner of the figure. After Porporato et al. (2003a).

nitrogen is thus necessary to maintain a C/N ratio equal to 8. The rate of net immobilization is determined by Eq. (10.15). The inequality in Eq. (10.19) is satisfied because the resulting immobilization rate is lower than the maximum potential one ($IMM_{max} = 1$), and φ is equal to one. The third case presents an example of nitrogen-poor litter, $(C/N)_l = 40$, where the required net immobilization rate (equal to 1.5) is greater than $IMM_{max} = 1$. This means that the inequality in Eq. (10.19) is not satisfied and the decomposition rate must be reduced by imposing $\varphi = \frac{1}{1.5} = 0.67$, as obtained from Eq. (10.20). With this value of φ, the C/N ratio of the fluxes into the biomass pool is equal to 8, and the immobilization rate is equal to IMM_{max}.

10.2.5   The mineral nitrogen in the soil

The balance of ammonium and nitrate in the soil can be modeled after Porporato et al. (2003a) as

$$\frac{dN^+}{dt} = MIN - IMM^+ - NIT - LE^+ - UP^+, \tag{10.21}$$

and

$$\frac{dN^-}{dt} = NIT - IMM^- - LE^- - UP^-, \tag{10.22}$$

in which the rates of mineralization and immobilization have already been described. The nitrification rate can be modeled as first-order kinetics, i.e.,

$$NIT = f_n(s)\, k_n\, C_b\, N^+, \tag{10.23}$$

where C_b expresses the dependence of nitrification on microbial activity. The constant k_n defines the rate of nitrification, while $f_n(s)$ (nondimensional) accounts for the soil moisture effects on nitrification. As seen in Figure 10.7, the optimum conditions for nitrification are very similar to those for decomposition, with the difference that nitrification tends to zero at soil saturation (Linn and Doran, 1984; Skopp et al., 1990). This behavior is reproduced by a linear increase up to field capacity followed by a linear decrease to soil saturation (see Figure 10.10),

$$f_n(s) = \begin{cases} \dfrac{s}{s_{fc}} & s \leq s_{fc} \\[2mm] \dfrac{1-s}{1-s_{fc}} & s > s_{fc}. \end{cases} \tag{10.24}$$

For both ammonia and nitrate, leaching occurs when nitrogen in the soil solution percolates below the root zone. It is thus simply proportional to the leakage term $L(s)$ modeled in Section 2.1.6,

$$LE^{\pm} = a^{\pm} \frac{L(s)}{s\, nZ_r}\, N^{\pm}, \tag{10.25}$$

where $L(s)$ is divided by the volume of water per unit area, $s\, nZ_r$, so that the term $\frac{L(s)}{s\, nZ_r}$ assumes the dimension of the inverse of time. The nondimensional coefficients $a^{\pm}$, $0 \leq a^{\pm} \leq 1$, are the fraction of dissolved ammonium and nitrate, respectively, and are related to the corresponding solubility coefficients. Since nitrate is a mobile ion, a^- can be taken equal to one, while a^+ is much lower, because a large fraction of ammonium is absorbed by the soil matrix. This is why leaching of ammonium is seldom important. At very low soil moisture levels, when liquid water becomes a highly disconnected system, the coefficients $a^\pm$ may be reduced by evaporation and become soil moisture dependent, $a^\pm = a^\pm(s)$. Although this is certainly not important for leaching, it may modify the water uptake at low soil moisture levels. The last terms of Eqs. (10.21) and (10.22), UP^+ and UP^-, are the rates of ammonium and nitrate plant uptake, which are modeled as described below.

10.2.6 Nitrogen uptake by plants

The process of nitrogen uptake by plants is complex and not yet completely understood. For this reason, its simplified modeling is rather delicate and requires a detailed justification. As many aspects of nitrogen uptake are similar for nitrate and ammonium, we will discuss them together, pointing out the differences when necessary.

For both ammonium and nitrate, passive and active uptake can be regarded as additive processes (Wild, 1988), i.e.,

$$UP^\pm = UP_p^\pm + UP_a^\pm. \tag{10.26}$$

The passive uptake can be assumed to be proportional to the transpiration rate, $E(s)$, and to the nitrogen concentration in the soil solution, i.e.,

$$UP_p^\pm = a^\pm \, \frac{E(s)}{s \, nZ_r} \, N^\pm, \tag{10.27}$$

where, as already pointed out for leaching, $E(s)$ must be divided by $s \, nZ_r$ for dimensional reasons and $a^\pm$ represents the fraction of dissolved inorganic nitrogen. As noticed above, $a^\pm$ may be reduced at very low soil moisture levels, $a^\pm = a^\pm(s)$, but very little information seems to be available in this regard. Given the behavior of the transpiration rate (linearly increasing up to s^* and then constant with s), the passive uptake has the form shown in light gray in Figure 10.12. The reason for the decrease above s^* is the dilution of the soil solution.

The active uptake mechanism is closely associated with the plant metabolic processes (Wild, 1988; Engels and Marschner, 1995; Larcher, 1995). It is accomplished by establishing a concentration gradient between the root surface and the soil, which triggers a diffusion flux of the nitrogen ions. The intensity of the flux is limited by the gradient itself which, in part, is controlled by the plant on account of its nitrogen demand and rate of passive uptake and by the diffusion coefficient. Accordingly, we assume that the plant tries to compensate for the deficit with the active mechanism of uptake only if the

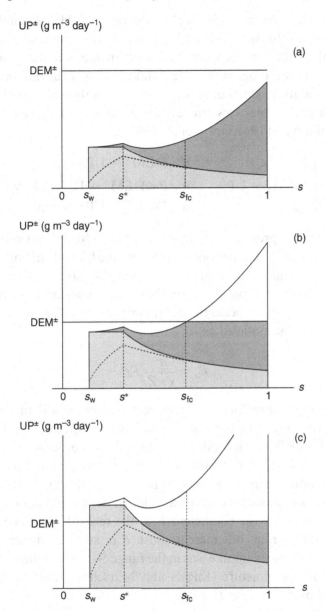

Figure 10.12 Plant nitrogen uptake as a function of the relative soil moisture, s for increasing concentrations of mineral nitrogen from (a) to (c). The light gray regions represent the passive component of the uptake, the dark gray regions the active component. The dashed lines refer to nonlinear reduction of the solubility coefficients, $a^{\pm}$, with soil moisture which lowers the passive uptake. After Porporato et al. (2003a).

passive uptake is lower than a given plant demand, $DEM^{\pm}$. When the diffusion of nitrogen ions into the soil is limiting, the active uptake is assumed to be proportional to the nitrogen concentration in the soil through a suitable diffusion coefficient, otherwise active uptake is simply the difference between the demand and the passive component (i.e., the total uptake satisfies the plant demand). Three possible cases can be assumed to occur in the representation of the active component (Porporato et al. 2003a),

$$
UP_a^{\pm} = \begin{cases} 0 & \text{if } DEM^{\pm} - UP_p^{\pm} \leq 0 \\ DEM^{\pm} - UP_p^{\pm} & \text{if } 0 < DEM^{\pm} - UP_p^{\pm} \leq k_u N^{\pm} \\ k_u N^{\pm} & \text{if } DEM^{\pm} - UP_p^{\pm} > k_u N^{\pm}. \end{cases} \tag{10.28}
$$

The term $k_u N^{\pm}$ expresses the dependence of the diffusive flux on the gradient of concentration between the root surface and the bulk of the soil, that in first approximation is taken to be proportional to the nitrogen concentration in solution (i.e., the concentration within the roots is supposed to be nearly zero). The parameter $k_u(day^{-1})$ bears the dependence of the diffusion process on soil moisture and can be modeled as

$$
k_u = \frac{a^{\pm}}{s\, nZ_r} F s^d \tag{10.29}
$$

where the term $\frac{a^{\pm}}{s\, nZ_r}$ transforms the concentration in the soil into a concentration in solution, the term F is a rescaled diffusion coefficient (Wild, 1988), and d expresses the nonlinear dependence of the diffusion process on soil moisture. If one considers that the diffusion coefficient is often related to the product between soil moisture and a tortuosity factor (which in turn has a quadratic dependence on soil moisture), a typical value of d is around 3. Porporato et al. (2003a) did not relate the coefficient F to the real diffusion coefficients for ammonium and nitrate, but merely chose F in such a way that the active contribution to nitrogen uptake was in the range 50–80% of the total uptake at high values of soil moisture (Engels and Marschner, 1995; and references therein), as shown in Figure 10.12.

When the nitrogen concentration is low compared to the plant demand (Figure 10.12a), the active uptake is limited by diffusion and plants are under nitrogen deficit; on the other hand, when nitrogen concentration is high (Figure 10.12c), plant requirements are usually met, either by passive uptake alone at low soil moisture values (i.e., when nitrogen in solution is more concentrated) or by both active and passive uptake at higher soil moisture values. As already noted, the passive uptake may overtake the plant demand and, if this situation lasts for some time, the excessive nitrogen uptake may have harmful effects and decrease the plant growth rate (Wild, 1988). At low soil moisture the role of the "solubility coefficient" $a^{\pm}$ may also become important: in the case of nonlinear dependence on soil moisture, a reduction in the passive uptake at low soil moisture appears (dashed line in Figure 10.12a–c).

Although in the present mechanistic model the demand must be a well-defined value, it is clear that in reality DEM is an average value that is representative of the typical nitrogen requirement of a given species. Since it is much more difficult to distinguish between ammonium and nitrate, Porporato et al. (2003a) considered an overall nitrogen demand and then, according to the type of plant, split it proportionally in to DEM^+ and DEM^-. When an ecosystem has reached stable conditions, the amount of nitrogen taken up on annual time scales tends to balance that returned to the soil in the form of plant litter (e.g., Wild, 1988; Larcher, 1995). It is thus reasonable to assume that plant nitrogen demand is of the same order of magnitude (or eventually a little bit higher) as the average rate of nitrogen added with the litter. Such a linkage confers to both the rate of added litter and the plant nitrogen demand an essential regulative control for the rate of nitrogen recycling in the ecosystem, in which the physiological characteristics of plants and environmental conditions play an important role.

The present model of nitrogen uptake is in part a demand-driven one (e.g., Hansen et al., 1995), whose behavior is in good agreement with the results reported in the literature. It is usually accepted, in fact, that the dependence of nitrogen uptake on nitrogen concentration is described by Michaelis–Menten kinetics, corrected with a linear increase at high nitrogen concentration (e.g., Haynes, 1986; Engels and Marschner, 1995). Such behavior corresponds to an increasing rate of uptake up to a saturation concentration followed by a further increase at very high nitrogen concentrations. Figure 10.13 shows the total nitrogen uptake in our model as a function of the nitrogen concentration. It is clear that, for relatively high soil moisture values, the behavior is similar to that just discussed. The plateau where the uptake remains constant is found at DEM and its width depends on the soil moisture value. The following linear increase at high N concentrations, due to the passive uptake, presents a shallower slope than the first part. Lower soil moisture values give a similar behavior, but with a narrower plateau.

10.2.7 Summary of the model

The evolution of carbon and nitrogen in the soil is described by Eqs. (10.1), (10.4), (10.5), (10.8), (10.21), and (10.22), along with the conditions (10.12), (10.13), and (10.19) defining net mineralization and immobilization. The

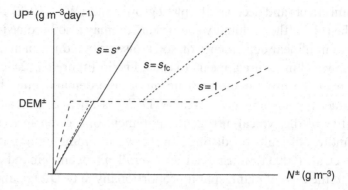

Figure 10.13 Nitrogen plant uptake as a function of nitrogen concentration for constant values of soil moisture. After Porporato et al. (2003a).

system is coupled with the soil moisture evolution equation, which accounts for the hydrologic forcing.

By summing up the three equations for the carbon and the five for the nitrogen, one obtains the corresponding differential equations for the total carbon and nitrogen in the system, $C_{tot} = C_l + C_h + C_b$ and $N_{tot} = N_l + N_h + N_b + N^+ + N^-$, respectively. In particular, from the sum of Eqs. (10.1), (10.5), and (10.8) one obtains

$$\frac{dC_{tot}}{dt} = ADD - r_r DEC_l - r_r DEC_h \tag{10.30}$$

and from the sum of Eqs. (10.4), (10.7), (10.9), (10.21), and (10.22)

$$\frac{dN_{tot}}{dt} = \frac{ADD}{(C/N)_{add}} - LE^+ - UP^+ - LE^- - UP^-. \tag{10.31}$$

The terms in Eqs. (10.30) and (10.31) represent the global gains and losses of the soil system. In this way the continuity of nitrogen and carbon in the soil system is ensured, closely following the scheme of Figure 10.9. In some aspects the structure of the model by Porporato et al. (2003a) is similar to that considered by Birkinshaw and Ewen (2000). In their paper, however, the reduction in the decomposition due to insufficient immobilization is not taken into account and, as a consequence, the closure of the carbon and nitrogen balance in their model is not complete.

10.3 An application to the Nylsvley savanna

In the present section, we present an application of the model of Section 10.2 to a broad-leafed savanna located in the Nylsvley region (South Africa), where the available data document the hydrologic and ecologic processes and allow for an adequate testing of the model (Scholes and Walker, 1993; see also Section 5.1). The properties of the nitrogen and carbon dynamics are simulated and studied in detail, analyzing in particular how the random fluctuations imposed by the stochastic forcing of precipitation propagate to the state variables of the system. The analysis follows the paper by D'Odorico et al. (2003). Although the investigation is based on the data of Nylsvley, the results can presumably be extended also to other water-controlled ecosystems.

10.3.1 Field data

As described in more detail in Section 5.1, the broad-leafed sites at Nylsvley are dominated by a vegetation composed mainly by *Eragostris pallens* (herbaceous) and *Burkea africana* (woody) with a canopy cover of about 30–40% and a discontinuous grass cover in the rest of the area. This vegetation is generally found on nutrient-poor, acidic soils, while fine-leafed savannas are typically observed on more fertile grounds (Scholes and Walker, 1993). Both in the fertile and in the unfertile sites, soils are sandy and about 1 m deep, though most of the organic matter and nutrients are concentrated in the top 80 cm.

Nitrogen may represent a limiting factor for productivity in the broad-leafed savanna, where the release of mineral nitrogen from decomposition of litter and plant residues is quite slow (average turnover time of about five years; Scholes and Walker, 1993). Most of the decomposition is due to soil microbes (compared to fire oxidation and termites) and is mostly controlled by soil water content. Figure 10.14a shows the mean annual carbon cycle in the broad-leafed sites at Nylsvley. The total carbon stock of the ecosystem is about $10\,000\,\text{gC m}^{-3}$ and most of it is in soil organic matter. Figure 10.14b shows the mean annual cycle of nitrogen; the total nitrogen stock is about $400\,\text{gN m}^{-3}$, and more than 99% is in the organic matter, with only a small percentage being available to plants as mineral nitrogen. Inorganic nitrogen is present in the soil as nitrate (about $1.0\,\text{gN m}^{-3}$), while ammonium is found in much smaller concentrations, indicating an almost instantaneous nitrification. This suggests that the nitrogen cycle is limited by the slow rate of decomposition (i.e., by the availability of substrate for nitrification), while the process of nitrification does not exert any constraint. From the data reported by Scholes and Walker (1993) (see Figure 10.14) it is possible to estimate the average values of the C/N ratios for the humus and biomass pools as well as the amount and the C/N ratio of the plant residues annually added to the soil as litterfall and dead roots (Table 10.2). For the other parameters characterizing climate, soil, vegetation, and nutrient cycles at Nylsvley see Section 5.1.

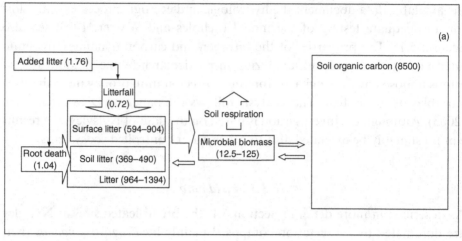

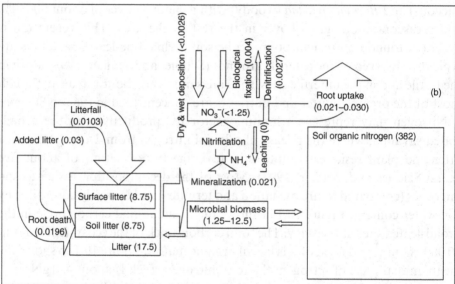

Figure 10.14 Mean carbon (a) and nitrogen (b) cycles in the broad-leafed savanna at Nylsvley (modified from Scholes and Walker, 1993); the concentrations of carbon are expressed as averages in an 80-cm-deep soil layer (in g m^{-3}), while the rates of the processes are expressed as daily means in a 242-day-long growing season (in g m^{-3}day^{-1}). The box sizes are approximately proportional to the pool sizes.

Soil microbes at Nylsvley tend to have short lifetimes and a strong sensitivity to water deficit. For this reason, microbial biomass experiences water stress earlier than starvation due to depletion of soil mineral nitrogen. As a consequence, apart from a few exceptions reported at the beginning of the growing

Table 10.2 *Parameters representative of the nutrient pools at Nylsvley (data from Scholes and Walker, 1993). Modified after D'Odorico et al. (2003).*

Parameter	Symbol	Units	Value
Added litter (avg. rate)	ADD	$(\text{gC m}^{-2} \text{ day}^{-1})$	1.5
Added litter (C/N ratio)	$(C/N)_{add}$	–	58.0
Microbial biomass (C/N ratio)	$(C/N)_b$	–	11.5
Humus (C/N ratio)	$(C/N)_h$	–	22.0

season, in normal conditions immobilization does not take place. The rate of nitrogen uptake, in the range of $5-7.25 \, \text{gN m}^{-3} \text{ year}^{-1}$, is in approximate equilibrium with the rate of mineralization. Considering also that the overall seasonal leaching losses are negligible, there is neither net seasonal accumulation, nor depletion of nitrate in the soil (Scholes and Walker, 1993).

Hypotheses and calibration of the model

Since at Nylsvley the nitrogen cycle and plant activity cease almost completely during the dry dormant season, nutrient dynamics was assumed to develop as a sequence of growing seasons in which the final condition at the end of a growing season becomes the initial condition of the following one (D'Odorico et al., 2003). For the sake of simplicity, litter was assumed to be incorporated at a constant rate throughout the growing season, leaving to future investigations the analysis of the possible effects of temporal patterns of temperature, litterfall, and plant uptake.

Consistently with the model of Section 10.2 and the values published by Scholes and Walker (1993), dry and wet deposition, nitrogen fixation, ammonia volatilization, and denitrification were neglected, representing only minor sources or sinks of nitrogen compared to the other components. As a result, the only source of NO_3^- (or NH_4^+) in the model is nitrification (mineralization), while the outputs of inorganic nitrogen are root uptake and, to a lesser extent, leaching and immobilization (Scholes and Walker, 1993).

The stochastic model of carbon and nitrogen cycles presented in Section 10.2 was calibrated by D'Odorico et al. (2003) to simulate the soil nutrient dynamics in the broad-leafed savanna at Nylsvley. All the model parameters having a clear physical link with the hydrogeochemical processes were directly estimated from the data on soil, climate, vegetation, and nutrients available at Nylsvley (Table 10.3 and Section 5.1), while the remaining parameters were determined by calibration to reproduce the observed average size of the soil nitrogen and carbon pools as well as the mean rates of decomposition, mineralization, and root uptake (Table 10.4).

Table 10.3 *Amount of carbon and nitrogen in the nutrient pools, and rates of decomposition, mineralization and root uptake observed at Nylsvley (data from Scholes and Walker, 1993). The carbon and nitrogen concentrations are expressed as the average of an 80-cm-deep soil layer, the rates as averages throughout a 242-day-long growing season. After D'Odorico et al. (2003).*

C in humus pool	$(gC\ m^{-3})$	8500
C in biomass pool	$(gC\ m^{-3})$	12.5–125
C in litter pool	$(gC\ m^{-3})$	960–1400
N in ammonium pool	$(gN\ m^{-3})$	≈ 0
N in nitrate pool	$(gN\ m^{-3})$	≤ 1.25
Average soil moisture	–	0.11
Rate of uptake	$(gN\ m^{-3}\ day^{-1})$	0.02–0.03
Rate of mineralization	$(gN\ m^{-3}\ day^{-1})$	0.021
Rate of litter decomp.	$(gC\ m^{-3}\ day^{-1})$	1.2

The values of the constants k_l, k_h, and k_b, for the first-order kinetics of litter and humus decomposition and death of microbial biomass (Table 10.4) are estimated using the steady-state solution of Eqs. (10.1), (10.5), and (10.8) along with the average litterfall rate (Table 10.2), and the carbon storage in litter (C_l), humus (C_h), and biomass (C_b) pools (Table 10.3). When in those equations the temporal derivatives are set to zero and the values of the other variables are assigned as reported in Tables 10.2 and 10.3 for the average conditions observed at Nylsvley, the solution of this linear system leads to the values of k_l, k_h, and k_b reported in Table 10.4. The nondimensional factor $f_d(s)$

Table 10.4 *Parameters of the model estimated by calibration for the case of the broad-leafed savanna at Nylsvley. After D'Odorico et al. (2003).*

Parameter	Units	Value
a^+	–	0.05
a^-	–	1
DEM^+	$(gN\ m^{-3}\ day^{-1})$	0.2
DEM^-	$(gN\ m^{-3}\ day^{-1})$	0.5
d	–	3
F	$(m\ day^{-1})$	0.1
k_d	(day^{-1})	8.5×10^{-3}
k_l	$(m^3\ day^{-1}\ gC^{-1})$	6.5×10^{-5}
k_i^+	$(m^3\ day^{-1}\ gC^{-1})$	1
k_i^-	$(m^3\ day^{-1}\ gC^{-1})$	1
k_h	$(m^3\ day^{-1}\ gN^{-1})$	2.5×10^{-6}
k_b	$(m^3\ day^{-1}\ gN^{-1})$	0.6
r_h	–	$\min\left[0.25, \frac{(C/N)_h}{(C/N)_l}\right]$
r_r	–	0.6

expressing the effect of soil water content on the rate of these first-order kinetics was estimated using Eq. (10.3) with an average soil moisture of 0.12.

The estimation of the value of the first-order constant of nitrification, k_n, (Eq. (10.23)) was facilitated by the lack of ammonium and by the high nitrification rates. Being limited by the supply of substrate (i.e., NH_4^+) rather than by the microbial activity, nitrification is not very sensitive to k_n, as long as an adequately large value was used. For the same reason, due to the limited amount of ammonium, the values of the parameters k_i^+ and k_i^- – expressing the different susceptibility of ammonium and nitrate to losses by immobilization (Eq. (10.16)) – are irrelevant to the overall dynamics since most of the immobilization is contributed by NO_3^-.

The parameters F and $DEM^{\pm}$ for the passive uptake (Eqs. (10.28) and (10.29)) were estimated so as to reproduce the observed rates of root uptake, while the tortuosity factor, d, was assumed to be equal to 3. The isohumic coefficient, r_h, representing the fraction of decomposing litter undergoing humification, was taken as the minimum value between 0.25 and the ratio $\frac{(C/N)_h}{(C/N)_l}$. Similarly, the constant r_r (i.e., the portion of decomposing carbon that is lost by respiration) was chosen equal to 0.6. Nitrate was assumed to be completely soluble in the soil solution only when soil moisture is above field capacity (i.e., $a^- = 1$), while below s_{fc} the solubility coefficient was assumed to decay according to a power law with exponent $g = 0.5$. The same behavior was also used for the solubility of ammonium, but a smaller value was selected for a^+ to account for the absorption of ammonium by the soil matrix and the consequent low mobility of these ions (Table 10.4). However, due to the low amounts of NH_4^+ in the soil and the consequently low rates of ammonium uptake and leaching, the particular value used for this parameter is irrelevant to the overall dynamics.

10.3.2 Results

Figure 10.15 shows examples of model-generated time series reported by D'Odorico et al. (2003) for some of the state variables relevant to the carbon and nitrogen cycles. Figure 10.16 shows the corresponding rates of decomposition, net mineralization, uptake, and leaching. In both cases, the average values observed at Nylsvley (broken lines) were found to be in general good agreement with the simulations.

Temporal dynamics

The input of precipitation was modeled by D'Odorico et al. (2003) as a marked Poisson process with Nylsvley parameters (see Section 5.1). As shown in

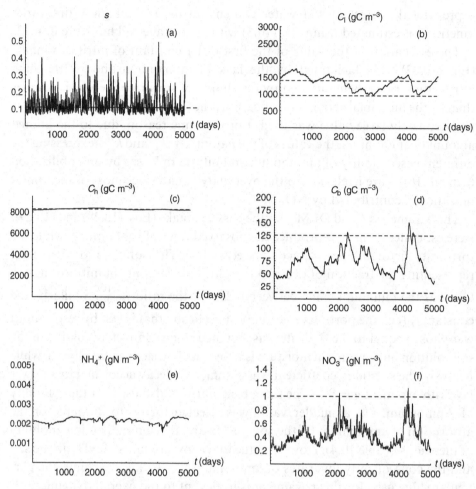

Figure 10.15 Temporal dynamics of relative soil moisture (a), carbon litter (b), carbon humus (c), carbon biomass (d), ammonium (e), and nitrate (f) simulated for the case of the broad-leafed savanna at Nylsvley. The broken lines represent the average values or the range of values observed at Nylsvley (Table 10.3) except in (f), where the broken line is an upper limit for nitrate concentration in the nitrogen-poor savanna. The broken lines are reported whenever observations are available. After D'Odorico et al. (2003).

Figure 10.15, depending on the inertia of the various pools and on the degree of dependence on soil moisture, the random fluctuations imposed by precipitation are filtered by the temporal dynamics of the state variables in a very interesting manner. Some variables (e.g., NO_3^-) preserve much of the high-frequency variability imposed by the random forcing of precipitation, while some others (C_h, C_l, and C_b) show smoother fluctuations (Figures 10.15 and 10.17).

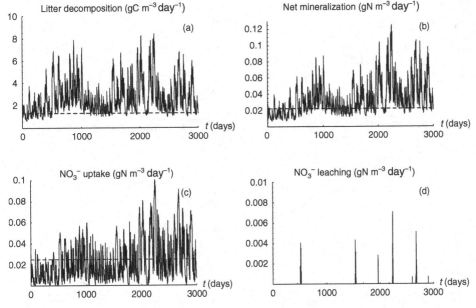

Figure 10.16 Simulated rates of litter decomposition (a), net mineralization (b), nitrate uptake (c) and nitrate leaching (d) in the broad-leafed savanna at Nylsvley. The broken lines represent the average values observed at Nylsvley (Table 10.3). The nitrate leaching (d) is reported to be negligible at Nylsvley (Scholes and Walker, 1993). After D'Odorico et al. (2003).

Nitrate dynamics is the final product of a number of intertwined processes in which both high- and low-frequency components interact. In particular, the high-frequency component of NO_3^- fluctuations (period of days to weeks) can be linked to the direct dependence of mineralization and nitrification on soil moisture (Figure 10.15e and f), which transfers the random fluctuations of the rainfall forcing to the budget of NO_3^-. On the other hand, the low-frequency variability (period of seasons to years) resembles the one of organic matter (in particular of microbial biomass; e.g., Figure 10.15d) and depends on the inertia imposed on the dynamics by the dimension of the soil carbon and nitrogen pools, which is very large compared to the fluxes (see Figure 10.14). Notice that the same low-frequency component also characterizes the litter dynamics, which is negatively correlated to C_b (see Figure 10.15b) as the growth of one of these pools occurs at the expense of the other one.

The different time scales occurring in the carbon and nitrogen cycles were analyzed by means of the power spectra of the different variables. Figure 10.18 shows the logarithmic plot of the normalized spectral density of s, C_b, and NO_3^-. As expected, the energy associated with the high frequencies is higher for soil moisture than for nitrate and microbial biomass, while the converse is true

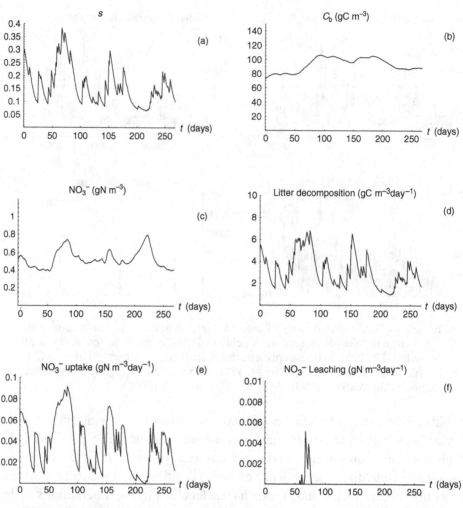

Figure 10.17 Temporal dynamics of soil moisture (a), carbon biomass (b), nitrate (c), litter decomposition (d), nitrate uptake (e), and nitrate leaching (f) simulated during a few growing seasons at Nylsvley (enlargement from Figures 10.15 and 10.16). After D'Odorico et al. (2003).

for the low frequencies. The crossing between the C_b (or NO_3^-) and s power spectra is located at frequencies corresponding to periods of seasons to years. The model thus reproduces the change in time scales occurring from the soil moisture dynamics to the nutrient dynamics observed by Schimel et al. (1997). This indicates that, while in semi-arid ecosystems soil moisture is not able by itself to provide memory to the system at scales larger than one year (due to the complete depletion of soil water content by the end of the dry season, e.g., Nicholson, 2000), nutrient and vegetation dynamics may have a much longer

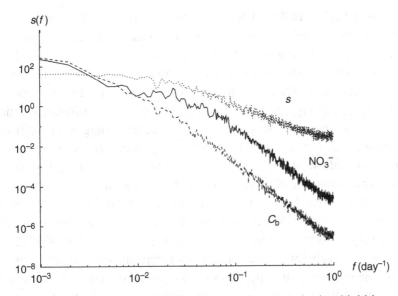

Figure 10.18 Power spectra of soil moisture, nitrate, and microbial biomass. After D'Odorico et al. (2003).

memory responsible for the interannual persistency observed both in hydro-climatic and ecosystem processes (Parton et al., 1988; Schimel et al., 1996, 1997).

Analysis of the mutual interactions among the different variables gives further insight into the dynamics of the process. From the analysis of the cross-correlation coefficients of the daily series (Table 10.5) it is seen that soil moisture, though being the driving force of the system, is poorly correlated at lag zero to all the other variables, due to the different time scales of the fluctuations. For similar reasons, there is a strong correlation between the amounts of nitrate and microbial biomass in the soil, while the mentioned negative correlation of C_l to C_b propagates also to the nitrate content. The correlation coefficients of the carbon humus are very low, because C_h tends to maintain a quasi-constant value with almost no interaction with the other

Table 10.5 *Cross-correlation coefficients among the state variables relevant to the soil carbon and nitrogen cycles. After D'Odorico et al. (2003).*

	Soil moisture	Carbon (litter)	Carbon (humus)	Carbon (biomass)	Nitrate
Soil moisture	1	−0.05	0.03	0.17	0.14
Carbon (litter)	−	1	0.17	−0.63	−0.64
Carbon (humus)	−	−	1	0.12	0.16
Carbon (biomass)	−	−	−	1	0.86

variables (see Figure 10.15). The correlation coefficients for ammonium are not reported because of their irrelevancy, due to the extremely low amounts of ammonium characterizing the nitrogen cycle at Nylsvley (see Figure 10.15e).

The low cross-correlation coefficient between s and NO_3^- is somewhat surprising, especially when compared to the results of Figures 10.15 and 10.17. In fact, although the resemblance of the time evolution of the two variables over short time scales (Figure 10.17) seems to suggest a certain degree of correlation, the s-independent low-frequency fluctuations of nitrate reduce the cross-correlation at lag zero (Table 10.5). As shown in Figure 10.19, however, the cross-correlation between soil moisture and nitrate slowly increases for positive lags, attaining its maximum at about 100–150 days. When the lowest soil moisture values are excluded from the computation, the maximum cross-correlation increases and moves to a lag of 15–20 days. Despite the relatively low values of cross-correlation, such regular behavior is a sign of the existence of a persistent (though infrequent) link between soil moisture and nitrate dynamics. A closer inspection of the time series of soil moisture and nitrate reveals that such a phenomenon is due to the presence of sudden flushes of nitrate, following a prolonged wet period after a drought (see Figure 10.20). In these conditions, in fact, first the dry soil hinders decomposition and favors soil organic matter (SOM) accumulation, then the subsequent wet period elicits biomass growth and enhances mineralization

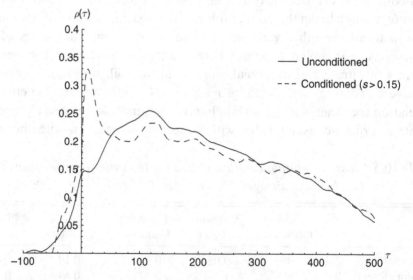

Figure 10.19 Cross-correlation function between soil moisture and nitrate concentration in the soil: considering all soil moisture data (continuous line) or soil moisture above 0.15 (dashed line). After D'Odorico et al. (2003).

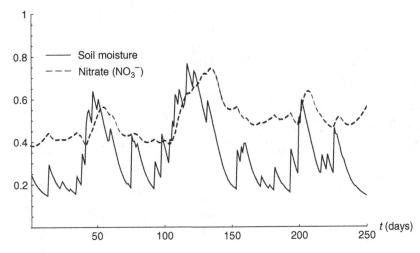

Figure 10.20 Response of the soil nitrate pool to fluctuations in the soil water content. After D'Odorico et al. (2003).

(e.g., Cui and Caldwell, 1997). Episodic changes in nitrate levels greatly influence plant growth, because a plant's response to increased availability of nitrogen tends to be very quick (e.g., Brady and Weil, 1996). These pulses of nitrate are therefore of considerable importance for natural ecosystems. Their modeling would not have been possible using a model based on monthly estimates of soil moisture: the daily temporal resolution is thus fundamental for capturing the impact of soil moisture on nutrient dynamics.

A further example of the importance of the stochastic characterization of the model is given in Figure 10.21, where an exceptionally long wet period is shown to cause a dramatic increase of the microbial biomass activity and nitrate and a consequent decrease of the substrate pools (C_l and, not shown, C_h). However, the high nitrate concentration is soon depleted by plant uptake and leaching and, in the long run, the event turns out to be unfavorable to plants in terms of nitrogen availability because of the marked depletion of the SOM substrate.

The rates of decomposition, mineralization, and root uptake, reported in Figures 10.16 and 10.17, also present interesting features. Owing to their direct dependence on soil water content (see Figures 10.7 and 10.10), they are subject to a high-frequency variability that is closely coupled to that of soil moisture (see Figures 10.16, 10.17, and the very high correlation coefficients in the first row of Table 10.6). These fluxes are also correlated to C_b, which induces in their dynamics a weak low-frequency component recognizable in Figure 10.16, and are negatively correlated to the soil litter content due to the negative correlation of the latter to C_b (see Table 10.5). When the fluxes are compared, one notices that the rate of net mineralization is subject to the same fluctuations as

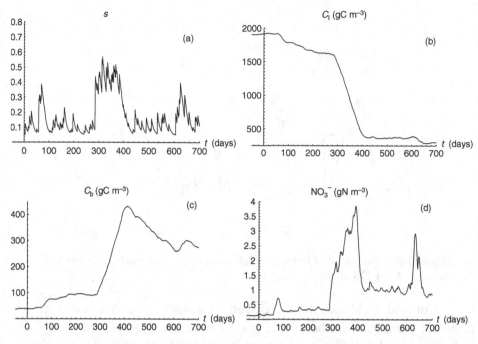

Figure 10.21 Effect of a prolonged wet period on the soil organic matter and nitrate dynamics. After D'Odorico et al. (2003).

decomposition (Figure 10.16), suggesting that there are no relevant losses of mineral nitrogen due to immobilization. The average seasonal rates of soil respiration (not shown) are qualitatively similar to the litter decomposition rate, with fluctuations around an average value of 5 gC m^{-3} day^{-1}, which is in good agreement with the Nylsvley observations (Scholes and Walker, 1993).

Since the prevailing arid conditions hamper diffusive transport of NO$_3^-$ in the soil, its total flux is mainly made up of the passive component, which in turn is mostly controlled by the interaction of the soil moisture and nitrate through the process of transpiration. Nitrogen concentration is always so low that plant nitrogen demand is never met, in agreement with the fact that the broad-leafed savanna at Nylsvley is a nitrogen-deficient ecosystem (Scholes and Walker, 1993). The dry conditions at Nylsvley are also responsible for the very infrequent occurrence of leaching (Figures 10.16 and 10.17).

10.3.3 Sensitivity to rainfall regime

D'Odorico et al. (2003) characterized the probabilistic structure of the long-term dynamics through the analysis of the probability distributions (pdf's) of the state variables and fluxes. The results are shown in Figures 10.22 and 10.23

Table 10.6 Cross-correlation coefficients between the state variables and the fluxes of the soil carbon and nitrogen cycles. After D'Odorico et al. (2003).

	Decomposition	Soil respiration	Net mineralization	NO_3^- uptake	Leaching
Soil moisture	0.81	0.80	0.74	0.78	0.45
Carbon (litter)	-0.10	-0.16	-0.43	-0.35	-0.06
Carbon (humus)	0.22	0.21	0.18	0.15	0
Carbon (biomass)	0.49	0.54	0.66	0.55	0.10
Nitrate	0.42	0.46	0.62	0.54	0.10

with continuous lines. All the pdf's are unimodal, but with relevant differences in shape: the positively skewed shape of the relative soil moisture pdf is transmitted to all the pdf's of the fluxes (Figure 10.23). In contrast, the probability distributions characterizing the state variables of the model are nearly symmetric (see also Figures 10.15 and 10.16). The fluctuations of microbial biomass and nitrate span a wide range of values (CV of the distribution $\simeq 0.4$), while the fluctuations are much smaller for C_l (CV $= 0.21$) and C_h (CV $= 0.02$).

In order to gain further insights into the system's sensitivity to climate forcing, the model was run with slight variations of the average rate of precipitation. Because the model does not include a closure of the system

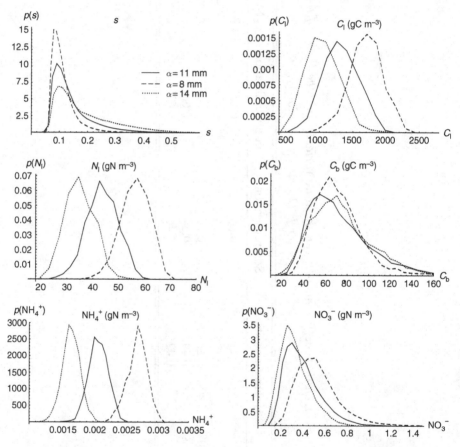

Figure 10.22 Probability density function of relative soil moisture, and of nitrogen and carbon content in the soil organic and inorganic pools with three different regimes of precipitation: $\alpha = 14.0$ mm, dotted line ("wet"); $\alpha = 11.0$ mm, solid line (Nylsvley); $\alpha = 8.0$ mm, broken line ("dry"). Soil, climate (e.g. λ) and vegetation parameters correspond with those of the broad-leaved savanna at Nylsvley. After D'Odorico et al. (2003).

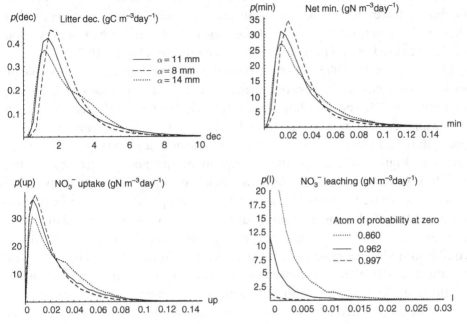

Figure 10.23 Probability density functions of the rates of decomposition, net mineralization, root uptake, and leaching with three different regimes of precipitation: $\alpha = 14.0$ mm, dotted line ("wet"); $\alpha = 11.0$ mm, solid line (Nylsvley); $\alpha = 8.0$ mm, broken line ("dry"). Soil, climate (e.g. λ) and vegetation parameters correspond with those of the broad-leaved savanna at Nylsvley. After D'Odorico et al. (2003).

with plant dynamics and productivity (see Section 10.2), it cannot account for variations in the rates of added litter, transpiration, and uptake with changes in the rainfall regime. This sensitivity analysis was thus only intended to investigate the hypothetical response of the carbon and nitrogen cycles to changes in precipitation conditions when the other parameters are kept fixed (D'Odorico et al., 2003).

Given that at Nylsvley the interannual variability of the mean storm depth, α, is larger than that of the mean storm frequency, λ, the variations in the rainfall regimes are attributed to changes around the mean value of α (11 mm), keeping λ constant. The results in Figures 10.22 and 10.23 show that a wet climate (dotted lines) tends to enhance decomposition (Figure 10.23), leading to a stronger depletion of litter. This explains the smaller amounts of nitrogen and carbon in the litter pool (Figure 10.22) and the slightly higher concentration of soil microbial biomass, which determines favorable conditions for the production of inorganic nitrogen and, in particular, of nitrate. Such findings are in agreement with those by Parsons et al. (1996, page 23 685), who report an accumulation of nitrate for wet soils in the absence of root uptake. Globally, however, the amount of nitrate accumulating in the soil is smaller than in dry conditions, because of the larger losses due to NO_3^- uptake and leaching caused by the more abundant precipitation (Figure 10.23).

D'Odorico et al. (2003) also compared the results obtained under stochastic hydrologic forcing with those obtained with constant continuous rain. As seen in Figure 10.24, the simulated average NO_3^- content still decreases with increasing average rainfall, since in both cases the effect of an increase in precipitation induces a faster cycling of carbon and nitrogen (the rates increase, Figure 10.23, and the dimension of the pools decreases, Figure 10.22). Interestingly, such a long-term response is quite different to its short-term counterpart (Figures 10.20 and 10.21): an episodic increase in the soil water content, following a relatively drier period, causes an increase of nitrate concentration, while persistent wet conditions drive the system towards an equilibrium characterized by smaller pools and larger fluxes. The stochastic forcing has the effect of continuously changing the equilibrium position, resulting in fluctuating temporal dynamics of carbon and nitrogen (Figure 10.15). In the absence of stochastic forcing, the system does not show inherent variability and reaches a steady-state characterized by values that, because of the nonlinearity of the system, are different from the corresponding mean values of the stochastic case (Figure 10.24).

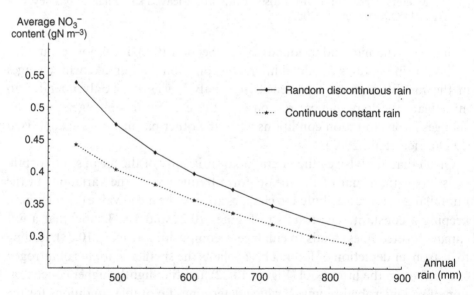

Figure 10.24 Decrease of the average nitrate content with the total rain, for the stochastic model presented in this paper and for the corresponding model with continuous constant rainfall. After D'Odorico et al. (2003).

From the previous results, it is clear that the high-frequency variability of soil moisture and the low-frequency variability of soil organic matter combine to produce a complex temporal dynamics of soil mineral nitrogen, while the nonlinear mutual interactions between the processes may either enhance or reduce the effect of changes in the climatic regime. In the case of Nylsvley, the larger losses of nitrate due to uptake and leaching over-compensate for the larger decomposition rate expected with the "wet" rainfall regime, leading to an overall diminution of the inorganic nitrogen in the soil. Moreover, the occurrences of exceptional hydrologic conditions, such as prolonged rainfall or drought, are capable of affecting the dynamics of the slow-varying pools inducing long-lasting effects.

In the cases considered up to now, the only variability of the system is due to the external hydrologic forcing. However, given the nonlinearity and degree of complexity of the whole system, one might also expect conditions in which the system shows inherent variability in the form of self-sustained oscillations. Figure 10.25 shows a preliminary example reported by D'Odorico et al. (2003) in which, under slower decomposition rates, the system approaches statistically steady conditions through damped oscillations having time scales of

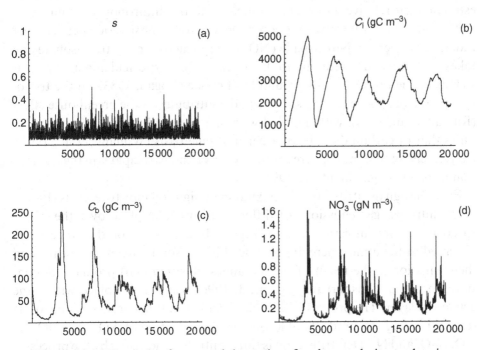

Figure 10.25 Example of temporal dynamics of carbon and nitrate showing damped oscillations towards equilibrium for a dry climate and slow decomposition rates. After D'Odorico et al. (2003).

several growing seasons and amplitudes considerably larger than those induced by the stochastic hydrologic forcing. Such behavior, that in part is reminiscent of the results by Thornley et al. (1995), would suggest that the soil nutrient system might show cases of richer (maybe chaotic) dynamics entangled with the stochastic variability directly induced by the stochastic hydrologic fluctuations. If this were confirmed, it would have important consequences for the understanding of the possible scenarios of ecosystem stability (e.g., Tilman and Wedin, 1991).

10.4 The problem of nitrogen oxide emissions

A very important environmental problem connected to the soil nitrogen cycle is that of nitrogen oxide emissions. In fact, as mentioned in Section 10.1.2, at high soil moisture values denitrification takes place with an ensuing emission of nitrogen oxides into the atmosphere. Although the magnitude of these fluxes is considerably smaller than that of the other exchanges in the soil nitrogen cycle, they may have a large environmental impact. As a consequence, biogenic emissions of nitrogen oxides from terrestrial ecosystems are intensely studied both to understand the processes responsible for their control and to estimate their relative importance compared to the anthropogenic sources.

Nitrogen oxides have strong consequences for atmospheric chemical and radiative properties. Nitric oxide (NO) is very reactive in the troposphere and takes part in reactions leading to the formation of nitric acid and acid rain as well as to the production of ground level ozone (Logan, 1983). In the troposphere, nitrous oxide (N_2O) is stable and contributes to the greenhouse effect (Ramanathan, 1988), with a global warming potential that is more than 200 times that of carbon dioxide on a per-molecule basis (Williams et al., 1992). Conversely, in the stratosphere, N_2O is reactive and participates in the destruction of ozone (e.g., Davidson, 1991).

Even though denitrification is maximum at high soil moisture levels (Figure 10.7), nitrogen gas emissions of different forms take place over the whole spectrum of soil moisture values (Figure 10.26). Most of the trace gases observed in the atmosphere are produced in the soil by microorganisms and their emission is the result of the dynamics of (microbial) production, consumption, and transport (e.g., Conrad, 1996) which takes place within the active soil layer. The process of microbial denitrification consists of anaerobic respiration (e.g., Conrad, 1996) leading to the reduction of NO_3^- and NO_2^- to NO, N_2O, and N_2. Denitrification is thus controlled indirectly by the processes affecting the rate of nitrification, the O_2 supply, and organic matter availability as well as directly by the activity of the denitrifying enzyme and the conditions of

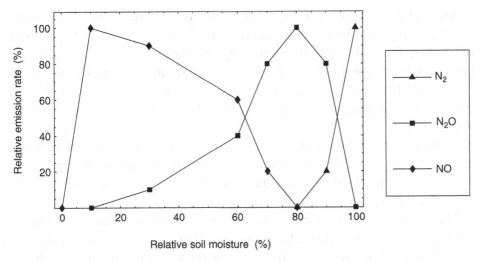

Figure 10.26 Normalized rate of nitrogen emissions as a function of relative soil moisture. Redrawn after Meixner and Eugster (1999).

soil aeration which are closely linked to soil moisture dynamics (Linn and Doran, 1984; Matson and Vitousek, 1990; Skopp et al., 1990; Potter et al., 1996).

Soil moisture affects the emission rate and its partitioning between NO, N_2O, and N_2, at least in three ways (Figure 10.26): (i) through the control of soil aeration (i.e., rate of nitrification/rate of denitrification), (ii) through the regulation of substrate diffusion (Skopp et al., 1990), and (iii) through the control of the physical transport of gases within the soil (Rosswall et al., 1989; Skopp et al., 1990; Davidson, 1991; Cardenas et al., 1993; Fowler et al., 1997; Yang and Meixner, 1997; Meixner and Eugster, 1999). There is general agreement that the nitrogen oxide emissions are non-monotonic functions of soil moisture: their rate increases with increasing values of soil water content up to a maximum value and then decreases (Cardenas et al., 1993; Yang and Meixner, 1997; Crill et al., 2000). Conversely, the rate of N_2 emission is an increasing function of s and occurs only with high soil moisture contents. The actual values of emissions can in general depend on other ecosystem properties (such as soil temperature and pH, amount and type of microbial biomass, land cover and use, etc.); however, these factors do not significantly affect the shape of the functions representing the dependence on soil water content.

It is thus clear that the analysis of the dynamics of biogenic nitrogen emissions should be linked to soil moisture dynamics as was done for the other nitrogen fluxes in Section 10.2. Potter et al. (1996) made a first step in this direction, although their estimation of the soil water content was performed through a monthly balance and then partitioned over the days of the month to capture the instantaneous nitrogen oxide emissions following daily storms.

Another analysis of the role of soil moisture in nitrogen emissions from soil was carried out by Ridolfi et al. (2003b). Using the soil moisture model of Chapter 2, coupled to a functional dependence between soil moisture and the normalized rates of emission[2] (Figure 10.26), they derived the probability distributions of the various fluxes from the pdf of soil moisture. Although their analysis is limited to the dependence on soil moisture in the absence of other controlling factors, Ridolfi et al. (2003b) provide a first indication of the relative importance of hydrologic processes on nitrogen oxide emissions at the daily time scale. Their results are in general good agreement with the field measurements in water-controlled ecosystems (e.g., Otter et al., 1999).

The results show a strong dependence of the nitrogen oxide fluxes on the climate, soil, and vegetation characteristics. Due to the relevant role of field capacity and of the hydraulic conductivity in the soil water balance at high soil water contents, fine-grain soils are found to emit at lower rates than coarse-grain soils and to favor the emission of N_2O both in dry and wet climates. Combining clayey soils and wet climates, the probability distribution of normalized NO emission becomes characterized by an atom of probability at zero as a consequence of the frequent occurrence of conditions close to saturation which preclude the supply of O_2 needed for the enzymatic activity of nitrifying bacteria. Figure 10.27 shows examples of the probability distribution of the normalized emission of NO and N_2O for different values of E_{max}. It is observed that higher values of maximum transpiration rates, which correspond to greater depletion of soil moisture, induce larger emissions of NO than N_2O, while for low E_{max} the situation is inverted. Figure 10.28 shows that only in extremely wet climates do clayey soils have high N_2 emissions while those of N_2O and NO are practically absent.

10.5 Towards a hierarchy of models

The results presented in Section 10.3 have shown that some of the fluxes and variables of the soil nitrogen cycle have very small fluctuations around their mean values (e.g., Figures 10.15 and 10.17). It is then natural to ask whether it is possible to develop simpler models for the short-term nitrogen balance in which such slow-varying variables are substituted by their mean values, thus reducing the order of the system and augmenting the possibilities for analytical developments without losing essential parts of the dynamics. Of course, the

---

[2] Ridolfi et al. (2003b) use φ to indicate the normalized rates (i.e., divided by the maximum rates) of gaseous emission of a generic nitrogen compound.

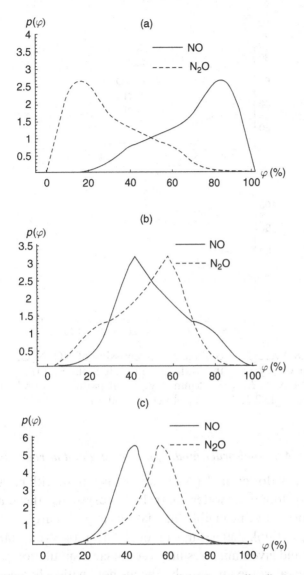

Figure 10.27 Probability density function of the (normalized) rate of emission of NO and N_2O for a 100-cm-deep loamy sand and different maximum transpiration rates: (a) $E_{max} = 5.0$ mm day^{-1}, (b) $E_{max} = 3.5$ mm day^{-1}, (c) $E_{max} = 2.0$ mm day^{-1}. Rainfall parameters: $\lambda = 0.2$ day^{-1}; $\alpha = 15.0$ mm; soil parameters and values of s_w and s^* as in Table 2.1. After Ridolfi et al. (2003b).

actual conditions for succesfully carrying out such simplifications must be carefully assessed case by case. Future research will hopefully provide useful criteria on this topic. Here we only hint at possible avenues of investigation along this line.

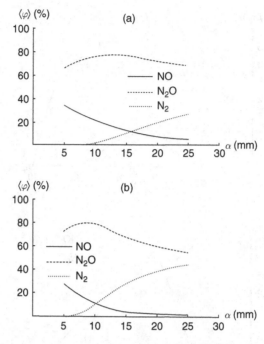

Figure 10.28 Average percentage rate of emission of NO, N_2O, and N_2 for a 40-cm-deep clayey soil and different rainfall regimes: (a) $\lambda = 0.2 \, day^{-1}$, (b) $\lambda = 0.2 \, day^{-1}$. $E_{max} = 3.5 \, mm \, day^{-1}$; soil parameters and values of s_w and s^* as in Table 2.1. After Ridolfi et al. (2003b).

10.5.1 *A second-order model for moisture and nitrogen in soils*

The results by D'Odorico et al. (2003) have shown that, with parameters of the type frequently found in water-controlled ecosystems, the humus and the ammonium pools have negligible fluctuations (e.g., Figure 10.15). Therefore, it is reasonable to think that a simpler model that employs constant amounts in those pools would give similar results over the same simulation periods. This is indeed the case, as some analyses by the authors with a five-equation model (two for litter, two for the microbial biomass, and one for the mineral nitrogen) have shown.

The simplification of the system can be brought to the extreme if one is only interested in a single growing season. Over these time scales, the results of Figure 10.17 suggest that the low-frequency variability of the SOM pools may be negligible so that only the high-frequency fluctuations directly connected with soil moisture are to be considered. Accordingly, the system may be reduced to a second-order model for relative soil moisture and mineral nitrogen dynamics of the form

$$\frac{ds(t)}{dt} = Y[s(t), t] - \rho[s(t)]$$

$$\frac{dN(t)}{dt} = \psi[s(t), N(t)],$$

(10.32)

where the first equation is the usual equation for soil moisture dynamics (Chapters 2 and 3), while the second describes the evolution of mineral nitrogen through the function ψ, which directly depends on s and N and accounts for the amount of biomass and litter available for decomposition through constant parameters.

The second-order stochastic system (Eq. (10.32)) is interesting from both the physical and the mathematical points of view. Using the same procedure of Section 2.2, it is possible to write the Chapman–Kolmogorov forward equation for the evolution of the joint probability of soil moisture and nitrogen, $p(s, N; t)$, as

$$\frac{\partial}{\partial t}p(s, N; t) = \frac{\partial}{\partial s}[\rho(s)p(s, N; t)] + \frac{\partial}{\partial N}[\psi(s, N)p(s, N; t)]$$

$$- \lambda'p(s, N; t) + \lambda' \int_{s_h}^{s} p(u, N; t)f_Y(s - u, u)\, du.$$

(10.33)

Unfortunately, even with extremely simple models for $\rho(s)$ and $\psi(s, N)$ and under steady-state conditions, the above Master equation does not seem to be easily solvable. Some insights into the dynamics of the system may be gathered by considering the temporal evolution of soil moisture and nitrogen dynamics, using linear losses for soil moisture dynamics (see the minimalistic models of Section 2.6.2) along with a very simple and continuous form of $\psi(s, N)$ that maintains the essential dependence of nitrogen fluxes on soil moisture and nitrogen concentration (e.g., Figure 10.7). Using a parabolic relationship for decomposition/mineralization minus a linear increasing function of both s and N to account for the combined leakage and uptake losses which both increase with N and s, one obtains

$$\frac{ds(t)}{dt} = Y[s(t), t] - \eta s$$

$$\frac{dN(t)}{dt} = bs(1 - s) - ksN,$$

(10.34)

where b and k are positive constant parameters. In particular, b accounts for both the rates of decomposition/mineralization and the amount of SOM substrate, while the simple quadratic dependence on soil moisture, $s(1 - s)$, qualitatively reproduces the soil moisture dependence of decomposition/

mineralization (e.g., Figure 10.7). The parameter k synthesizes the soil and plant characteristics controlling leaching and transpiration.

Although the approximations contained in Eq. (10.34) may lead to an overestimate of the mineralization at low soil moisture and to an underestimate of the losses of mineral nitrogen at high soil moisture, such a minimalistic model should reproduce, at least qualitatively, the dynamics of the complete system when the fluctuations in SOM are not important.

As follows from the second equation of the system (10.34), in the absence of soil moisture fluctuations, N would always tend toward an equilibrium value,

$$bs(1-s) - ksN_{eq} = 0 \quad \rightarrow \quad N_{eq} = \frac{b}{k}(1-s). \qquad (10.35)$$

In general, however, N_{eq} is time dependent through the soil moisture evolution and therefore N_{eq} is only an instantaneous point of attraction for $N(t)$. If the response time of the nitrogen pool is very short, the fluctuations of $N(t)$ are closely coupled to those of N_{eq} and the pdf of N (also because of the linear relationship between s and N_{eq}) turns out to be very similar to that of s. Conversely, if the nitrogen response is slow, the trace of N is more autocorrelated and its pdf tends to develop around a different equilibrium point. Such dynamics can be analyzed in the phase space defined by the variables $\{s, n\}$, where n is the normalized mineral nitrogen, $n = N/N_{max} = Nk/b$. Figure 10.29 shows the phase space trajectories of s and n for two different values of k corresponding to a rapid (left figure) and slow (right figure) response of n to s during phases of soil dry-down. Two different realizations of the stochastic bivariate process (obtained by integration of the system (10.34)) are shown in

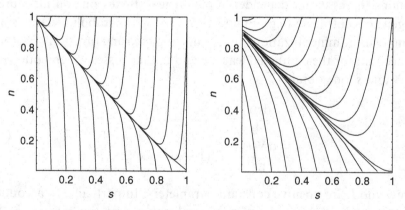

Figure 10.29 Phase space trajectories of s and n according to the model (10.34) for $\eta = 0.02$ and two different values of k. Left panel: $k = 1$ (rapid response of n); right panel: $k = 0.2$ (slow response of n).

Figure 10.30. The top figures refer to a high value of γ (i.e., shallow soil and large mean depth of rainfall events), in which the large fluctuations of soil moisture in turn induce large fluctuations of mineral nitrogen. The bottom figures refer to a low value of γ (i.e., deep soil and small mean depth of rainfall events) that, producing only a small soil moisture fluctuation, hardly affects the dynamics of mineral nitrogen. It is thus clear that, even with the same value of k (which depends mainly on temperature and on the amount of SOM present in the soil), the different soil moisture dynamics may produce a completely different evolution of n.

Unfortunately, the results of Figure 10.30 refer specifically to a given realization and are difficult to generalize. The solution of the probabilistic problem posed by Eq. (10.33) would be most interesting from the ecohydro-logical point of view, as it would synthesize the effects of the direct impact of soil moisture on the soil nitrogen dynamics. A possible line of attack in this direction could be the derivation of the equations for the mean nitrogen value following the method used in Section 8.1 for the transient and seasonal dynamics of mean soil moisture.

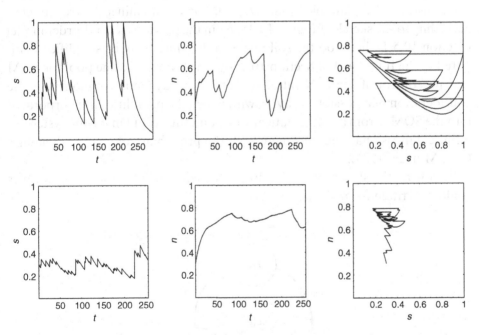

Figure 10.30 Examples of traces of soil moisture and mineral nitrogen and corresponding phase space evolution for different soil and mean rainfall depths. Top figures: shallow active soil and large mean depth of rainfall events; bottom figures: deep active soil and small mean depth of rainfall events. Both examples refer to the case with $k = 0.2$ shown in the right panel of Figure 10.29.

10.5.2 *Nonlinear dynamics of the soil–plant nitrogen cycle*

In the analyses of Sections 10.2 and 10.3 the plant characteristics were assumed to be invariant in time. Such a simplification was motivated by the extreme complexity of describing plant growth and by the interest in concentrating on the soil part of the nitrogen cycle. It should be clear, however, that important additional features may arise in the long-term evolution when plant dynamics is included in the system. Simple models of plant growth would certainly be useful for investigating these kinds of problems, such as the possible presence of self-sustained oscillations in the soil–plant system, in part already suggested by the results of Figure 10.25.

The pronounced internal cycle of nitrogen in the soil–plant system is almost closed by the sequence of uptake and plant biomass growth, litter production, SOM decomposition, and mineralization. Environmental conditions, soil moisture in particular, act as an external forcing through decomposition, uptake, and plant productivity (see Section 10.1). A very simple model to account for these main interactions could be represented by a system of four ordinary differential equations describing the temporal dynamics of relative soil moisture, $s(t)$, plant biomass, $b(t)$, SOM, $\sigma(t)$, and mineral nitrogen, $n(t)$, according to the sketch of Figure 10.31. As in the previous second-order model (Section 10.5.1), here too the soil nitrogen dynamics may be simplified compared to the full model of Section 10.2, by employing only one pool for SOM and no control of the C/N ratio. An important dynamic component is the consideration of a possible delay, τ, with which the plant litter is incorporated into the SOM. From a mathematical viewpoint this would make the system an infinite dimensional one thus increasing the probability of chaotic dynamics (e.g., Murray, 1989).

Following the sketch of Figure 10.31, the equations for the four variables could be written formally as

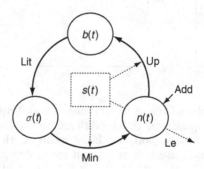

Figure 10.31 Scheme of a simple model for the soil–plant nitrogen dynamics.

$$nZ_r \frac{ds(t)}{dt} = R(t) - I[b(t), t] - Q[s(t), t] - E[s(t)] - T[s(t), b(t)] - L[s(t)]$$

$$\frac{db(t)}{dt} = \mathcal{F}\{Up[s(t), b(t), n(t)]; T[s(t), b(t)]\} - Lit[p(t)]$$

$$\frac{d\sigma(t)}{dt} = Lit[b(t - \tau)] - Min[s(t), \sigma(t)]$$

$$\frac{dn(t)}{dt} = Min[s(t), \sigma(t)] + Add - Up[s(t), b(t), n(t)] - Le[s(t), n(t)],$$

$$(10.36)$$

in which the dynamics of plant biomass, $b(t)$, depends on the rates of transpiration, T, and uptake, Up, that, in turn, are functions of the amount of biomass present. Interception, I, and litter input into the soil, Lit, also depend on b, the latter with a possible delay time, τ. $\mathcal{F}$ is a function describing plant growth dynamics, while Min, Up, and Le may be modeled as in Section 10.2. The term Add accounts for the nitrogen input due to wet and dry deposition and biological fixation, which are the necessary additions to compensate for the leaching, which is the only loss from the system when denitrification is neglected.

To proceed quantitatively, all the components of the above system could be modeled in a simple way, and the resulting system may be initially analyzed considering constant rainfall rate (i.e., no stochastic forcing). Once the stability scenario of the system is understood and the possible conditions, if any, of periodic or chaotic behavior are assessed, the interaction with the external hydrologic fluctuation provided by the stochastic forcing of soil moisture would provide the required realism to the analysis.

11

Hydrologic variability and ecosystem structure

Throughout the book we have described various ways in which plants respond to hydrologic forcing: in this last chapter, these elements are put together to analyze the role of hydrologic variability on the space–time ecosystem structure. Vegetation has been observed to exhibit a degree of spatial organization in a number of ecosystems, including rainforests (Sole and Manrubia, 1995), semi-arid grasslands (Milne, 1992), temperate forests (Mladenoff et al.,1993), and savannas (Couteron and Kokou, 1997; Scholes and Archer, 1997; see also Figure 1.10). The emergence of these organized patterns is attributable to a range of temporal and spatial variations in the interaction among species, localized dispersal abilities, and disturbance regimes as well as in the climate and soil characteristics.

Understanding the temporal and spatial patterns of vegetation in semi-arid ecosystems is perhaps one of the most fascinating aspects of ecohydrology. It is a rapidly evolving topic in which a great deal of research is currently being done. As a consequence, what follows is not to be intended as an exhaustive review of the subject, but rather as a discussion of results arising from simplified models with the same explorative spirit of the previous chapters. We just mention other studies that, though not directly dealing with fluctuating hydrologic conditions, might be of interest here (e.g., Klausmeier, 1999; Hellerislambers et al., 2001; Rietkerk et al., 2002; and references therein). Many of these studies have proposed models in which the pattern-forming component is an activation-inhibition mechanism that generates a Turing-type of instability (e.g., Murray, 1989). However, to justify spatial patterns such models always imply the presence of a diffusion mechanism which, depending on the variable being considered (e.g., soil moisture, plant dispersal, or surface runoff), may limit the validity of the approach to very small spatial scales (e.g., tens of centimeters in the case of soil moisture) or is quite difficult to justify from a physical point of view (e.g., surface runoff is not driven by diffusion). In any case, whether the temporal and spatial variability of the soil and climatic forcing changes the main conclusions of these investigations constitutes an important avenue of research.

In the following discussion on ecosystem structure we will often refer to tree–grass coexistence in savannas, where soil water availability is generally considered critical for tree density (Bourliere and Hadley, 1970; Belsky, 1990; Scholes and Archer, 1997; Fensham and Holman, 1999; see Section 5.2) and is the key factor in establishing the function, spatial pattern, and individual structure of vegetation (Scholes, 1997). A savanna is not an ecologic middle ground between forests and grasslands, but a system with its own characteristics, including a remarkably stable coexistence of trees and grasses. The explanation for their existence has been the subject of much research throughout the years. Perhaps the most common explanation is the so-called Walter hypothesis, which is based on the rooting-depth separation with respect to water access by trees and grasses (Walter, 1971). If trees are assumed to have roots mostly in the deeper soil layers and grasses only in the surface layer, trees then have preferential access to subsoil water and grasses have more effective access to topsoil water. In effect, the Walter hypothesis proposes that two different resources regulate trees and grasses in savannas so that the competitive-exclusion principle[1] can be bypassed.

However, although a number of models have applied the Walter hypothesis to study the codominance of trees and grasses under different conditions (e.g., Walker et al., 1981; Eagleson, 1982; Eagleson and Segarra, 1985), detailed field observations do not support the Walter hypothesis in many of the world's savannas (Scholes and Walker, 1993; Scholes and Archer, 1997). This does not mean that such a hypothesis is not a realistic representation for some cases, but it implies that it cannot be used as a general mechanism which fully explains the structure of savannas. The variety of soils and rainfall regimes on which savannas are found hints at searching for an explanation through more general mechanisms than the Walter hypothesis or that interact with this hypothesis. In Chapters 5 and 7 we saw how different physiological and soil characteristics may differently impact the site water balance, often producing similar levels of stress for both trees and grasses, thus justifying different forms of adaptation for various species. As a consequence, hydrologic fluctuations (Chapter 8), changes in soil texture (Section 7.2) or topographic conditions (Section 9.2),

---

[1] This principle, when generalized to n species, states that they cannot coexist with fewer than n resources or limiting factors (MacArthur and Levins, 1964). In the case of trees and grasses in savannas, only one species will dominate in the presence of a single resource. As pointed out by Lehman and Tilman (1997), however, this view of competition does not include any consideration of spatial structures or temporal niches in plant communities.

when associated with different strategies of water use (Chapters 5 and 6), become possible sources of temporal and spatial niches for the coexistence of different types of plants and the starting point for the developments of vegetation patterns.

The models that we discuss in this chapter investigate the inter-relationship between soil moisture dynamics and the space–time dynamics of vegetation. The first analysis deals with a spatially explicit model of tree–grass competition, in which the competition for water is achieved through an optimization of the water stress (Rodríguez-Iturbe et al., 1999b, 1999c); although no evolution of vegetation is considered, the spatial hydrology variability and the competition for water are found to lead to tree–grass compositions that are in close agreement with those present in southern Texas and South Africa. The second model, by Van Wijk and Rodríguez-Iturbe (2002), is a cellular automaton that describes the spatial and temporal evolution of trees and grasses whose characteristics of colonization and mortality are based on the dynamic water stress developed in Chapter 4. The issue of the interannual rainfall variability and tree–grass competition is further developed in the models by Fernandez-Illescas and Rodríguez-Iturbe (2003, 2004). In the first paper the temporal structure of a hierarchical model of competition between trees and grass driven by the dynamic water stress is considered. Finally, in Fernandez-Illescas and Rodríguez-Iturbe (2004) the model is extended to include the role of the spatial dynamics.

11.1 Tree–grass coexistence and optimization of water stress

Rodríguez-Iturbe et al. (1999b, 1999c) analyzed the coexistence of trees and grasses in savannas using a model based on the optimization of plant water stress for different spatial distributions of trees and grasses. The definition of water stress is similar to the one used for static water stress (Section 4.2), but is applied to the long-term mean soil moisture, i.e.,

$$\xi = \begin{cases} \left[\frac{s^* - \langle s \rangle}{s^* - s_{\mathrm{w}}}\right]^q & \text{for } s < s^* \\ 0 & \text{otherwise,} \end{cases} \tag{11.1}$$

where q represents the nonlinear effect of water deficit on the plant response and the mean soil moisture $\langle s \rangle$ is calculated using the simplified model of Section 2.6.1. Although the representation of water stress is simpler than the one provided by the dynamic water stress (Section 4.3), it embeds the general effect of soil moisture dynamics on vegetation and represents the magnitude of the impact of randomly varying water deficit on the vegetation at the site.

The model considers a region heterogeneous in soil and vegetation represented by a grid with square cells which typically correspond to the canopy coverage of the average woody vegetation (e.g., 5 m × 5 m or 10 m × 10 m). The area covered by roots of the individual plants is assumed to be larger than the individual cell to allow for competition among neighboring vegetation. The whole region is characterized by the same statistically steady climatic conditions.

From the definition of mean water stress (Eq. (11.1)), one can visualize that different kinds of vegetation undergo different stress conditions under any prescribed climatic characteristics. An example of this is shown in Figure 11.1 for a typical case of trees and grasses. The different values of s^* and s_w for trees and grasses lead to stress–soil water curves that intersect regardless of the particular exponent, q, used in Eq. (11.1). When a site is considered in isolation from its neighbors and no spatial interaction takes place, soil moisture values above A favor trees while soil moisture values below A favor the presence of grasses.

Competition for soil moisture is assumed to take place locally and to depend on stresses among neighboring plants. The spatial dynamics controlling the competition for soil moisture is assumed to globally minimize the vegetation stress resulting from deficits of soil water content. No attempt is made to model the temporal evolution of vegetation and the different spatial vegetation

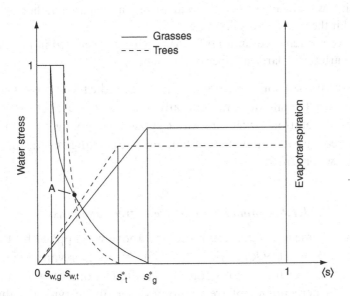

Figure 11.1 Stress diagram used to model for plant responses to water stress for mean typical parameter values of trees and grasses. After Rodriguez-Iturbe et al. (1999c).

fields (e.g., different percentages of trees versus grasses in savannas) are assumed to be in equilibrium with the soil moisture structure corresponding to a minimum global stress over the region. Notice that, even though the goal is a global optimum (e.g., minimum stress), the competition for soil moisture is always local. In nature, where competition frequently leads to associations benefiting many partners, competition is always between individual organisms rather than species, which are only indirectly competing entities (Bonner, 1988).

Initially, soil moisture conditions at each site are computed assuming no spatial interaction. Then, the dynamics develops according to the following rules (Rodríguez-Iturbe et al., 1999c):

(1) Choose at random a site i and one of its neighbors n.
(2) The neighbor subtracts an amount of soil moisture, η, from the site i according to

$$\eta = \begin{cases} (1 - \xi_i)(s_i - s_{w_n})\epsilon & \text{for } s_i \geq s_{w_n} \\ 0 & \text{otherwise,} \end{cases} \tag{11.2}$$

where ξ_i is the vegetation stress in i and ϵ is a random variable uniform in [0,1].
(3) The new stresses ξ_i and ξ_n are computed: if the global stress, $\sum_i \xi_i$, decreases, the above dynamics is accepted and the procedure is repeated in a new site chosen at random; if instead the global stress increases, the changes in soil moisture do not take place and the process is restarted.
(4) If the site i was already subtracting water from the neighbor n, then the transfer is stopped if the global stress decreases.
(5) The process is repeated until no further decrease is detected in the global stress after an arbitrarily large number of interactions.

Since the transfers of soil moisture go directly into the receiving vegetation, the soil at the receiving site does not actually increase its water content although the existing vegetation lowers its water stress. The giving site, on the other hand, will decrease its soil water content and the existing vegetation will likely increase its water stress.

11.1.1 Optimal coexistence of trees and grasses

The spatial scheme of interaction presented above was applied by Rodríguez-Iturbe et al. (1999b, 1999c) to the case of the Nylsvley savanna (Section 5.1). The model was run for various random configurations of trees and grasses with different percentages of tree canopy cover, using typical values of the parameters for the main species of plants found in the broad-leafed savannas at Nylsvley (Section 5.1). All sites are assumed to be covered by vegetation.

Several stress-function exponents, q, ranging between 1 and 3 and different grid sizes were used in Eq. (11.1), without significant changes in the results.

Figure 11.2 shows the dependence of the total (i.e., global) vegetation stress as a function of the percentage of tree canopy cover for the Nylsvley case when $q = 1$. The minimum-stress situation is found to correspond to a tree canopy density near 35% which matches quite well the values of 30–40% observed in the field (Scholes and Walker, 1993). Figure 11.2 also shows the dependence of average soil moisture at the tree and grass sites as a function of tree canopy density. As expected, tree sites show a higher average soil water content than grass sites and values agree well with those observed in the region (as inferred from Figure 6.6 in Scholes and Walker, 1993). Notice in Figure 11.2 that when there are no spatial interactions, the optimal condition is a field 100% covered with grass. Spatial dynamics leads to mixtures of trees and grasses that reduce global vegetation stress.

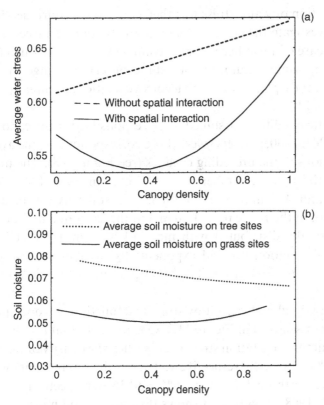

Figure 11.2 (a) Average water stress and (b) average soil moisture in tree and grass sites as a function of tree canopy cover for broad-leafed savannas at Nylsvley. After Rodriguez-Iturbe et al. (1999c).

The effects of clustering of trees on the total stress were investigated by implementing the same dynamics on a grid where trees are spatially clustered according to a Poisson spatial process with cluster centers chosen at random and the size of the tree cluster is a random variable independent of spatial position. The results (Rodríguez-Iturbe et al., 1999c; Figure 10) show that clustering is not effective in reducing the total vegetation stress. In natural savannas, the degree of clustering of the woody vegetation varies widely and, in some cases, evenly spaced plants can result from competitive interactions. In any case, clustering does not seem to be due to competition for soil moisture (which works against it) but rather results from autogenic site modification. Deposition of seeds, together with local enrichment of soil nutrients and alteration of the microclimate, are conducive to clustering of woody vegetation.

11.1.2 Vegetation dynamics and climate fluctuations

The optimization of water stress was then employed (Rodríguez-Iturbe et al., 1999c) to investigate the effect of climatic fluctuations on tree–grass coexistence for the case of a southern Texas savanna (Section 5.2). Such fluctuations are a very important natural factor behind observed changes in the relative densities of woody plants and herbaceous vegetation, as already discussed in Sections 5.2 and 7.1.

The climatic conditions assumed to be responsible for the canopy densities existing in 1941, 1960, and 1983 were those corresponding to the average of the growing seasons in the preceding decades (see Figure 5.8). The mean rainfall from May through September was 395 mm in 1931–1941, 339 mm in 1950–1960, and 412 mm in 1973–1983. The soil was assumed to have a random depth in the range 0.70–1.10 m and the saturated hydraulic conductivity log-normally distributed with mean of 82.2 cm day^{-1} (USDA, 1979), coefficient of variation 0.1 and exponentially decaying spatial correlation, $\rho(l) = e^{-(l/I)}$, with $I = 3$ pixels. The other parameters are described in detail in Section 5.2.

Minimizing the global vegetation stress as a function of woody plant density results in stresses shown in Figure 11.3. One observes that for 1941, 1960, and 1983 the minimum vegetation stress occurs when the density of woody plants is near 15%, 0%, and 35%, respectively. These values agree quite well with the observed ones for those dates (13%, 8%, and 36%, respectively, as reported in Archer et al., 1988). Figure 11.4 shows the average soil moisture in tree sites and grass sites as a function of tree canopy density for the growing seasons of the decades preceding 1960 and 1983. As expected, the average soil water

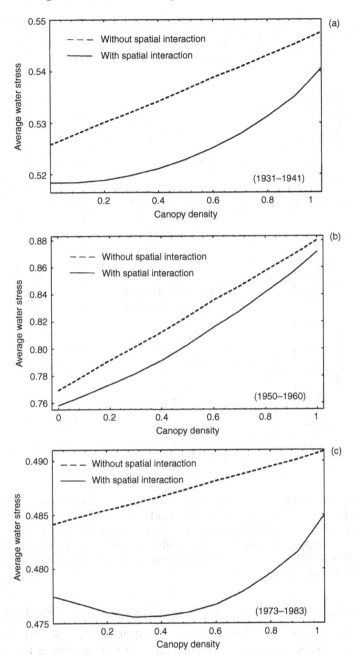

Figure 11.3 Average water stress of Eq. (11.1) as a function of tree canopy coverage for vegetation at La Copita, Texas (see Section 5.2). Rainfall conditions refer to months from May through September in Alice during (a) 1931–1941 (average rainfall 395 mm), (b) 1950–1960 (average rainfall 339 mm), and (c) 1973–1983 (average rainfall 412 mm). After Rodriguez-Iturbe et al. (1999c).

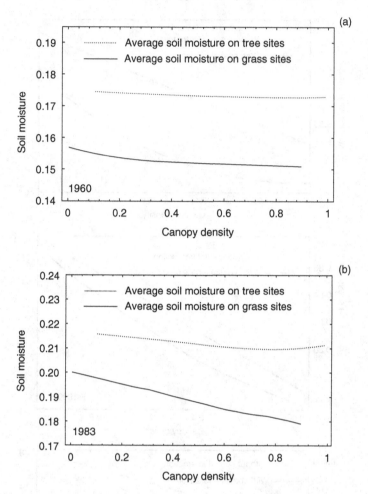

Figure 11.4 Average soil moisture in tree and grass sites after competition as a function of tree canopy density at La Copita (Texas). The values correspond to the condition of minimum global stress with spatial interactions under different canopy coverages. Rainfall conditions refer to the months from May through September during the decades preceding (a) 1960 and (b) 1983. After Rodriguez-Iturbe et al. (1999c).

content in the years leading to 1983 is considerably higher for both kinds of plants. The dependence of evapotranspiration on canopy density also is quite different for the decades preceding 1960 and 1983. As shown in Figure 11.5, in the wet case of 1983, evapotranspiration increases monotonically with tree canopy density, reflecting the greater availability of water for the woody vegetation which, on average, transpires more than the herbaceous vegetation. For 1960, the extreme scarcity of water leads to spatial interactions conducive to an average evapotranspiration that declines as a function of increasing tree

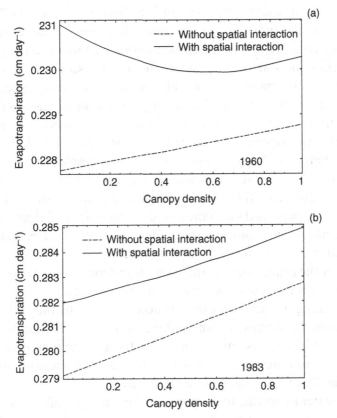

Figure 11.5 Average evapotranspiration after competition as a function of tree canopy density at La Copita (Texas). The values correspond to the condition of minimum global stress under different canopy coverages. Rainfall conditions refer to the months from May through September in Alice, Texas, during the decade preceding 1960 (a) and the decade preceding 1983 (b). After Rodriguez-Iturbe et al. (1999c).

canopy density for low densities but which tends to increase for high values of woody plant coverage. It is important to note that Figures 11.4 and 11.5 represent soil moisture and evapotranspiration after a condition of minimum global water stress has been reached for a given canopy coverage.

11.1.3 The role of bare soil

The same model of competition can be applied without changes to cases where there are more than two functional dependences of evapotranspiration on available soil moisture, as in the case of bare soil. On the one hand, a bare site increases access of neighboring sites to moisture because there is no competition with site vegetation. For a bare site, Eq. (11.2) becomes $\eta = (s_i - s_{w_n})\epsilon$. Because there is no vegetation stress at bare-soil sites, all comparisons regarding global stress are normalized by the actual number of vegetated sites and one then considers the average stress per plant over the region. On the other hand, the available soil water at bare-soil sites may be on average much less than for the vegetated sites. The main reasons for this are: (i) in the presence of vegetation, root penetration induces the presence of macropores and preferential infiltration paths; and (ii) lack of vegetation frequently leads to sealing of the soil surface as the result of raindrop impacts. When the soil dries, a hard crust may favor surface runoff from subsequent precipitation. Such runoff has been suggested to be an important factor in the development of patterned vegetation, such as the so-called tiger bush formed in some semi-arid environments (Bromley et al., 1997). Depending on the climate and soil characteristics, the presence of bare soil may be locally beneficial for the nearby vegetation in the short term but sealing of the surface soon occurs with the accompanying detrimental effects.

Surface sealing on bare-soil sites was accounted for by the reduction of the mean depth of the storms, assuming that, on average, the rain that falls and infiltrates on a bare site is only a fraction of the average rain on the vegetated sites. The bare-soil sites were assumed not to cluster into regions large enough for appreciable surface runoff to be generated.

The soil water losses due to evaporation from a bare-soil site were modeled as linearly increasing from 0 at $s = 0$ up to a maximum rate E^* at $s = 1$ (Rodríguez-Iturbe et al., 1999c). This scheme is obviously a crude simplification; however, when combined with leakage, it reproduces the general observation that for bare soil the curvature of the evaporation versus soil moisture relationship is opposite to that in the vegetated case (Philip, 1957). The value of E^* is generally lower than the maximum evapotranspiration occurring on vegetated sites, but the bare soil generally has less moisture than that found under vegetation.

According to the previous modeling scheme, the role of bare soil was studied under two conditions. First, bare-soil sites were assumed to have negligible infiltration due to the superficial crust they develop. Second, the infiltration was assumed to be reduced (but not eliminated). No evolution in time was allowed for the bare soil. Figure 11.6 shows a typical result of the experiments for Nylsvley, in which bare soil is explicitly included. In this example, the vegetated sites have a fixed proportion of 40% trees and 60% grasses. When bare crust prevents infiltration, we observe that – as expected – the total stress on the vegetated sites increases monotonically with increasing bare soil. When bare-soil sites allow a reduced infiltration (modeled through the assumption that on average they receive one-third of the rain received by vegetated sites)

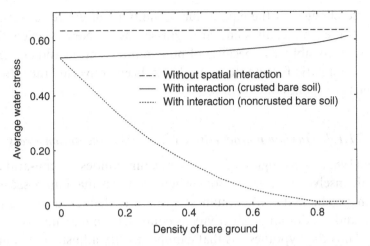

Figure 11.6 Average water stress in vegetated sites as a function of the percentage of bare soil for a savanna where 40% of the vegetated soil is occupied by trees. The parameters correspond to the case of the broad-leafed savannas at Nylsvley. After Rodriguez-Iturbe et al. (1999c).

the global stress on vegetated sites decreases with increasing bare soil. The second situation may represent what happens in areas that have been plowed.

These results show the sensitivity of savannas and grasslands to external interactions either of natural or human origins. In arid and semi-arid climates, a cover of low grasses prevents drop-splash erosion and reduces overland flow more than do the canopies of typical woody vegetation in this region (McAuliffe, 1995). A change from perennial grasses to a more patchy cover of shrubs most frequently carries profound consequences for the ecological equilibrium of a region. Loss of grass during drought periods leads to compaction of soils and reduced infiltration rates. Locally augmented runoff increases the rate of soil erosion and landscape incision on the random fluctuations of the topography (Schlesinger et al., 1990) and in turn induces an increased transport of water, nitrogen, and other nutrients across the landscape. As pointed out by Schlesinger et al. (1990), the net effect of these fluxes is to reduce the availability of soil moisture and nutrients and to increase the heterogeneity of their spatial distribution. Through landscape incision and natural topographic fluctuations, water may accumulate locally due to the increase of local runoff. This facilitates deeper infiltration, which favors the existence of shrubs. More generally, the greater spatial heterogeneity of water and nutrients may produce islands of fertility and patchy distributions of shrub vegetation, like those presently seen in the La Copita area (see Figure 1.10). In some extreme cases, the differences between bare soil and vegetation and their interactions with the site water balance and nutrient cycles may lead to dramatic changes in the equilibrium conditions of entire regions. This is the case, for example, in some arid grasslands in New Mexico that (see Figure 1.11), possibly also because of human intervention and change in the interannual climatic forcing, have gradually been converted into shrublands with extensive patches of bare soil.

11.1.4 Maximum productivity vs minimum vegetation stress

Eagleson (1982, 1994) suggested that plant communities in semi-arid climates arrange themselves to operate somewhere between maximum security and maximum productivity. Maximum security was defined as maximum soil moisture, and maximum productivity as equivalent to maximum evapotranspiration. Thus the hypothesis is that canopy density adjusts to a value in the range whereby the community is between maximum soil moisture and maximum evapotranspiration.

Using the same scheme presented before, Rodríguez-Iturbe at al. (1999c) also tested an interaction scheme that subtracts moisture from a site by a randomly chosen neighbor to verify if it leads to an increase in global transpiration (surrogate of productivity). Figure 11.7 shows the average global transpiration and the relative soil moisture for the Nylsvley region when the global transpiration is explicitly maximized for different values of tree canopy coverage (e.g., woody plant density) and there is no bare soil over the region.

With no spatial interactions, the global transpiration decreases monotonically with increasing percentage of woody plants. The response to maximization of productivity is quite different from the previous simulations (Figure 11.2). Global transpiration increases as the tree coverage increases to about 40%, it then remains at the maximum value of 0.25 cm day^{-1} until canopy values of near 60% and then decreases with the increase of woody plants. The canopy densities obtained through maximization of transpiration are only slightly larger than those obtained as a result of minimizing the global water stress of the vegetation. This qualitatively agrees with the hypothesis of Eagleson (1982), further studied by Salvucci and Eagleson (1992), that the optimal canopy density is to be found between maximum security and maximum productivity. The average values of soil moisture for tree and grass sites agree well with those resulting from the minimization of vegetation water stress.

11.2 A water-stress-based cellular automaton for tree–grass evolution

The previous analysis provides a static analysis of vegetation under given environmental conditions. To account for the long-term fluctuations in

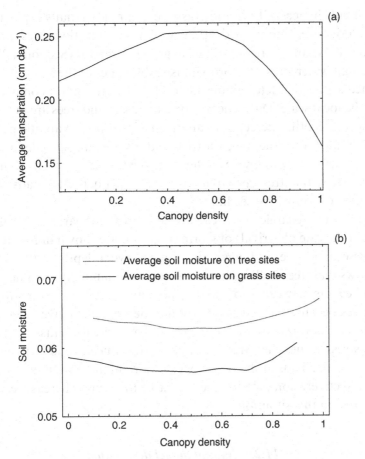

Figure 11.7 Average transpiration resulting from a spatial dynamics that maximizes transpiration under prescribed tree canopy coverages. The parameters are those corresponding to the Nylsvley case and $q = 1$. (a) Average global transpiration after competition as a function of canopy density; (b) average soil moisture in the tree and grass sites. After Rodriguez-Iturbe et al. (1999c).

rainfall in a spatially explicit model describing savanna functioning, Van Wijk and Rodríguez-Iturbe (2002) analyzed a cellular automaton based on water stress for tree–grass evolution and applied it to study the dynamics of the savanna in southern Texas. The site and all the parameters are described in Section 5.2. Their goal was to develop a simple model to investigate what kind of temporal and spatial behavior the interannual rainfall fluctuation could induce on tree–grass coexistence. This section is taken from Van Wijk and Rodríguez-Iturbe (2002).

The dynamic water stress function (Section 4.3) was taken by Van Wijk and Rodríguez-Iturbe (2002) as the basis for describing the death and establishment chances of trees and grasses in a simple spatially explicit scheme. In their model, tree and grass death as well as colonization rates are directly linked to the dynamic stress during the growing season (Section 4.3). In this way, through water stress, the probabilistic dynamics of the soil water content is assumed to be the determining factor for the tree–grass competition for space at the local scale. Differences between grasses and trees are incorporated in the rooting depth, interception, and transpiration. The abiotic parameters of the point model are the same for trees and tree seedlings, while to incorporate a higher sensitivity to severe drought for the tree seedlings, as compared to the mature trees, the stress sensitivity parameter k (Eq. (4.21)) is adjusted from 1.0 for trees to 0.5 for tree seedlings.

The values of dynamic stress calculated with these parameter values as a function of the rate of arrivals of storm events, λ, are shown in Figure 11.8 for trees, grasses, and tree seedlings, using a mean event depth, α, of 1.5 cm. The figure shows that tree seedlings have the highest dynamic stress values. At rates of rainfall events between 0.15 and 0.25 per day, trees have lower stress values than grasses because of their larger rooting depth, which gives them a larger water buffer than grasses. At very low λ, however, the grasses have lower dynamic stress values than trees because of the concentration of roots in the upper soil layers. Thus at the occurrence of a rainfall event grasses react very fast and effectively compete for the available moisture whereas trees are not very effective in this situation.

11.2.1 Spatial model description

Van Wijk and Rodríguez-Iturbe (2002) define a cellular automaton on a grid of 100 by 100 cells in which the status of each cell is updated at each time

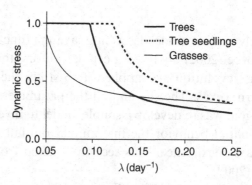

Figure 11.8 Simulated dynamic stress values as a function of $\lambda(\alpha = 1.5\,\mathrm{cm})$. After Van Wijk and Rodriguez-Iturbe (2002).

depending on the status of its neighbors. The cells are considered to have the canopy size of one individual mature tree; periodic boundary conditions are assumed throughout the simulation in order to minimize boundary effects. A cell can be empty, be occupied by a tree, by tree seedlings (i.e., trees in the age of one to five years) or by grasses. No mixed occupation is possible in a cell. Tree seedlings cannot reproduce, and after five years become mature trees.

The spatial model incorporates no spatial heterogeneity in the abiotic components of the model: rainfall is evenly distributed over the total grid, all soil physical parameters used in the point model are the same, and no spatial redistribution of runoff water takes place. Therefore, any patterns that may arise in the spatial vegetation output of the model are determined by the biotic part of the model, i.e., death and reproduction.

Death and reproduction of both trees and grasses are modeled very simply. The chances of death for trees, tree seedlings, and grasses are directly linked to the water stress experienced by the vegetation through the value of dynamic stress minus a threshold value. By incorporating a threshold, a minimum of dynamic stress will not immediately lead to an increase in death chances, which would be a gross over-estimation of the sensitivity of trees and grasses to water stress. The various mechanisms of drought avoidance are implicitly incorporated in the model by the use of a maximal death probability, which is much smaller than one. The threshold parameters are deduced from the maximum values of tree and seedling deaths, occurring at severe water stress, based on the values presented in the study of Fensham and Holman (1999). Also a minimum death chance is used, in the absence of "damaging" water stress (i.e., the value of the dynamic water stress is below the threshold value). Death occurs randomly in space, with no sheltering effects or any other spatial dynamics on death chances being taken into account.

Settlement of tree seedlings or grasses can only take place in empty cells. Of an empty cell the 24 nearest neighbor cells are assumed to be potential colonizers (the first ring of 8 neighbors, and the next ring of 16 neighbors). From this neighborhood one of the cells occupied by either trees or grasses is selected at random. If a tree occupies the cell, there is a possibility of settlement of a tree seedling in the empty cell depending on the value of dynamic water stress of trees for that particular growing season. If the cell is still empty after this calculation or if the randomly chosen cell was not occupied by a tree, there is the possibility of grass settlement in the empty cell according to the value of the dynamic water stress of grasses. Therefore, for grasses it is not strictly necessary to be present in the neighborhood of an empty cell in order to colonize the space. The probabilities of settlement of trees and grasses as a function of the dynamic stress minus the threshold parameters as defined for the death chances (i.e., which have different values for grasses and trees) are given in Figure 11.9. The optimal values of tree and grass settlement are based on the values of Jeltsch et al. (1996). Figure 11.9 shows the difference in sensitivity and maximum probabilities of settlement for trees and grasses. If the stress threshold is not exceeded, trees have the highest probability of settlement because of their massive local seed distribution. However, with the occurrence of stress, their settlement probability decreases much faster than that of grasses, because of the higher drought sensitivity of both seed production and settlement of trees versus grasses. The difference in drought stress sensitivity was estimated from measurements in southern Texas (Archer et al., 1988; see also Section 5.2). The spatial model of Van Wijk and Rodríguez-Iturbe (2002) therefore consists of three threshold parameters for the dynamic stress function, four propagation and settlement parameters (two for trees and two for grasses), and three minimum values of death chance, for a total of ten parameters. Depending on the occupancy of a certain cell, whether tree, tree seedling or grass, the parameters are assigned to that cell. No further spatial interactions, resulting in adjusted parameter values affected by neighborhood cell values, were included in the model.

During a model run, each time step corresponds to a growing season whose values of λ and α are sampled from distributions obtained from historical rainfall data in southern Texas (Section 5.2). From these values the dynamic water stress experienced by trees, tree seedlings, and grasses is then calculated, and from there the chances of death and settlement are computed. In the spatial grid the status of the cells is updated depending on the occupancy of the neighboring cells, and the chances of death and settlement. After this, for the next time step, new values of dynamic water stress are calculated, death and settlement chances are computed and cell status updated.

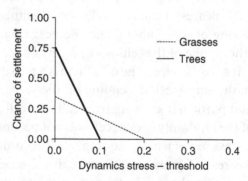

Figure 11.9 Probability of settlement in the spatial model of trees and grasses as a function of dynamic stress minus the threshold value (for trees 0.60 and for grasses 0.23). After Van Wijk and Rodriguez-Iturbe (2002).

11.2.2 *Rainfall sensitivity*

The sensitivity of tree and grass abundances to rainfall fluctuations was analyzed by Van Wijk and Rodríguez-Iturbe (2002) considering several values of standard deviations of λ and α.

The combined effects of increasing rainfall and year-to-year rainfall variability are shown in Figure 11.10. When there is no interannual variability in the rainfall parameters, the tree population (including tree seedlings) becomes sustainable for values of $\lambda > 0.11$ day^{-1} and shows a sharp increase from a cover of near 0% to a cover close to 50%. For λ values between 0.115 and 0.12 day^{-1} the tree population shows a small decrease in density. This is caused by the fact that at these λ's, trees do not experience major water stress (the value of the dynamic stress is below their threshold), whereas the tree seedlings are still under high stress conditions (see Figure 11.8). Because the water stress of the grasses decreases in this range of λ values, the density of grasses increases, and as only limited space is available it does so at the expense of seedlings. For λ values above 0.12 day^{-1}, the dynamic stress of the tree seedlings also decreases, and the density of the trees increases sharply again.

The above effect totally disappears when year-to-year variation is included in the rainfall amounts. When interannual rainfall variation is present, trees are absent in the lower range of mean λ values, and then increase rapidly in density over a relatively small interval of mean λ. The mean value of λ at which the tree population becomes sustainable becomes higher with increasing year-to-year variability. Also visible in Figure 11.10 is the increase in the standard deviations of tree and grass densities when the year-to-year variation in rainfall amounts increases. With increasing rainfall variability, very dry years occur more often, resulting in high tree and grass kill-off, and thereby prohibiting the occurrence of only trees in the model at higher λ values. The increasing year-to-year variability therefore shows a tendency to increase the interval of tree–grass coexistence, in which over a broader range of λ values the grasses are present in significant amounts, especially when considering the increased standard deviation. Values of the year-to-year variations in rainfall larger than those shown in Figure 11.10 were not considered because the populations of trees and grasses are then still in transient behavior after 3000 years, and the random rainfall characteristics have a large effect on the mean value of the last 1000 years. At the highest coefficients of variations (CV) applied in the analysis (CV of $\lambda = 0.2$ and CV of $\alpha = 0.1$) there is a stable coexistence between trees and grasses at a λ interval of from 0.144 to about 0.17 day^{-1} (e.g., between 320 mm and 380 mm of mean rainfall per growing season).

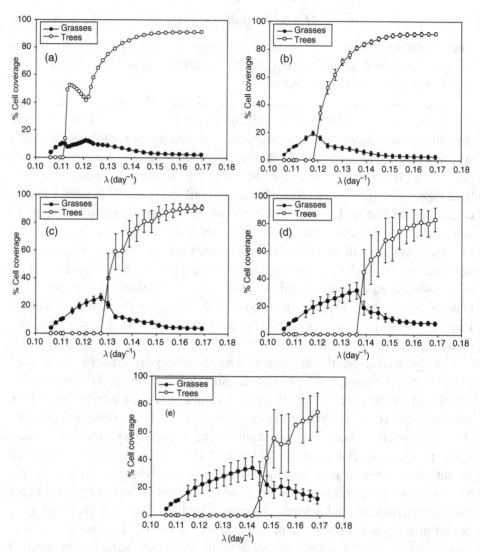

Figure 11.10 The mean and standard deviation (bars) of cell coverage with trees (including tree seedlings) and grasses as a function of the mean value of λ (mean value of $\alpha = 1.5\,\mathrm{cm}$). Computations are for the last 1000 years of a 3000-year simulation. (a) $\mathrm{CV}(\lambda) = 0.000$ and $\mathrm{CV}(\alpha) = 0.00$; (b) $\mathrm{CV}(\lambda) = 0.025$ and $\mathrm{CV}(\alpha) = 0.05$; (c) $\mathrm{CV}(\lambda) = 0.050$ and $\mathrm{CV}(\alpha) = 0.10$; (d) $\mathrm{CV}(\lambda) = 0.075$ and $\mathrm{CV}(\alpha) = 0.15$; (e) $\mathrm{CV}(\lambda) = 0.100$ and $\mathrm{CV}(\alpha) = 0.20$. After Van Wijk and Rodriguez-Iturbe (2002).

Coexistence between grasses and trees in the model occurs at rainfall values at which the death rate of trees is high enough to prevent a total dominance and low enough to keep the tree population sustainable. Coexistence of numerous plant species competing for a single limiting resource can be

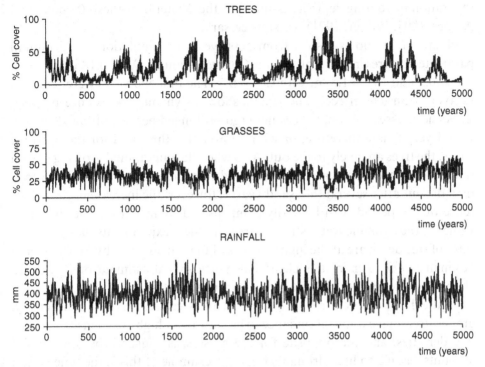

Figure 11.11 Model input (shown as a 10-year running average of rainfall) and output (percentage coverage of trees and grasses) for 5000 years of simulation. Rainfall parameters are $\langle \lambda \rangle = 0.18$ day^{-1}, $CV(\lambda) = 0.2$, and $\langle \alpha \rangle = 1.5$ cm, $CV(\alpha) = 0.3$. After Van Wijk and Rodriguez-Iturbe (2002).

accounted for in classical ecological models that are spatially explicit if a trade-off exists in colonization, competition, and longevity (Tilman, 1994). Disturbance is a key factor for this tradeoff, because it assures the availability of free places where settlement of one or the other species can occur. Periods of severe water stress cause the death of trees and especially tree seedlings, thereby giving grasses the opportunity for settlement. In periods with high rainfall amounts the trees take over because of their lower overall death rate and their higher settlement chance (see Figure 11.9) at low water stress compared to grasses.

Although different rooting depths based on Texas data were used for trees and grasses in the point model (Van Wijk and Rodríguez-Iturbe, 2002), this was not a key factor determining coexistence because no competition for water was incorporated into the model. Better estimates of the death and colonization parameters for both grasses and trees could help to clarify whether hydrological disturbances are a determining factor for tree–grass coexistence in savanna ecosystems as suggested by this model, or whether a niche separation by rooting depth as assumed by the Walter hypothesis (Scholes and Archer, 1997; Walter, 1971) is also necessary.

Figure 11.11 shows results from 5000 years of simulation using rainfall parameters representative of the La Copita savanna, $\langle \lambda \rangle = 0.18$ day^{-1} with CV $= 0.2$ and $\langle \alpha \rangle = 1.5$ cm with CV $= 0.3$. The first 500 years were removed to avoid transient effects. The results show a dynamic coexistence between trees and grasses, although trees undergo prolonged periods of low densities. In a dryer climate therefore, or with a lower tree threshold for the dynamic stress, the trees are likely to die out. Also when the interannual growing season rainfall variance is increased, grasses tend to overcome the trees, which may then die out after several thousand years. However, the increase in the occurrence of dry periods can be easily compensated by increasing the threshold value for the dynamic water stress of trees. In these experiments the maximum rates of tree death are at the higher values of the ranges given by Fensham and Holman (1999). Again it is clear that a field-based estimate of tree death chances is essential for reliable model parameterization.

Figure 11.11 also clearly shows that if trees/grasses have a high density, grasses/trees have a low abundance. The shift from high tree densities to low tree densities, and the opposite for the grasses, is triggered by low rainfall amounts leading to high drought stress. An example of this is the time period between 1300 and 1500 years in Figure 11.11. During that period a severe drought stress occurs leading to a high mortality of both grasses and trees. The grasses, however, recover much faster than the trees from this extreme event, and thereby fill up a large amount of the cells of the model. These cells are then unavailable for settlement of tree seedlings, and the tree density undergoes a prolonged period of low values. If, however, the amount of rain is relatively high, the trees can profit from increased amounts of settlement, and slowly recover from earlier severe drought mortality.

11.2.3 Time spectra

The temporal series of the rainfall input and the corresponding vegetation output were studied by Van Wijk and Rodríguez-Iturbe (2002) through their power spectra. Although the spectrum of the rainfall input is flat over the whole frequency range, the spectra of model outputs show considerable structure over all frequencies (Figure 11.12). Moreover, the power spectra of the density coverage for grasses and bare soil show a power law relation over an extended frequency range. Thus local interactions based on the water stress present at each site lead to temporal structures at many frequencies when the system is driven by simple white noise. Power law power spectra in time are

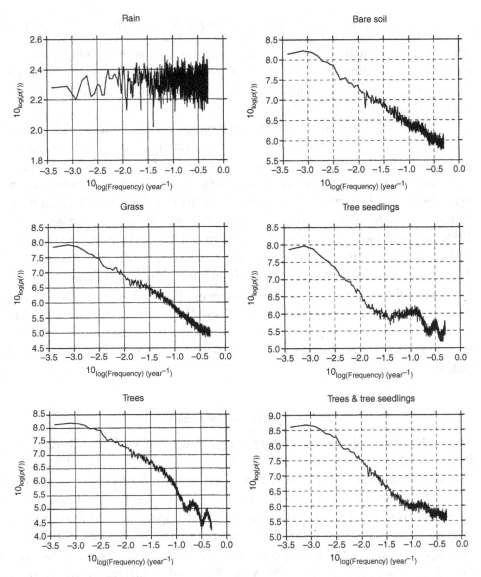

Figure 11.12 Power spectral density plots of the rainfall input and the corresponding proportional abundance of trees, seedlings, grasses, and bare soil for a long-term simulation (the rainfall characteristics are the same as in Figure 11.11). After Van Wijk and Rodriguez-Iturbe (2002).

a typical signature of temporal fractal signals, which display long-term dependence. More concretely, a spectrum $S_x(f) \propto 1/(f^D)$ indicates a self-affine structure where the process remains statistically unchanged when rescaled by different factors along the different axes, e.g., time and density coverage. Thus,

$$X(t + \gamma \Delta t) - X(t) \overset{d}{=} \gamma^H [X(t + \Delta t) - X(t)], \qquad (11.3)$$

where d means equality in probability distribution, γ is the arbitrary scaling factor along the time axis, H is the Hurst coefficient, and γ^H is the rescaling factor along the axis of density coverage. The power spectrum and H are related by $D = 2H + 1$ with $0 < H < 1$.

The power law structure of the spectrum indicates that the ecosystem avoids getting locked in any dominant frequency and displays a wide range of dynamic states with a well-defined structure far from complete disorder (e.g., flat spectrum) or from much order (e.g., strong cyclic components). In many biological systems this is a sign of healthy dynamics which avoid both extremes of order and disorder (e.g., heart rate variations, Goldberger et al., 1985; Sole and Goodwin, 2000). The spectra of the density coverage for trees and tree seedlings show departures from power law dependence of frequencies. The trees show peaks at frequencies of about $1/2.5$ years and $1/5$ years, whereas the tree seedlings show peaks at frequencies of about $1/3$ years and $1/8$ years. These peaks are likely to result from dynamic effects arising from the five-year interval chosen as seedling lifetime. Notice that the spectrum for trees and tree seedlings together does not show the above trend, and approaches a power law behavior.

11.2.4 Spatial pattern analysis

The spatial fields of the model were analyzed by Van Wijk and Rodríguez-Iturbe (2002) using two statistics. First, for a consecutive set of 1500 years the spatial distribution of trees was tested for spatial clustering by applying Ripley's K-function (Ripley, 1976; Haase, 1995; Wiegand et al., 1998). Field measurements indicate that both random and strongly clustered distributions of trees may exist in the field. The univariate form of Ripley's K-function investigates whether the model can represent both situations for prolonged periods of time. The analysis by Van Wijk and Rodríguez-Iturbe (2002) proceeds as follows. For each spatial field an inner plot of 70 by 70 cells is selected, to which the univariate form of Ripley's K-function is applied and, in the univariate case, the clustering or hyperdispersion of a set of points in a circular area is estimated through comparison with the expected values for a randomly distributed field. The approximately unbiased estimator for $K(l)$, where l is the radius of the circular area (also called the length scale), is

$$\hat{K}(l) = n^{-2} A \sum_{i \neq j} w_{ij}^{-1} I_l(u_{ij}), \qquad (11.4)$$

where *n* is the number of events (trees) in the analyzed field, *A* is the area of the field, I_l is a counter variable, u_{ij} is the distance between the events *i* and *j*, and w_{ij} is a weighting factor to correct for edge effects (Ripley, 1976). For all events where $u_{ij} \leq l$, the counter variable I_l is set to one, otherwise it is set to 0. With an inner field of 70 by 70 cells, one may apply a maximum radius of the circular area of 15 cells without using edge corrections (Haase, 1995). To test whether the trees in the fields simulated by the model show significant departures from a random pattern, Van Wijk and Rodríguez-Iturbe (2002) estimated the 95% confidence interval around the expected values for *K* in the case of a random pattern, using randomly filled fields with the same density of trees as in the field being tested (Haase, 1995). By comparing the maximum and minimum *K*-value of the randomization procedure with the values obtained for the field under analysis, the significance of the departure from the null hypothesis of random distribution could then be tested (Haase, 1995; Wiegand et al., 1998).

Van Wijk and Rodríguez-Iturbe (2002) calculated the Ripley's *K*-function and tested whether there was significant clustering present in the spatial tree distributions. These tests did not include tree seedlings because given the pattern of seedling settlement, i.e., close to the mature trees, the spatial distribution both of trees and tree seedlings would always show significant clustering, and therefore would not be very illuminating. The results of the tree spatial distributions are shown in Figure 11.13, where the initial 500 years were deleted from the model outcome to remove the effects of model initialization.

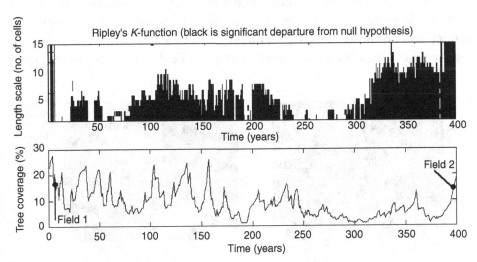

Figure 11.13 Results of Ripley's *K*-function together with percentage tree cover (the rainfall characteristics are the same as in Figure 11.11). After Van Wijk and Rodriguez-Iturbe (2002).

In the upper graph the significant departures from the random spatial distributions are shown for length scales of 2 to 15 cells. Clearly visible are prolonged periods in which significant clustering occurs at several length scales. These periods end when a major tree die back is caused by severe drought. Since tree death occurs in a spatially uncorrelated pattern, this leads to a breakdown of the clusters. The clustering increases in length scales at periods in which there is an increase in tree density. This may be explained by the fact that tree seedling establishment can only occur in the neighborhood of mature trees. When the seedlings become mature trees, the spatial distribution of these is strongly clustered around the trees whose seeds led to the original establishment.

In Figure 11.14 two fields are shown, which have about the same tree density ($\approx 15\%$) but are totally different in their cluster characteristics (Field 1 and Field 2 in Figure 11.13). In Field 2 significant clustering is present at all 15 length scales of Ripley's K-function. Clearly visible are isolated patches of trees separated by grass vegetation and bare soil. In Field 1 the spatial tree distribution is not significantly different from a random one. As shown in Figure 11.13, Field 1 occurs just after a severe tree die back (and also grass die back) and thus the density of bare-soil cells is much higher than in Field 2, which occurs in a period in which tree densities are increasing. The model can yield totally different spatial distributions even at the same level of vegetation density and may thus account for prolonged periods of both spatially random tree distributions, and spatially clustered tree distributions. Both types of spatial patterns have been documented with field data (Jeltsch et al., 1998). The above result does not mean that soil heterogeneity, redistribution of surface water or processes such as animal dispersion of seeds are not important for the spatial distribution of trees found in the field. It only means that from

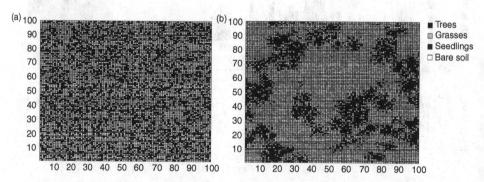

Figure 11.14 Spatial structures of Fields 1 and 2 – (a) and (b), respectively – of Figure 11.13 (see text for explanation). After Van Wijk and Rodriguez-Iturbe (2002).

the point of view of dynamics driven by hydrologic fluctuations such processes are not essential for explaining differences in tree distribution, although they may be very important in the complex evolution of real ecosystems.

Van Wijk and Rodríguez-Iturbe (2002) also studied the probability distributions of clusters of trees of a large number of modeled spatial fields. The eight direct neighbors of a certain tree cell may belong to the same cluster in case they are also occupied by a tree. Additionally, the probability distribution of cluster sizes was also studied separately for two types of simulated fields: those that have an increment in tree density with respect to the previous time step, and those that show a decrease in tree density. In this way the effects of the totally random occurrence of tree death on the cluster size distribution can be tested.

Figure 11.15a shows the frequency of occurrence of clusters of different sizes for trees and seedlings. There is an approximate power law distribution of cluster sizes, except for very large clusters. The significant deviation for very large sizes is likely the result of the small sample size as well as the finite size effects of the domain which with 10 000 cells is obviously small for clusters beyond a few hundred connected trees. The steepness of the power law for the trees-only analysis is larger than for the case of both trees and seedlings. This is caused by the fact that in the latter case the number of large clusters is much higher than in the trees-only analysis, and the number of small clusters is lower.

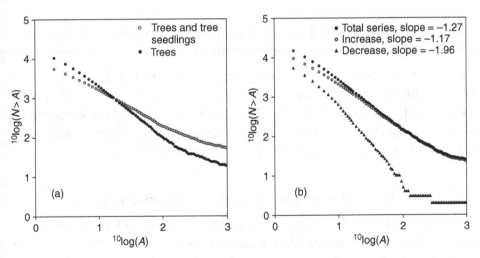

Figure 11.15 (a) Cluster-size distributions of 200 fields of model output taken at random throughout the simulation (the vertical axis gives the logarithm of the number of clusters with size larger than A). (b) Tree cluster-size distribution of model output separated into fields obtained at temporally increasing and temporally decreasing tree coverage. After Van Wijk and Rodriguez-Iturbe (2002).

In Figure 11.15b the cluster-size distribution for trees is plotted, distinguishing between fields that show an increase of trees in time and fields that show a decrease of trees in time. The number of fields obtained when trees decrease with respect to the previous year is smaller than the number of fields obtained with an increase of trees in time. This implies that most of the time, as can also be seen in Figure 11.11, trees were increasing in density, and decreases in tree density occurred in relatively short time spans of severe drought. The distribution of the "decrease" fields shows a higher slope than that of the "increase" fields: the "decrease" fields therefore have relatively more small clusters, caused by the spatially uncorrelated tree deaths occurring at the moment the field is analyzed.

As clustering of trees is observed on a mosaic of grasses and bare soil, the distribution of cluster sizes is an important criterion in the quantification of the observed patterns. Moreover, the possible existence of power laws in these distributions suggests the presence of fractal structures in spatial vegetation patterns. The temporal and spatial fractal structures identified in the model output point toward the possible absence of dominant frequencies – or scales – in the space-time dynamics of water-controlled ecosystems driven by hydrologic fluctuations. This in turn would suggest that these systems tend to self-organize through the interaction of many different processes operating on different time scales (Van Wijk and Rodríguez-Iturbe, 2002).

11.3　A hierarchical model for tree–grass competition driven by interannual rainfall variability

This section studies vegetation competition in water-limited ecosystems by means of hierarchical competition-colonization models that have been extensively used in ecological research. The models are driven by the stress conditions resulting from the annual fluctuations in rainfall. The presentation is taken from Fernandez-Illescas and Rodríguez-Iturbe (2003).

11.3.1　Hierarchical competition-colonization model

Levins and Culver (1971), Nee and May (1992), Tilman (1994), among others, describe and discuss hierarchical competition-colonization models for site occupancy dynamics. Such n-species models divide the landscape into sites, each corresponding to the area potentially occupied by an adult. A strict competitive hierarchy between the species allows them to replace each other and/or colonize empty sites while the mortality of the species causes sites to become unoccupied. The ranking of the species is done according to their competitive abilities, with species 1 being the best competitor and species n being the worst competitor. Resource competition is not explicitly incorporated into the model but the ranking of species from superior to inferior competitor reproduces qualitatively the features of resource competition (Tilman, 1994). Superior competitors landing on sites occupied by inferior competitors immediately displace the inferior species. Inferior competitors cannot invade sites occupied by superior competitors. The previous assumptions lead to the following equation of site occupancy

$$\frac{dp_i}{dt} = c_i p_i \left(1 - \sum_{j=1}^{n} p_j\right) + c_i p_i \sum_{j=i+1}^{n} p_j - p_i \sum_{j=1}^{i-1} c_j p_j - m_i p_i, \qquad (11.5)$$

where i indicates the rank of the species, n is the number of species, p_i is the fraction of sites occupied by the ith species, c_i is the colonization rate or average number of propagules per individual of species i per unit time, and m_i is the mortality rate of the ith species (Levins, 1969; Hastings, 1980; Tilman, 1994). In a spatially discrete scheme the colonization rate is equivalent to the average number of cells successfully colonized per cell of species i per unit time. Propagule dispersal is random across sites and the rate of production of propagules per species i is given by $c_i p_i$. The first term in Eq. (11.5) represents the rate of production of newly colonized sites, while the second gives the colonization of sites occupied by inferior competitors. The third term accounts for loss of sites due to the arrival of superior competitors, while the last term indicates mortality losses or the rate at which occupied sites become vacant.

In the case where $n = 2$, the corresponding equations are

$$\frac{dp_1}{dt} = c_1 p_1 (1 - p_1) - m_1 p_1 \qquad (11.6)$$

$$\frac{dp_2}{dt} = c_2 p_2 (1 - p_1 - p_2) - c_1 p_1 p_2 - m_2 p_2, \qquad (11.7)$$

where m_1 and c_1 are the mortality and colonization rates of the superior competitor whose evolution is unaffected by the inferior competitor which can only colonize empty sites and has parameters m_2 and c_2.

Equilibrium takes place when $\frac{dp_i}{dt} = 0$, which for two species yields the equilibrium abundances as

$$\widehat{p}_1 = 1 - \frac{m_1}{c_1} \qquad (11.8)$$

and

$$\widehat{p}_2 = 1 - \frac{m_2}{c_2} - \widehat{p}_1 \left(1 + \frac{c_1}{c_2} \right). \tag{11.9}$$

The necessary and sufficient conditions for stable coexistence of a superior and inferior competitor in a subdivided habitat (i.e., $\widehat{p}_1, \widehat{p}_2 > 0$) are given by Tilman (1994) as

$$c_1 > m_1 \tag{11.10}$$

and

$$c_2 > \frac{c_1(c_1 + m_2 - m_1)}{m_1}. \tag{11.11}$$

For the case where the two species present have some mortality and a finite colonization rate, it is impossible for them to occupy the entire habitat. Therefore, at equilibrium, some amount of bare soil is always present in the system (Tilman, 1994).

For the case of constant mortality across species, i.e., $m_1 = m_2 = m$, a common assumption in applications of this model (Kinzig et al., 1999), Eqs. (11.10) and (11.11) simplify to

$$c_1 > m \tag{11.12}$$

and

$$c_2 > \frac{c_1^2}{m}. \tag{11.13}$$

Equations (11.12) and (11.13) give the necessary and sufficient conditions for stable coexistence of two competing species of similar mortality in a subdivided habitat under no interannual resource variability and with constant c_i and m_i. Therefore, in order for the stable coexistence of the two species to occur, the colonization rates of the two species should be greater than their mortality rates. Also, there should be an inverse relationship between competitive ability and colonization ability so that the colonization rates per unit time (c_i) increase with species rank. Equations (11.12) and (11.13) imply that $c_2 > c_1$, because $c_1 > m$ and c_1/m is thus larger than 1. This inverse relationship is called the competition-colonization tradeoff and it allows inferior competitors to survive in sites not occupied by superior competitors. Field experiments by Gleeson and Tilman (1990) confirm strong tradeoffs between root and reproductive allocation, species with greater allocation to roots being able to reduce the soil concentration of key nutrients to significantly lower levels. This allocation towards a higher use of resources corresponds with a lower allocation to reproduction and slower dispersal. Thus superior competitors (e.g., $i = 1$) are not, in general, the best colonizers and this provides inferior competitors with the opportunity to persist. Other experimental evidence comes from studies like those of Grubb (1986) and Hanski and Ranta (1983). The competition-colonization tradeoff is thus crucial in permitting stable coexistence between competing species.

The model described by Eq. (11.5) has been extensively used and studied in ecology literature, where it has been a most useful tool in the analysis of competition and biodiversity in spatially structured habitats (Tilman, 1994). In particular, in the analysis by Fernandez-Illescas and Rodríguez-Iturbe (2003) the key resource is soil moisture and the competitive ranking of a species is determined by its use of the soil moisture at the site. Since soil water availability changes stochastically throughout the growing season and also from year to year, rankings based on the use of such a resource will also vary throughout time.

11.3.2 Hydrologic forcing of the model

Fernandez-Illescas and Rodríguez-Iturbe (2003) assign the ranking of the species, as better or worse competitors for water, based on comparison of their dynamic water stresses (Section 4.3) which in turn depends on the hydrologic conditions throughout the years; the species with the lowest $\bar{\theta}$ will be the superior competitor (i.e., rank 1) while the species with the highest $\bar{\theta}$ will be the inferior competitor (i.e., rank 2). As explained before, rank determines the type of equation for the dynamics of site occupancy. Since the ranking of the species may change from year to year as a result of the interannual rainfall variability, so may the corresponding equation describing the dynamics of site occupancy for the given species.

Seed germination has been shown to be positively influenced by rainfall (Marone et al., 2000) but the relationship between rainfall amount and seed germination varies greatly among species. Marone et al. (2000) report different seed germination values for grasses and forbs for seasons with the same total precipitation amount. In the model described by Eq. (11.5), these relative differences are assumed to be a function of ranking of the species in the competitive hierarchy (Fernandez-Illescas and Rodríguez-Iturbe, 2003). For a given level of $\bar{\theta}$, a plant colonizes more if it is the inferior competitor for the resource than if it is the superior competitor. This reflects the tradeoff between root and reproductive allocation and results in two different colonization functions for each species, one applicable if the species is the superior competitor and the other one applicable if the species is the inferior competitor. Since rainfall amount and seed germination are positively related and because for given vegetation and soil characteristics, increasing growing season rainfall corresponds to decreasing dynamic water stress, $\bar{\theta}$, the colonization functions are assumed to decrease linearly with $\bar{\theta}$, reaching zero as $\bar{\theta}$ equals 1, i.e., at the onset of permanent damage to the plant.

Assuming two species, X and Y, species X would have two colonization functions given by

$$c_{X_{sup}} = R(1 - \bar{\theta}_X) \tag{11.14}$$

or

$$c_{X_{inf}} = S(1 - \bar{\theta}_X), \tag{11.15}$$

where $R < S$, $\bar{\theta}_X$ is the dynamic water stress value corresponding to species X and $c_{X_{sup}}$ and $c_{X_{inf}}$ are the colonization parameters of species X given that it is the superior or inferior competitor for water, respectively.

Similarly, species Y would have two colonization functions of the form

$$c_{Y_{sup}} = T(1 - \bar{\theta}_Y) \tag{11.16}$$

or

$$c_{Y_{inf}} = U(1 - \bar{\theta}_Y), \tag{11.17}$$

where $T < U$, and $\bar{\theta}_Y$ is the dynamic water stress value corresponding to species Y. For simplicity, Fernandez-Illescas and Rodríguez-Iturbe (2003) assume mortality rates in Eq. (11.5) as independent of the interannual rainfall fluctuation.

Model inputs

Fernandez-Illescas and Rodríguez-Iturbe (2003) implemented the model using vegetation and climatic data obtained from the Nylsvley savanna in South Africa and southern Texas savanna of La Copita (see Sections 5.1 and 5.2). They further analyzed the available rainfall data of the two sites and the historical statistics were used to generate 10 000-year time series of growing-season rainfall amounts using a Markovian-log normal model

$$y_i = \mu_y + \rho_y(y_{i-1} - \mu_y) + e_i, \tag{11.18}$$

where y_i and y_{i-1} are the generated natural logarithms of the growing-season total rainfall at year i and year $i-1$ respectively, μ_y is the mean, and ρ_y is the lag -1 correlation of the logarithms of the historical rainfall values. The random component e_i follows a log-normal distribution with mean zero and variance $\sigma_y^2(1 - \rho_y^2)$, where σ_y^2 is the variance of the logarithms of the historical precipitation values. The parameters μ_y, σ_y^2, and ρ_y are estimated preserving the statistics of the historical data (Matalas, 1967). The time series of growing-season total rainfall, x_i, are then obtained from the generated logarithms, i.e., $x_i = \exp(y_i)$. Table 11.1 gives the parameters used in the generation of rainfall at both sites.

The rainfall traces generated from such a model are statistically indistinguishable from the historical growing-season rainfall records and are thus used to drive the hierarchical competition-colonization model. In order to account for the dependence of the rate of storm arrivals, λ, on the growing-season total rainfall, Fernandez-Illescas and Rodríguez-Iturbe (2003) divided the historical record of growing-season rainfall into two categories, corresponding to wetter or drier than average conditions.

Based on whether the generated growing-season total rainfall is above or below average, λ is then sampled from a log-normal distribution with either the wet or dry conditional λ statistics as parameters. Subsequently, for every year the growing-season mean depth of rainfall events (e.g., mean daily rainfall during wet days), α, is calculated as

$$\alpha = \frac{x_i}{T_{\text{seas}}\lambda},\qquad(11.19)$$

where x_i corresponds to the generated growing-season total rainfall for each particular year, T_{seas} corresponds to the length of the growing season, and λ corresponds to the generated rate of storm arrivals for the particular year.

The generated time series of α and λ are then used with the vegetation parameters of Nylsvley and La Copita to simulate 10 000-year time series of dynamic water stress values, $\bar{\theta}$, for *Paspalum setaceum* and *Prosopis glandulosa* in La Copita and for *Burkea africana* and *Eragostris pallens* in Nylsvley. The hydrologically driven hierarchical competition-colonization model translates the dynamic water stress values, $\bar{\theta}$, of the two species at La Copita and of the two species at Nylsvley into proportional abundances of the corresponding species. In this manner, Fernandez-Illescas and Rodríguez-Iturbe (2003) study the impact of interannual rainfall variability not just on the overall condition of a single species, i.e., $\bar{\theta}$, but on ecosystem structure.

In general, colonization rates for woody and herbaceous plants are difficult to obtain and are not available at either the La Copita or the Nylsvley site.

Table 11.1 *Rainfall statistics used in the generation of rainfall at La Copita, Texas and at Nylsvley, South Africa. After Fernandez-Illescas and Rodriguez-Iturbe (2003).*

Site	Years	Growing season	μ (mm)	σ (mm)	$\rho(1)$	$\overline{\lambda_{dry}}(day^{-1})$	$\overline{\lambda_{wet}}(day^{-1})$	$\sigma_{\lambda dry}$ (mm)	$\sigma_{\lambda wet}$ (mm)
La Copita	1911–2000	May–Sept	387	163	0.177	0.153	0.214	0.036	0.05
Nylsvley	1917–2000	Sept–April	585	130	0.0	0.21	0.256	0.048	0.045

Even when measured at a given site, large variability in seed germination values among species has been reported (Marone et al., 2000). However, measurements of the mean proportional abundances of trees and grasses are available at these two sites. At La Copita, Archer et al. (1988) report an average tree coverage of 20% while at Nylsvley, Scholes and Walker (1993) indicate that trees and grasses have areal coverages of 32% and 30% respectively. The colonization intercepts in the colonization-dynamic stress functions were chosen so that the modeled average proportional abundances match observations at the two sites (Fernandez-Illescas and Rodríguez-Iturbe, 2003). They are shown in Table 11.2 for all the cases considered here.

As mentioned before, applications of hierarchical competition-colonization models frequently assume constant mortality rates across species (Kinzig et al., 1999). For simplicity, the mortality of the two species at each of the two sites is assumed equal to 0.1, since this is a commonly used value in the literature (Tilman, 1994).

11.3.3 Impact of stochasticity of climate on species abundances

In order to investigate the effect of precipitation fluctuations on vegetation composition at La Copita and at Nylsvley, Fernandez-Illescas and Rodríguez-Iturbe (2003a) also implemented the previous model with interannual and without interannual rainfall variability. Interannual rainfall variability is represented by inputting into the model the generated time series of dynamic water stress, $\overline{\theta}$, for each species. The lack of interannual rainfall variability is simulated by inputting into the model a temporally constant $\overline{\theta}$ value for each species computed from the mean observed rainfall descriptors at each site

Table 11.2 *Colonization intercepts characterizing the two species at La Copita, Texas, and the two species at Nylsvley, South Africa. After Fernandez-Illescas and Rodriguez-Iturbe (2003).*

Species	Colonization intercept (year^{-1})
Prosopis glandulosa (rank 1)	0.5
Paspalum setaceum (rank 1)	0.6
Prosopis glandulosa (rank 2)	7.6
Paspalum setaceum (rank 2)	7.9
Burkea africana (rank 1)	0.3
Eragostris pallens (rank 1)	0.64
Burkea africana (rank 2)	0.67
Eragostris pallens (rank 2)	0.7

(i.e., $\alpha = 1.42\,\text{cm}$, $\lambda = 0.178\,\text{day}^{-1}$ at La Copita and $\alpha = 1.93\,\text{cm}$, $\lambda = 0.23$ day^{-1} at Nylsvley).

Figure 11.16a shows the modeled proportional abundances of *Prosopis glandulosa* and *Paspalum setaceum* with and without interannual rainfall variability for 10 000 years. Figure 11.16b shows the same for *Burkea africana* and *Eragostris pallens* (Fernandez-Illescas and Rodríguez-Iturbe, 2003). It is clear that interannual rainfall variability has a profound effect on the time series of proportional abundances for both species at both sites. The effect is greater at La Copita since this site has larger interannual rainfall fluctuations than Nylsvley (i.e., coefficient of variation of growing-season rainfall of 0.42 vs 0.22). The larger colonization intercepts at La Copita (and thus the greater

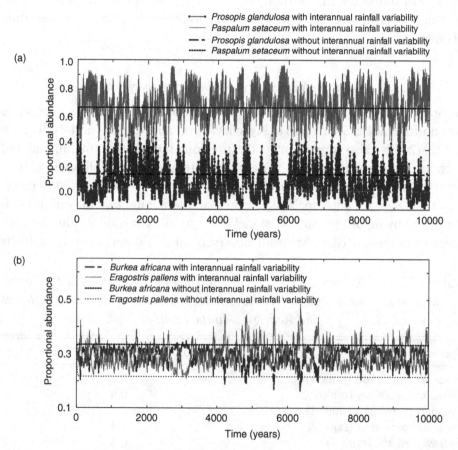

Figure 11.16 Modeled proportional abundances with and without interannual rainfall variability for (a) *Paspalum setaceum* and *Prosopis glandulosa* and (b) *Eragostris pallens* and *Burkea africana*. After Fernandez-Illescas and Rodriguez-Iturbe (2003).

slopes of the colonization functions at this site) are also responsible for the enhanced sensitivity to interannual rainfall fluctuations.

The type of coexistence predicted by the hydrologically driven hierarchical competition-colonization model in Figure 11.16a is in accordance with observations made at La Copita, where the proportions of the two species at this site have been recorded to be highly sensitive to abnormally dry and wet periods (Archer et al., 1988). The mean modeled proportional abundance of *Prosopis glandulosa* is 18%, which matches well the observed areal coverage reported by Archer et al. (1988). Although a different set of colonization intercepts for a given site results in proportional abundances with mean values different from the observed ones, the qualitative effects of interannual rainfall fluctuations were found to be robust to changes in colonization parameters. The modeled proportional abundances of *Burkea africana* and *Eragostris pallens* shown in Figure 11.16b have mean values of 30% and 29% respectively, matching well the areal coverage for both species reported in Scholes and Walker (1993). For both La Copita and Nylsvley, and for the two functionally different species considered in each case, the mean proportional abundances under interannually fluctuating rainfall conditions are significantly different from those resulting under constant hydrologic inputs. This results from the highly nonlinear dynamics operating in the hierarchical competition-colonization model.

Impact of rainfall statistics on species abundances

Figures 11.17, 11.18a, and 11.19a show quartiles for the proportional abundances for *Paspalum setaceum* and *Prosopis glandulosa* calculated for a range of mean growing-season total rainfall μ, standard deviation σ, and lag -1 coefficients $\rho(1)$, respectively. Figures 11.18b and 11.19b show those of *Eragostris pallens* and *Burkea africana* calculated for a range of standard deviations, σ, and lag -1 coefficients, $\rho(1)$, respectively. Each parameter is independently varied while the other two parameters are fixed at the observed values. For each combination of μ, σ, and $\rho(1)$, the values of growing-season total rainfall, λ, α, and $\bar{\theta}$ were generated by Fernandez-Illescas and Rodríguez-Iturbe (2003) as previously described.

As expected, relative levels of proportional abundances for trees and grasses are highly sensitive to mean growing-season total rainfall amounts. Figure 11.17 shows that for amounts below 300 mm of mean growing-season total rainfall, trees fail to maintain a significant presence at the site throughout the 10 000-year simulation. However, tree abundances rise sharply for rainfall amounts above 350 mm (at the expense of grasses) and tend to dominate the vegetative cover at the site when the rainfall is above 475 mm. Thus, when maintaining the observed rainfall variance and correlation, there appears to

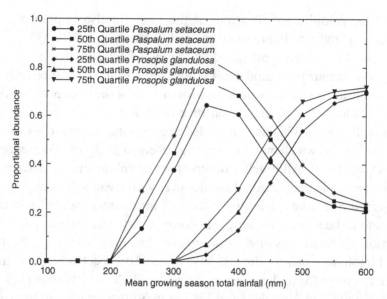

Figure 11.17 The 25th, 50th, and 75th quartiles of the proportional abundances of *Paspalum setaceum* and *Prosopis glandulosa* for a range of mean values of growing-season total rainfall. The rainfall interannual standard deviation and the lag − 1 correlation coefficient are kept constant. After Fernandez-Illescas and Rodriguez-Iturbe (2003).

exist a relatively sharp transition towards the disappearance of trees (e.g., *Prosopis glandulosa*) for mean growing-season rainfall values smaller than those observed at this site (e.g., 387 mm). Although not shown, very similar results are obtained for the proportional abundances of *Burkea africana* and *Eragostris pallens* in the Nylsvley savanna.

Quite interesting also is the observed sensitivity to interannual rainfall standard deviation (Figure 11.18a, b). Grasses benefit from the introduction of realistic amounts of interannual rainfall variability at both sites. Raising the standard deviation of the interannual rainfall amounts from 0 (i.e., no variation in annual rainfall) to the observed value of 130 mm in Nylsvley or 163 mm in La Copita leads to a reduction in the proportional abundances of *Burkea africana* and *Prosopis glandulosa* and a corresponding increase in the abundances of *Eragostris pallens* and *Paspalum setaceum*. This effect, however, is more pronounced in Nylsvley where the transition between tree and grass domination takes place around the observed rainfall standard deviation. Not surprisingly, increasing interannual rainfall variability also increases the observed range of proportional abundances for both species at both sites. This effect is more pronounced in La Copita than in Nylsvley due to the higher colonization intercepts used at this site which result in an enhanced sensitivity

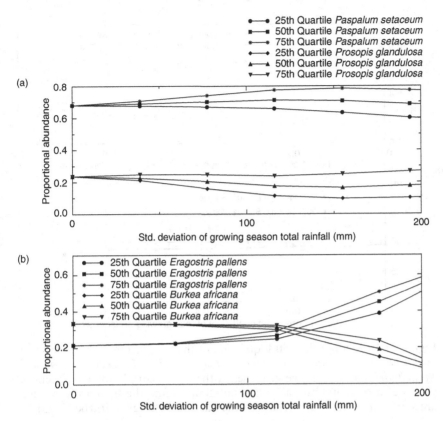

Figure 11.18 The 25th, 50th, and 75th quartiles of the proportional abundances of (a) *Paspalum setaceum* and *Prosopis glandulosa* and (b) *Eragostris pallens* and *Burkea africana* for a range of values of the standard deviation of growing-season total rainfall. The rainfall mean and the lag − 1 correlation coefficient are kept constant. After Fernandez-Illescas and Rodriguez-Iturbe (2003).

of the proportional abundances to interannual rainfall fluctuations. Also, low standard deviation values drive the proportional abundances of the two species at the two sites towards the equilibrium abundance values predicted by Eqs. (11.8) and (11.9), i.e., those corresponding to a hierarchical competition-colonization model without interannual rainfall variability.

Changes in the lag − 1 correlation coefficient at the La Copita site suggest that trees benefit from larger autocorrelation in growing-season total rainfall amounts. Figure 11.19a shows that median *Prosopis glandulosa* abundances associated with an uncorrelated white noise model for growing-season total rainfall are lower than those simulated for a Markovian model with larger lag − 1 correlation coefficients. In contrast, proportional abundances at the

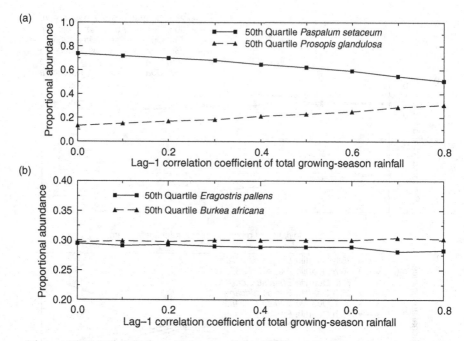

Figure 11.19 Median (i.e., 50th quartile) of the proportional abundances of (a) *Paspalum setaceum* and *Prosopis glandulosa* and (b) *Eragostris pallens* and *Burkea africana* for a range of values of the lag−1 correlation coefficient of growing-season total rainfall. The rainfall mean and the standard deviation are kept constant. After Fernandez-Illescas and Rodriguez-Iturbe (2003).

Nylsvley site (Figure 11.19b) show little sensitivity to changes in the lag − 1 correlation coefficient. This is likely to be due to smaller rainfall fluctuations at this site and thus a decreased sensitivity to the manner in which these fluctuations are correlated in time.

Dynamics of temporal evolution

Figure 11.20a, b shows spectral analyses of the temporal evolution of the proportions of *Paspalum setaceum* and *Prosopis glandulosa* and of the proportions of *Burkea africana* and *Eragostris pallens*. To avoid transient effects, the first 2000 years of the simulated time series were discarded from the analysis (Fernandez-Illescas and Rodríguez-Iturbe, 2003). The power spectral densities of the proportional abundances of the four species display a power law structure over an extended frequency range. This type of spectral structure, i.e., $S_x(f) \propto 1/f^D$, is a classical signature of self-affinity (see Section 11.2.3). The power spectra of the model inputs at both sites, namely growing-season total rainfall and the corresponding dynamic water stress time series, do not

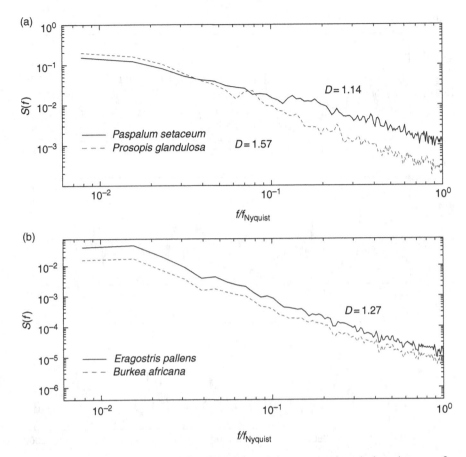

Figure 11.20 Power spectral density plots of the proportional abundances of (a) *Paspalum setaceum* and *Prosopis glandulosa* and (b) *Eragostris pallens* and *Burkea africana*. After Fernandez-Illescas and Rodriguez-Iturbe (2003).

show a power law relation in their spectra (Fernandez-Illescas and Rodríguez-Iturbe, 2003). Thus, the observed fractal temporal structure of the species' abundances results from the internal competition-colonization dynamics driven by randomly fluctuating hydrologic inputs and not from the structure of the inputs themselves.

Self-affinity means statistical (or deterministic) scale invariance (see Eq. (11.3) and following discussion). A self-affine process remains statistically unchanged when rescaled by different scaling factors along different axes, e.g., time and proportional abundance. Figure 11.21a shows 400 years of the original modeled time series of *Prosopis glandulosa* at La Copita. The Hurst exponent H for *Prosopis glandulosa* at this site is obtained from Figure 11.20a as 0.289. These 400 years are rescaled by $\gamma = 2$ along the time axis and by

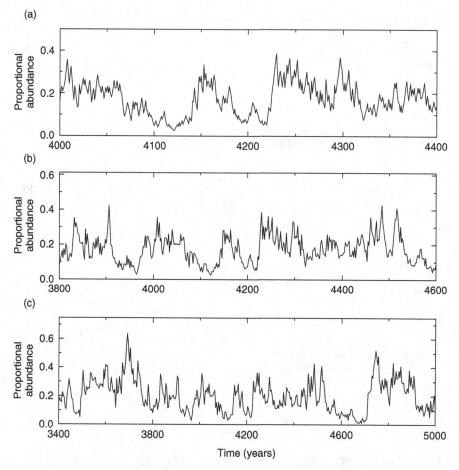

Figure 11.21 (a) 400 years of the original modeled proportional abundances of *Prosopis glandulosa*. (b) Properly rescaled 800 years of the modeled proportional abundances of *Prosopis glandulosa*. (c) Properly rescaled 1600 years of the modeled proportional abundances of *Prosopis glandulosa* at La Copita, Texas. Hurst exponent $H = 0.289$. After Fernandez-Illescas and Rodriguez-Iturbe (2003).

$\gamma^H = 1.22$ along the proportional abundance axis and shown in Figure 11.21b. The original 400 years are rescaled by $\gamma = 4$ along the time axis and by $\gamma^H = 1.49$ along the proportional abundance axis and shown in 11.21c. The 400, 800 and 1600 years shown in Figure 11.21a–c highlight the self-affinity of the time series of proportional abundances of *Prosopis glandulosa* at this site. The series are in fact statistically identical. As observed by Fernandez-Illescas and Rodríguez-Iturbe (2003), this is a model prediction that would need to be checked through analysis of long-term vegetation data which will need to be built by using proxy variables. For time scales of decades to millennia, pollen data are the main source of information that can be used to reconstruct records of past vegetation (Webb III et al., 1978; Clark, 1996). As already noted by Van Wijk and Rodríguez-Iturbe (2002), the fact that power law spectra contain a broad range of frequencies without any dominant cyclic component means that the ecosystem avoids getting locked into any dominant frequency. Similar to other biologic dynamics (e.g., heart rates), too much order, like that displayed in rhythmic patterns, is a sign of danger, as will be a complete disorder in which all frequencies are equally strong (e.g., flat power spectrum). As pointed out by Sole and Goodwin (2000): "The state of health is the normal biological attractor that combines both order and chaos." Although the hierarchical competition-colonization model is an extremely simplistic representation of very complicated dynamics, the output it yields when driven by fluctuating hydrologic conditions reflects, through the power spectra of the proportional abundances, the interaction of many different processes operating on different time scales.

The question may be asked if a single species under fluctuating hydrologic conditions will also display a power law spectrum in its proportional abundance. To investigate this, Fernandez-Illescas and Rodríguez-Iturbe (2003) ran the competition-colonization model for trees and grasses, separately, for both La Copita and Nylsvley. The colonization functions used in each case correspond to those described before for the two-species model when the species in question has rank 1. Since only one species is evolving, no change in the rank throughout time is present. Colonization only takes place on the bare soil of the region and the growing-season rainfall simulation is implemented with the historic parameters at each site. The results were similar in both cases and showed that a sizeable range of frequencies does not follow power law behavior, indicating the importance of species interactions in the spectra shown in Figure 11.20a,b. This is important because it suggests that species diversity is a crucial factor in the normal biological attractor that "combines order and chaos" (Sole and Goodwin, 2000). In the case of the single species, at least half of the spectrum is flat reflecting a more disordered type of evolutionary dynamics.

The dynamical evolution of *Paspalum setaceum* and *Prosopis glandulosa* as well as that of *Eragostris pallens* and *Burkea africana* exhibit high and negative cross-correlation coefficients (Figure 11.22a,b). Figure 11.23a,b shows the coherence function for the two species at La Copita and the two species at Nylsvley, respectively. Any stationary time series may be decomposed into uncorrelated frequency components, and the coherence between two series identifies the proportion of the power at frequency f in either series

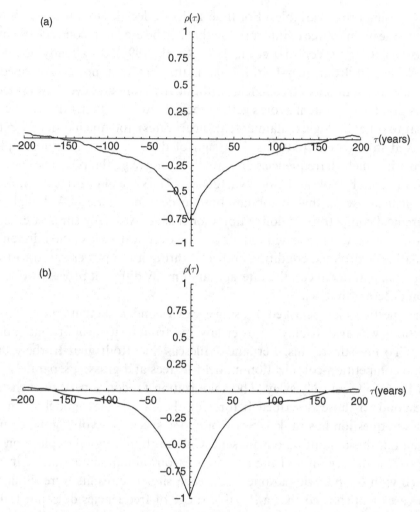

Figure 11.22 (a) Cross-correlation function of the proportional abundances of (a) *Prosopis glandulosa* and *Paspalum setaceum*; and (b) *Eragostris pallens* and *Burkea africana*. After Fernandez-Illescas and Rodriguez-Iturbe (2003).

that can be explained by its linear regression on the component of the other series centered at the same frequency f (Koopmans, 1995). It can be interpreted as a correlation coefficient between the frequency components centered at the same frequency, f, and provides a powerful tool for finding out which frequencies are responsible for any observed correlation between two series. The calculated coherence between the abundances of *Paspalum setaceum* and *Prosopis glandulosa* is small at high frequencies but rises sharply below 0.05 year^{-1} (see Figure 11.23a). This indicates that the majority of observed

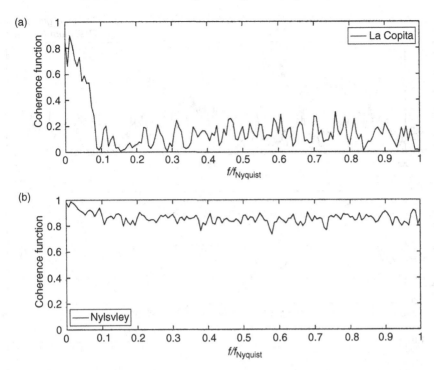

Figure 11.23 Coherence function of the proportional abundances of (a) *Prosopis glandulosa* and *Paspalum setaceum* and (b) *Eragostris pallens* and *Burkea africana*. After Fernandez-Illescas and Rodriguez-Iturbe (2003).

correlation between the abundance time series results from their close coupling at low frequencies. On the other hand, the calculated coherence between *Eragostris pallens* and *Burkea africana* shown in Figure 11.23b is high at all frequencies.

Figure 11.24a–c displays the coherence results in La Copita in a more intuitive manner. Figure 11.24a shows 1000 years of the proportional abundances of trees and grasses and Figure 11.24b shows a 150-year portion of the same series. The evident cross-correlation of the longer series is not as apparent in the shorter one and indeed comparison of the cross-correlation functions for both time series confirms this (not shown). In addition, filtering out high-frequency fluctuations isolates the highly cross-correlated low-frequency components of both time series. For this, a symmetric running average of 125 years was applied to both series by Fernandez-Illescas and Rodríguez-Iturbe (2003). The results in Figure 11.24c show the extremely high correlation among frequency components centered below 0.05 year^{-1}. Qualitatively similar results were obtained for different colonization functions.

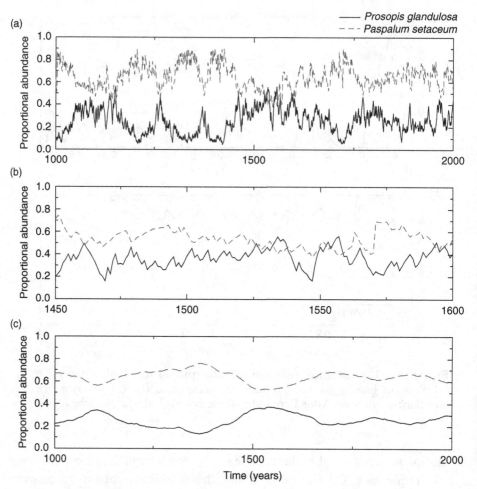

Figure 11.24 Proportional abundances of *Prosopis glandulosa* and *Paspalum setaceum* (a) for 1000 years of the simulated 10 000 years, (b) for 150 years of the simulated 10 000 years, (c) after a symmetric running average of 125 years is applied to the series shown in Figure 11.24a. After Fernandez-Illescas and Rodriguez-Iturbe (2003).

11.3.4 Is there a preferred state of proportional abundances?

Fernandez-Illescas and Rodríguez-Iturbe (2003) used information entropy to link the fluctuations in vegetation structure of a savanna ecosystem to the corresponding climatic forcings. The Shannon entropy H is defined as

$$H = -\sum_i \sum_j p(x_i, y_j) \log[p(x_i, y_j)], \tag{11.20}$$

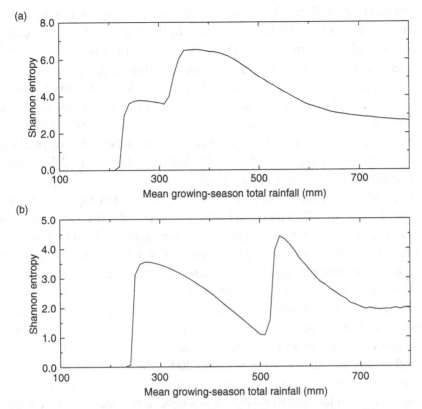

Figure 11.25 Change in Shannon entropy with amount of growing-season total rainfall for (a) La Copita, Texas and (b) Nylsvley, South Africa. The rainfall standard deviation and the lag−1 correlation coefficient are kept constant. After Fernandez-Illescas and Rodriguez-Iturbe (2003).

where (x_i, y_j) are the proportional abundance pairs of grasses and trees respectively and $p(x_i, y_j)$ is their joint probability. The subscripts i and j indicate that the range of the proportional abundances, $0 \leq x \leq 1$ and $0 \leq y \leq 1$, has been partitioned to a given resolution. H is a measure of the order-disorder present in the interannual transitions of the tree–grass composition when the system is driven by year-to-year changing rainfall.

For various levels of mean growing-season rainfall, Figure 11.25a,b shows the Shannon entropy of the binary grass–tree La Copita and Nylsvley ecosystems. Other rainfall statistics (i.e., standard deviation and lag−1 correlation coefficient) were maintained at levels estimated from observed rainfall at both sites. For very small rainfall amounts one expects the tree component to disappear followed by a corresponding decrease of grasses. This makes the entropy equal to zero for a mean annual growing-season rainfall below a critical threshold corresponding to the minimum rainfall required to sustain grass growth (e.g., 225 mm in the two cases considered by Fernandez-Illescas and Rodríguez-Iturbe, 2003). As shown in Figure 11.25a,b the entropy increases sharply once rainfall is above the critical threshold up to a value that reflects a maximum diversity in the possible abundance of grasses. As the rainfall increases above that corresponding to this local entropy maximum, there is a tendency for more and more grasses to be present in the system. As a consequence, the dynamic diversity in the composition decreases and so does the entropy. This is especially marked for the case of Nylsvley throughout a range of growing-season rainfall amounts between 300 mm and 500 mm per year. For rainfall above a certain value the entropy increases again until it reaches a global maximum at approximately the mean growing-season rainfall observed at each site. This increase is due to the proliferation of trees, which add dynamic diversity to the composition of the system. The previous feature suggests that under interannual rainfall fluctuations, water-controlled ecosystems tend to self-organize in a manner that reflects a maximum in the richness of possible dynamical responses. It is not clear whether the above feature results from an adaptation that maximizes the system's capacity for survival or is a generic result from a certain type of interaction dictated by the hierarchical competition-colonization model when driven by a random hydrologic input (Fernandez-Illescas and Rodríguez-Iturbe, 2003). More research is needed to corroborate the possible correspondence between maximum entropy and mean growing-season rainfall and to verify if it is a general feature of water-controlled ecosystems. Similar types of entropy maximization have been observed to operate in other ecological systems (e.g., Sole and Miramontes, 1995) and may help to clarify the role of stochastic climate in determining vegetation cover within water-limited ecosystems (Fernandez-Illescas and Rodríguez-Iturbe, 2003).

11.4 Impact of interannual rainfall variability on temporal and spatial vegetation patterns

The hydrologically driven hierarchical competition-colonization model of the previous section was extended by Fernandez-Illescas and Rodríguez-Iturbe (2004) to investigate in a spatially explicit manner the effect of temporal (i.e., interannual rainfall fluctuations) and spatial (i.e., local dispersal) processes on vegetation patterns. They implemented a cellular automaton version of the hierarchical competition-colonization model and applied it to the case of La Copita in southern Texas. The habitat is considered as a square domain consisting of 256 by 256 cells, each being the size of the area occupied by an adult tree. Each site can contain one individual tree, be occupied by grasses, or be empty. An individual can disperse to a square set of neighbors, of side

$2r + 1$ (where r is dispersal range), centered on itself. Depending on the dispersal range chosen for each species, the set of neighboring sites can be as small as the immediate surrounding square of eight sites ($r = 1$) or as large as the entire habitat (global dispersal). Seed dispersal is assumed to be the result of passive agents, primarily wind. Grasses are assumed to disperse globally because of their more diffuse seed dispersal, while trees are only allowed to disperse locally.

As in Fernandez-Illescas and Rodríguez-Iturbe (2003), comparison of $\bar{\theta}$ values among species establishes the species' competitive ranking. For the two-species case, the species with the lower $\bar{\theta}$ is assumed to be the superior competitor and the species with the higher $\bar{\theta}$ becomes the inferior competitor. So as to preserve the colonization–competition tradeoff, colonization values are defined through two colonization-dynamic stress functions for each species. As in Section 11.3, these functions have the form $c_x = a(1 - \bar{\theta}_x)$, with a value of the parameter a for the case where the species is the superior competitor, i.e., $x = 1$, and a different value of a for the case where the species is the inferior competitor, i.e., $x = 2$.

As discussed by Fernandez-Illescas and Rodríguez-Iturbe (2004), the cellular automaton version of Eqs. (11.6) and (11.7) can be easily described by a set of rules defining the site transitions from one time step to the next. In a spatially explicit context, the colonization rate is equivalent to the average number of cells that are successfully colonized per cell of species x per unit time. At each time step, an individual sends out a set of colonists to randomly chosen neighboring sites within its characteristic dispersal area. If the sites are empty or occupied by an inferior competitor, the propagules establish, otherwise they die. If the colonization rate c_x is less than one, the individual plant sends out a single colonist with probability equal to the colonization rate c_x. If the colonization rate c_x is greater than one, the individual plants send out n colonists where n is defined as the greatest integer less than c_x. An additional colonist is then dispersed with probability $(c_x - n)$. Death is modeled as a random process in space with probability at each time step equal to the mortality rate m_x. Periodic boundary conditions are assumed throughout the simulation to avoid edge effects (Fernandez-Illescas and Rodríguez-Iturbe, 2004).

11.4.1 Dynamics of species abundances

Temporal evolution of species

To investigate the effect of the *Prosopis glandulosa* dispersal range on the competitive interaction between this plant and *Paspalum setaceum* at La Copita, Fernandez-Illescas and Rodríguez-Iturbe (2004) ran the spatially explicit competition model with and without interannual rainfall fluctuations. Interannual rainfall variability is represented by forcing the competition-colonization model with the 10 000-year time series of dynamic water stress, $\bar{\theta}$, for each species. Homogeneous rainfall is simulated by using as model input a temporally constant $\bar{\theta}$ value for each species derived from the long-term mean observed rainfall at the site (i.e., $\alpha = 1.42\,\text{cm}$, $\lambda = 0.178\,\text{day}^{-1}$).

Figure 11.26 shows the change in mean proportional abundance obtained by Fernandez-Illescas and Rodríguez-Iturbe (2004) during 10 000-year simulations

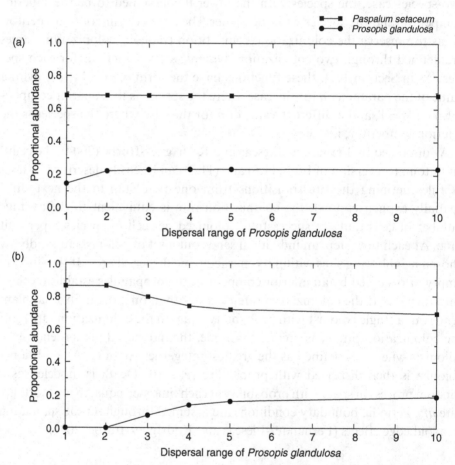

Figure 11.26 Mean proportional abundances for *Paspalum setaceum* and *Prosopis glandulosa* for different dispersal ranges of *Prosopis glandulosa* (a) without interannual rainfall variability and (b) with interannual rainfall variability. Interannual rainfall variability is generated by a Markovian log-normal model with parameters indicated in Table 11.1 for the case of La Copita. The lack of interannual rainfall variability is characterized by a time-invariant mean depth of rainfall events. After Fernandez-Illescas and Rodriguez-Iturbe (2004).

of both *Prosopis glandulosa* and *Paspalum setaceum* for different *Prosopis glandulosa* seedling dispersal ranges. *Paspalum setaceum* is assumed to disperse globally in both graphs, and the first 2000 years of the modeled time series of vegetation abundances have been removed from the analysis to avoid transient effects. Comparison of Figures 11.26a and 11.26b suggests that the impact of local dispersion is sensitive to the temporal variability of rainfall used to drive the competition-colonization model. In contrast to the homogeneous rainfall case, the inclusion of interannual rainfall variability causes *Prosopis glandulosa* to become extinct for dispersal lengths less than two. Furthermore, in the absence of interannual rainfall variability, dispersal ranges greater than one yield results virtually indistinguishable from those of the global dispersion case. The interannual rainfall variability case, however, shows sensitivity for dispersal ranges up to ten. This suggests (Fernandez-Illescas and Rodríguez-Iturbe, 2004) that the impact of local spatial processes (i.e., local dispersion) is limited to very strong local interactions for the homogeneous rainfall case (i.e., dispersal ranges of one or two) but is enhanced by the presence of rainfall variability. Moreover, if climate stochasticity is included, the proportional abundance of the globally dispersing species – *Paspalum setaceum* – is impacted by the dispersal properties of *Prosopis glandulosa* (Figure 11.26b). This occurs because, unlike the homogeneous rainfall case, where *Paspalum setaceum* is always dominant and therefore insensitive to the presence of *Prosopis glandulosa*, interannual rainfall variability produces abnormally wet years where *Prosopis glandulosa* has a competitive advantage and can colonize sites occupied by *Paspalum setaceum*. This switching of competitive rankings on a yearly basis effectively couples the fates of both species to each other.

The dominance of *Prosopis glandulosa* in wet years is accompanied by a reduction in colonization ability in accordance with the competition–colonization tradeoff mechanism (Tilman, 1994). Because dominance is an advantage only when one species is attempting to colonize a location occupied by an inferior competitor, and the probability of such cross-species colonization attempts is reduced by the clustering associated with local dispersal, the advantages of being dominant (i.e., ability to colonize inferior competitors) are outweighed by the disadvantages (i.e., lower colonization ability). Consequently, the introduction of interannual rainfall variability is, on average, detrimental to trees at the site, and the negative impact of reducing dispersal ranges from global to local distances is enhanced when the reduction is accompanied by interannual rainfall fluctuations (Fernandez-Illescas and Rodríguez-Iturbe, 2004).

Since under conditions of interannual rainfall variability *Prosopis glandulosa* needs to maintain a dispersal range of at least three to survive

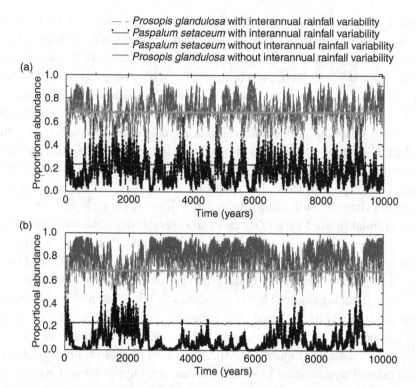

Figure 11.27 Modeled proportional abundances for *Prosopis glandulosa* (gray lines) and *Paspalum setaceum* (black lines) with and without (horizontal lines) interannual rainfall variability if (a) both species disperse globally and if (b) *Paspalum setaceum* disperses globally and *Prosopis glandulosa* disperses locally with dispersal range of three. After Fernandez-Illescas and Rodriguez-Iturbe (2004).

(Figure 11.26b), such a range was assigned to *Prosopis glandulosa* unless otherwise specified. Figure 11.27a, b provides a visual description of the impacts described above by plotting the 10 000-year time series of proportional abundances of *Prosopis glandulosa* and *Paspalum setaceum* with and without interannual rainfall fluctuations for the case where both disperse globally (Figure 11.27a) and for the case where *Prosopis glandulosa* disperses locally (i.e., dispersal distance of three) and *Paspalum setaceum* disperses globally (Figure 11.27b). *Prosopis glandulosa* has less of a presence in the area when its dispersal ability is local and this allows *Paspalum setaceum* to increase its abundance at La Copita.

Spatial evolution of species

Local dispersal is also associated with spatial patterns in site occupation that differ significantly from those patterns resulting from hierarchical

(a)

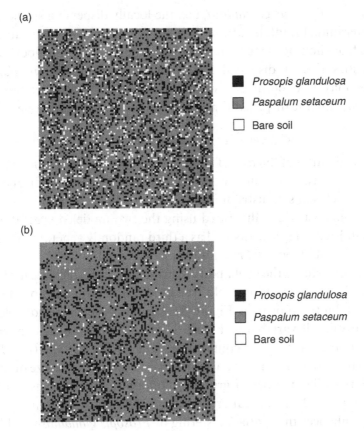

Figure 11.28 Modeled vegetation field at the same time step, (a) without interannual rainfall variability and (b) with interannual rainfall variability. In both cases the proportional abundance of *Prosopis glandulosa* is 23%. For clarity, only a portion of the 256 by 256 field is shown. *Paspalum setaceum* disperses globally and *Prosopis glandulosa* disperses locally with a dispersal range of three. After Fernandez-Illescas and Rodriguez-Iturbe (2004).

competition-colonization models lacking local spatial interaction between cells. To investigate the effect of interannual rainfall variability on the formation of these patterns, Fernandez-Illescas and Rodríguez-Iturbe (2004) focus on the spatial patterns of *Prosopis glandulosa* with a local dispersal range of three.

Figure 11.28a,b shows typical snapshots of two modeled vegetation fields derived without and with interannual rainfall variability. The particular case shown in Figure 11.28b corresponds to simulations having the same proportional abundance of *Prosopis glandulosa*, namely 23% (Fernandez-Illescas and Rodríguez-Iturbe, 2004). In this way, the effect of interannual rainfall variability in the observed spatial patterns can be isolated from any density effects. Visual inspection of Figures 11.28a and 11.28b shows significant clustering in the patterns of *Prosopis glandulosa*, i.e., the locally dispersing species, for the case of interannual rainfall variability. Clustering is not apparent in the field generated from homogeneous rainfall forcings despite the same local dispersal characteristics (i.e., a dispersal length of three) for *Prosopis glandulosa*. Clustering may also occur in the homogeneous rainfall case; however, it requires reducing the dispersal range of *Prosopis glandulosa* to one.

Cluster size probability distribution

A further evaluation of the role of rainfall variability on the spatial patterns of vegetation was carried out by Fernandez-Illescas and Rodríguez-Iturbe (2003b) using cluster size distribution, variance scaling, and multi-scale mosaic structure. The results are illustrated using the two modeled vegetation fields presented in Figure 11.28a,b as well as a third randomly generated field with a proportional abundance of *Prosopis glandulosa* equal to 23%.

Figure 11.29a shows the probability distribution of cluster sizes for the three fields considered: the random field, the modeled field under no interannual rainfall variability (shown in Figure 11.28a) and the modeled field under interannual rainfall variability (Figure 11.28b). Two cells are considered to be in the same cluster if they contain the same landcover type (i.e., *Prosopis glandulosa* in this case) and share a common grid cell side. The size or mass of a cluster refers to the number of grid cells contained within a cluster (Stauffer and Aharony, 1992). It is clear from Figure 11.29a that interannual rainfall variability enhances the spatial clustering of *Prosopis glandulosa* and leads to generally larger clusters relative to the random case. The impact of localized dispersal on cluster size distribution is significantly less profound without the presence of interannual rainfall fluctuations. As shown in Figure 11.29b when there is no interannual rainfall variability, clustering becomes more significant if the dispersal range of *Prosopis glandulosa* is decreased to one. The effect of interannual rainfall variability is thus to make a dispersal range of three produce spatial organization that is consistent with the clustering observed in the homogeneous rainfall case for a dispersal length of one. Results presented in Figure 11.29 are typical for all those time steps in Figure 11.27b where *Prosopis glandulosa* has the same proportional abundance for both cases; namely, with and without interannual rainfall variability.

Self-similarity of vegetation patterns

The scaling of higher-order spatial statistical moments was used by Fernandez-Illescas and Rodríguez-Iturbe (2004) as a measure of statistical spatial organization. To examine this in the generated vegetation patterns, nonoverlapping windows of different sizes were superimposed on the three

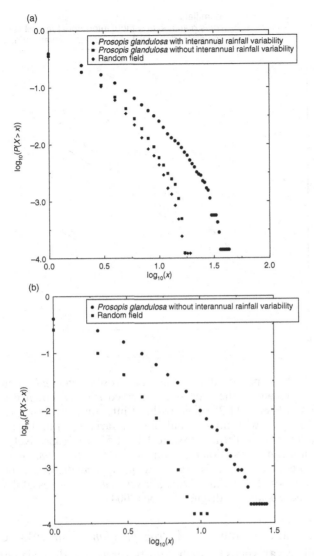

Figure 11.29 (a) Probability distribution of cluster sizes of *Prosopis glandulosa* for the modeled vegetation field without interannual rainfall variability (shown in Figure 11.28a), for the modeled vegetation field with interannual rainfall variability (shown in Figure 11.28b), and for a random field. All three fields have a mean proportional abundance of *Prosopis glandulosa* equal to 23%. For the modeled field, *Prosopis glandulosa* disperses locally with a dispersal range of three while *Paspalum setaceum* disperses globally. (b) Probability distribution of cluster sizes for *Prosopis glandulosa* for a modeled vegetation field without interannual rainfall variability and for a random field. Both fields have a mean proportional abundance of *Prosopis glandulosa* equal to 15%. For the modeled field, *Prosopis glandulosa* disperses locally with a dispersal range of one while *Paspalum setaceum* disperses globally. After Fernandez-Illescas and Rodriguez-Iturbe (2004).

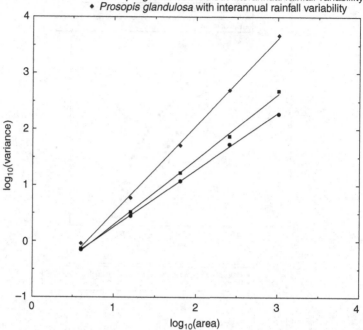

Figure 11.30 Variance of the number of *Prosopis glandulosa* individuals versus area over which the process is observed for the modeled vegetation fields shown in Figure 11.28a (i.e., without interannual rainfall variability), Figure 11.28b (i.e., with interannual rainfall variability) and for a random field. The slopes of the fitting lines are 1.11, 1.55, 1.0 respectively. All three fields have a mean proportional abundance of *Prosopis glandulosa* equal to 23%. For the modeled fields, *Prosopis glandulosa* disperses locally with dispersal range of three while *Paspalum setaceum* disperses globally. After Fernandez-Illescas and Rodriguez-Iturbe (2004).

fields considered and the number of *Prosopis glandulosa* within each window was counted. The variances of such counts versus window size are plotted in Figure 11.30. For each scale, edge effects were removed by applying a buffer zone of appropriate width within which plants are not analyzed. Results in Figure 11.30 indicate a power law relationship over three logarithm scales for the variance of the number of trees versus scale (i.e., window size) for each of the three cases (i.e., the random case, the case without interannual rainfall variability and the case with interannual rainfall variability). As expected, a purely random field has a slope of one. The slopes of the other two lines are 1.11 and 1.55 respectively. Both rainfall cases demonstrate power law behavior and therefore second-order self-similarity in the number of *Prosopis glandulosa* present in the field with a slope greater than the slope of the random case. Like the cluster size distribution results presented before, the deviation from a random case is much more pronounced in the case of a field with interannual rainfall variability than in the homogeneous rainfall case (slope of 1.55 versus slope of 1.11). Interannual rainfall variability greatly enhances the spatial scaling that differs significantly from the random case.

Multi-scale mosaic structure

As another indicator of vegetation spatial structure, following the analysis of Milne (1992), Fernandez-Illescas and Rodríguez-Iturbe (2004) analyzed the scaling of the number of *Prosopis glandulosa* individuals in windows of side L centered on a given pixel occupied by *Prosopis glandulosa*. Figure 11.31 shows the probability density function of the number of *Prosopis glandulosa* individuals, m, found within a window of length L, $p(m, L)$, for the three vegetation fields considered, i.e., random, with and without interannual rainfall

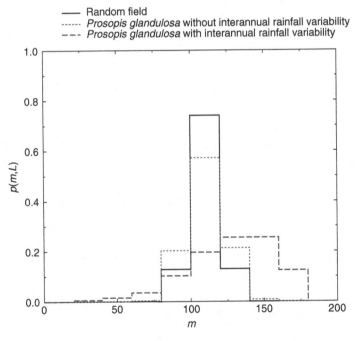

Figure 11.31 Probability density function of the number of *Prosopis glandulosa* individuals, m, within a window of length $L = 15$ for the vegetation fields shown in Figure 11.28a (i.e., without interannual rainfall variability), Figure 11.28b (i.e., with interannual rainfall variability) and for a random field. All three fields have a mean proportional abundance of *Prosopis glandulosa* equal to 23%. For the modeled fields, *Prosopis glandulosa* has a dispersal range of three while *Paspalum setaceum* disperses globally. After Fernandez-Illescas and Rodriguez-Iturbe (2004).

variability. It is clear from comparing the three probability density functions in Figure 11.31 that interannual rainfall fluctuations result in a much greater diversity of m values, as well as a greater mean and variance. This means that the patterns of *Prosopis glandulosa* resulting from the inclusion of interannual rainfall variability in the spatially explicit model of vegetation competition depict a greater diversity of pixel arrangements and clusters compared to a random map. Such enhanced diversity of spatial arrangements is less pronounced in the case of no interannual rainfall variability. Vegetation patterns generated with interannual rainfall variability are more spatially complex suggesting, once more, that temporal rainfall variability plays an important role in driving spatial diversity within water-limited ecosystems.

References

Aber, J. D., and C. T. Driscoll, 1997. Effects of land use, climate variation, and N deposition on N cycling and C storage in northern hardwood forests, *Global Biogeochemical Cycles*, **11**(4), 639–648.

Aber, J. D., and J. M. Melillo, 2001. *Terrestrial Ecosystems*, 2nd edn., San Diego, Calif., Academic Press.

Aber, J. D., J. M. Melillo, K. J. Nadelhoffer, J. Pastor, and R. D. Boone, 1991. Factors controlling nitrogen cycling and nitrogen saturation in northern temperate forest ecosystems, *Ecological Applications*, **1**(3), 303–315.

Abramowitz, M., and I. A. Stegun, 1964. *Handbook of Mathematical Functions*, New York, Dover.

Anderson, M. G. (ed.), 1988. *Modelling Geomorphological Systems*, New York, Wiley.

Anderson, M. G., and T. P. Burt (ed.), 1990. *Process Studies in Hillslope Hydrology*, 450 pp., New York, Wiley.

Annegarn, H. J., S. Cole, J. T. Suttles, and R. Swap, 2001. SAFARI 2000 dry-season airborne campaign, *The Earth Observer*, **12**(5), 18–22.

Archer, S., 1989. Have southern Texas savannas been converted to woodlands in recent history?, *American Naturalist*, **134**(4), 545–561.

Archer, S., 1994. Woody plant encroachment into southwestern grasslands and savannas: rates, patterns, and proximate causes, in *Ecological Implications of Livestock Herbivory in the West*, M. Vavra, W. A. Laycock, and R. D. Pieper, (ed.), pp. 13–68, Denver, Colo., Society for Range Management.

Archer, S., C. Scifres, C. R. Bassham, and R. Maggio, 1988. Autogenic succession in a subtropical savanna: conversion of grassland to thorn woodland, *Ecological Monographs*, **58**, 90–102.

Balakrishnan, V., C. Van den Broeck, and P. Hanggi, 1988. First-passage times of non-Markovian processes: the case of a reflecting boundary, *Physical Review A*, **38**, 4213–4222.

Baldocchi, D., and T. Meyers, 1998. On using eco-physiological, micrometeorological and biogeochemical theory to evaluate carbon dioxide, water vapor and trace gas fluxes over vegetation: a perspective, *Agricultural and Forest Meteorology*, **90**(1–2), 1–25.

Baldocchi, D. D., L. Xu, and N. Kiang, 2004. How plant functional-type, weather, seasonal drought, and soil physical properties alter water and energy fluxes of an oak-grass savanna and an annual grassland, *Agricultural and Forest Meteorology*, **123**, 13–39.

417

Ball, J. T., I. E. Woodrow, and J. A. Berry, 1987. A model predicting stomatal conductance and its contribution to the control of photosynthesis under different environmental conditions, in *Progress in Photosynthesis Research*, Vol. IV, pp. 221–224, Dordrecht, Martinus Nijhoff Publishers.

Band, L. E., 1991. Distributed parametrization of complex terrain, in *Landsurface–Atmosphere Interaction: Observations, Models and Analysis*, E. Wood, (ed.), pp. 249–270, Norwell, Mass., Kluwer Academic.

Beldring, S., L. Gottschalk, A. Rodhe, and L. M. Tallaksen, 2000. Kinematic wave approximations to hillslope hydrological processes in tills, *Hydrological Processes*, **14**, 727–745.

Belsky, A. J., 1990. Tree/grass ratios in East African savanna: a comparison of existing models, *Journal of Biogeography*, **17**, 483–489.

Benjamin, J. R., and C. A. Cornell, 1970. *Probability, Statistics, and Decision for Civil Engineers*, New York, McGraw-Hill.

Beven, K., 1989. Changing ideas in hydrology – the case of physically-based models, *Journal of Hydrology*, **105**, 157–172.

Beven, K., and P. Germann, 1982. Macropores and water flow in soils, *Water Resources Research*, **18**, 1311–1325.

Birkinshaw, S. J., and J. Ewen, 2000. Nitrogen transformation component for SHETRAN catchment nitrate transport modelling, *Journal of Hydrology*, **230**, 1–17.

Black, A. R., and A. Werritty, 1997. Seasonality of flooding: a case study of north Britain, *Journal of Hydrology*, **195**(1–4), 1–25.

Bloom, A. J., F. S. Chapin III, and H. A. Mooney, 1985. Resource limitation in plants: an economic analogy, *Annual Review of Ecology and Systematics*, **16**, 363–392.

Bolker, B. J., S. W. Pakala, and W. J. Parton, 1998. Linear analysis of soil decomposition: insights from the century model, *Ecological Applications*, **8**(2), 425–439.

Bonan, G. B., 1995. Land-atmosphere CO_2 exchange simulated by a land surface process model coupled to an atmospheric general circulation model, *Journal of Geophysical Research*, **100**, 2817–2831.

Bonan, G. B., 2002. *Ecological Climatology. Concepts and Applications*, Cambridge, Cambridge University Press.

Bonner, J. T., 1988. *The Evolution of Complexity*, Princeton, N. J., Princeton University Press.

Bourliere, F., and M. Hadley, 1970. The ecology of tropical savannas, *Annual Review of Ecology and Systematics*, **1**, 125–152.

Boyer, J. S., 1982. Plant productivity and environment, *Science*, **218**, 443–448.

Bradford, K. J., and T. C. Hsiao, 1982. Physiological responses to moderate water stress, in *Physiological Plant Ecology II, Water Relations and Carbon Assimilation*, O. L. Lange, P. S. Nobel, C. B. Osmond, and H. Ziegler, (ed.), pp. 279–286, New York, Springer-Verlag.

Brady, N. C., and R. R. Weil, 1996. *The Nature and Properties of Soils*, 11th edn., Upper Saddle River, N. J., Prentice Hall.

Branson, F. A., R. F. Miller, and I. S. Mc Queen, 1970. Plant communities and associated soil and water factors on shale-derived soils in northeastern Montana, *Ecology*, **51**(3), 391–407.

Branson, F. A., R. F. Miller, and I. S. Mc Queen, 1976. Moisture relationships in twelve northern desert shrub communities near Gran Junction, Colorado, *Ecology*, **57**(6), 1104–1124.

Bromley, J., J. Brouwer, A. P. Barker, S. R. Gaze, and C. Valentin, 1997. The role of surface water redistribution in an area of patterned vegetation in a semiarid environment, South-West Niger, *Journal of Hydrology*, **198**, 1–29.

Bronstert, A., 1999. Capabilities and limitations of detailed hillslope hydrological modelling, *Hydrological Processes*, **13**, 21–48.

Bronstert, A., and E. J. Plate, 1997. Modelling of runoff generation and soil moisture dynamics for hillslopes and micro-catchments, *Journal of Hydrology*, **198**, 177–195.

Brown, J. R., and S. Archer, 1990. Water relations of a perennial grass and seedling vs adult woody plants in a subtropical savanna, Texas, *Oikos*, **57**, 366–374.

Brutsaert, W., 1982. *Evaporation into the Atmosphere: Theory, History, and Applications*, Dordrecht, D. Reidel Publishing Company.

Brutsaert, W., and D. Chen, 1995. Desorption and the two stages of drying of natural tallgrass prairie, *Water Resources Research*, **31**(5), 1305–1313.

Brutsaert, W., and D. Chen, 1996. Diurnal variation of surface fluxes during thorough drying (or severe drought) of natural prairie, *Water Resources Research*, **32**(7), 2013–2019.

Budyko, M. I., 1974. *Climate and Life*, San Diego, Calif., Academic Press.

Burgess, T. L., 1995. Desert grassland, mixed shrub savanna, shrub steppe, or semidesert shrub? The dilemma of coexisting growth forms, in *The Desert Grassland*, M. P. McClaran and T. R. Van Devender, (ed.), pp. 31–67, Tucson, Ariz., The University of Arizona Press.

Burke, I. C., W. K. Lauenroth, R. Riggle, P. Brannen, B. Madigan, and S. Beard, 1999. Spatial variability of soil properties in the shortgrass steppe: the relative importance of topography, grazing, microsite and plant species in controlling spatial patterns, *Ecosystems*, **2**, 422–438.

Cabon, F., G. Girard, and E. Ledoux, 1991. Modelling of the nitrogen cycle in farm land areas, *Fertilizer Research*, **27**, 161–169.

Campbell, S. G., and J. M. Norman, 1998. *An Introduction to Environmental Biology*, New York, Springer-Verlag.

Canadell, J., R. B. Jackson, J. R. Ehleringer, H. A. Mooney, O. E. Sala, and E. D. Schulze, 1996. Maximum rooting depth of vegetation types at the global scale, *Oecologia*, **108**, 583–595.

Cardenas, L., A. Rondon, C. Johansson, and E. Sanhueza, 1993. Effects of soil moisture, temperature and inorganic nitrogen on nitric oxide emission from acidic tropical savanna soils, *Journal of Geophysical Research*, **98**(D8), 14 783–14 790.

Cardon, J. E., and J. Letey, 1992. Plant water uptake term evaluated for soil water and solute movement models, *Soil Science Society of America Journal*, **32**, 1876–1880.

Carson, M. A., and M. J. Kirkby, 1972. *Hillslope Form and Process*, Cambridge, Cambridge University Press.

Casanova, M., I. Messing, and A. Joel, 2000. Influence of aspect and slope gradient on hydraulic conductivity measured by tension infiltrometer, *Hydrological Processes*, **14**, 155–164.

Caylor, K. K., H. H. Shugart, and I. Rodríguez-Iturbe, 2004. Tree canopy effects on simulated water stress in southern African savannas. *Ecosystems*, (in press).

Celia, M. A., E. F. Bouloutas, and R. L. Zarba, 1990. A general mass-conservative numerical solution for the unsaturated flow equation, *Water Resources Research*, **26**(7), 1483–1496.

Cerda, A., 1997. Seasonal changes of the infiltration rates in a Mediterranean scrubland on limestone, *Journal of Hydrology*, **198**, 209–225.

Chapin, III F. S., 1991. Integrated responses of plants to stress, *BioScience*, **41**(1), 29–36.

Chapin, III F. S., A. J. Bloom, C. B. Field, and R. H. Waring, 1987. Plant response to multiple environmental factors, *BioScience*, **37**(1), 49–57.

Chapman, S. B., 1976. *Methods in Plant Ecology*, Oxford, Blackwell.

Chow, V. T. (ed.), 1964. *Handbook of Applied Hydrology*, New York, McGraw-Hill.

Clapp, R. B., and G. N. Hornberger, 1978. Empirical equations for some soil hydraulic properties, *Water Resources Research*, **14**, 601–604.

Clark, J. S., 1996. Testing disturbance theory with long-term data: alternative life-history solutions to the distribution of events, *American Naturalist*, **148**, 976–996.

Cody, M. L., 1986. Structural niches in plant communities, in *Community Ecology*, J. Diamond and T. Case, (ed.), New York, Harper & Row.

Collatz, J. G., J. T. Ball, C. Grivet, and J. A. Berry, 1991. Physiological and environmental regulation of stomatal conductance, photosynthesis and transpiration: a model that includes a laminar boundary layer, *Agricultural and Forest Meteorology*, **54**, 107–136.

Collatz, J. G., M. Ribas-Carbo, and J. A. Berry, 1992. Coupled photosynthesis-stomatal conductance model for leaves of C_4 plants, *Australian Journal of Plant Physiology*, **19**, 519–538.

Conrad, R., 1996. Soil microorganisms as controllers of atmospheric trace gases (H_2, CO, CH_4, OCS, N_2O and NO), *Microbiology Review*, 609–640.

Cordova, J. R., and R. L. Bras, 1981. Physically based probabilistic models of infiltration, soil moisture, and actual evapotranspiration, *Water Resources Research*, **17**(1), 93–106.

Corradini, C., R. Morbidelli, and F. Melone, 1998. On the interaction between infiltration and Hortonian runoff, *Journal of Hydrology*, **204**, 52–67.

Corradini, C., F. Melone, and R. E. Smith, 2000. Modeling local infiltration for a two-layered soil under complex rainfall patterns, *Journal of Hydrology*, **337**, 58–73.

Cosby, B. J., R. B. Hornberger, R. B. Clapp, and T. R. Ginn, 1984. A statistical exploration of the relationships of soil moisture characteristics to the physical properties of soils, *Water Resources Research*, **20**(6), 682–690.

Coupland, R. T., 1950. Ecology of mixed prairie in Canada, *Ecological Monographs*, **20**, 271–315.

Couteron, P., and K. Koukou, 1997. Woody vegetation spatial patterns in a semi-arid savanna of Burkina Fasso, West Africa, *Plant Ecology*, **132**, 211–227.

Cowan, I. R., 1965. Transport of water in the soil-plant-atmosphere system, *Journal of Applied Ecology*, **2**(1), 221–239.

Cowan, I. R., 1986. Economics of carbon fixation, in *On the Economy of Plant Form and Function*, T. J. Givnish, (ed.), pp. 133–170, Cambridge, Cambridge University Press.

Cox, D. R., and V. Isham, 1986. The virtual waiting-time and related processes, *Advances in Applied Probability*, **18**, 558–573.

Cox, D. R., and H. D. Miller, 1965. *The Theory of Stochastic Processes*, London, Methuen.

Crave, A., and C. Gascuel-Odoux, 1997. The influence of topography on time and space distribution of soil surface water content, *Hydrological Processes*, **11**, 203–210.

Crill, P. M., M. Keller, A. Weitz, B. Grauel, and E. Veldkamp, 2000. Intensive field measurements of nitrous oxide emissions from tropical agricultural soil, *Global Biogeochemical Cycles*, **14**(1), 85–95.

Cui, M., and M. M. Caldwell, 1997. A large ephemeral release of nitrogen upon wetting of dry soil and corresponding root responses in the field, *Plant and Soil*, **191**, 291–299.

Cuomo, C. J., R. J. Ansley, P. W. Jacoby, and R. E. Sosebee, 1992. Honey mesquite transpiration along a vertical site gradient, *Journal of Range Management*, **45**(4), 334–338.

Daly, E., A. Porporato, and I. Rodríguez-Iturbe, 2004a. Modeling photosynthesis, transpiration, and soil water balance: hourly dynamics during interstorm periods. *Journal of Hydrometeorology*, **5**, 546–558.

Daly, E., A. Porporato, and I. Rodríguez-Iturbe, 2004b. Ecohydrological significance of the coupled dynamics of photosynthesis, transpiration, and soil water balance. *Journal of Hydrometeorology*, **5**, 559–566.

Daubenmire, R., 1968. Soil moisture in relation to vegetation distribution in the mountains of northern Idaho, *Ecology*, **49**(3), 431–438.

Davidson, E. A., 1991. Fluxes of nitrous oxide and nitric oxide from terrestrial ecosystems, in *Microbial Production and Consumption of Greenhouse Gases: Methane, Nitrogen Oxides and Halomethanes*, J. E. Rogers, and W. B. Whitman, (ed.), pp. 219–235, Washington D.C., American Society of Microbiology.

Davidson, J. M., D. R. Nielsen, and J. W. Biggar, 1963. The measurement and description of water flow through Columbia silt loam and Hisperia silt loam, *Hilgardia*, **34**, 601–616.

De Wiest, J. M., 1969. *Flow through Porous Media*, New York, Academic Press.

Dewar, R. C., 1995. Interpretation of an empirical model for stomatal conductance in terms of guard cell function, *Plant, Cell and Environment*, **18**, 365–372.

Diamond, D. D., D. H. Riskand, and S. L. Orzell, 1987. A framework for plant community classification and conservation in Texas, *Texas Journal of Science*, **39**, 203–221.

Dickinson, R. E., A. Henderson-Sellers, P. J. Kennedy, and M. F. Wilson, 1986. *Biosphere-Atmosphere Transfer Scheme (BATS) for the NCAR CCM*, Boulder, Colo., NCAR/TN-275-STR, National Center for Atmosphere Research,

Dickinson, R. E., J. A. Berry, G. B. Bonan, G. J. Collatz, C. B. Field, I. Y. Fung, M. Goulden, W. A. Hoffman, R. B. Jackson, R. Myneni, P. J. Sellers, and M. Shaikh, 2002. Nitrogen control on climate model evapotranspiration, *Journal of Climate*, **15**, 278–295.

Dingman, S. L., 1994. *Physical Hydrology*, Englewood Cliffs, N. J., Prentice-Hall.

D'Odorico, P., L. Ridolfi, A. Porporato, and I. Rodríguez-Iturbe, 2000. Preferential states of seasonal soil moisture: the impact of climatic fluctuations, *Water Resources Research*, **36**(8), 2209–2220.

D'Odorico, P., F. Laio, A. Porporato, and I. Rodríguez-Iturbe, 2003. Hydrologic controls on soil carbon and nitrogen cycles. II. A case study, *Advances in Water Resources*, **26**, 59–70.

Dowty, P., P. Frost, P. Lasolle, G. Midgley, M. Mukrlabai, L. Otter, J. Privette, J. Ramontsho, S. Ringrose, R. J. Scholes, and Y. Wang, 2001. Summary of the SAFARI 2000 wet season field campaign along the Kalahari transect, *The Earth Observer*, **12**(3), 29–34.

Eagleson, P. S., 1978a. Climate, soil, and vegetation. 1. Introduction to water balance dynamics, *Water Resources Research*, **14**(5), 705–712.

Eagleson, P. S., 1978b. Climate, soil, and vegetation. 3. A simplified model of soil moisture movement in the liquid phase, *Water Resources Research*, **14**(5), 722–730.

Eagleson, P. S., 1982. Ecological optimality in water-limited natural soil-vegetation systems. 1. Theory and hypothesis, *Water Resources Research*, **18**(2), 325–340.

Eagleson, P. S., 1994. The evolution of modern hydrology (from watershed to continent in 30 years), *Advances in Water Resources*, **17**, 3–18.

Eagleson, P. S., 2000. Interview with Peter Hanneberg, in *Our Struggle for Water*, Stockholm, Stockholm International Water Institute.

Eagleson, P. S., 2002. *Ecohydrology: Darwinian Expression of Vegetation Form and Function*, Cambridge, Cambridge University Press.

Eagleson, P. S., and R. I. Segarra, 1985. Water-limited equilibrium of savanna vegetation systems, *Water Resources Research*, **21**(10), 1483–1493.

Ehleringer, J. R., and R. K. Monson, 1993. Evolutionary and ecological aspects of photosynthetic pathway variation, *Annual Review of Ecology and Systematics*, **24**, 411–439.

Eltahir, E. A. B., 1996. The role of vegetation in sustaining large-scale atmospheric circulations in the tropics, *Journal of Geophysical Research*, **101**(D2), 4255–4268.

Eltahir, E. A. B., 1998. A soil moisture rainfall feedback mechanism. 1. Theory and observations, *Water Resources Research*, **34**(4), 765–776.

Engels, C., and H. Marschner, 1995. Plant uptake and utilization of nitrogen, in *Nitrogen Fertilization in the Environment*, P. E. Bacon, (ed.), New York, Marcel Dekker Inc.

Entekhabi, D., G. R. Asrar, A. K. Betts, K. J. Beven, R. L. Bras, C. J. Duffy, T. Dunne, R. D. Koster, D. P. Lettenmaier, D. B. McLaughlin, W. J. Shuttleworth, M. T. van Genuchten, M. Y. Wei, and E. F. Wood, 1999. An agenda for land surface hydrology research and a call for the second international hydrological decade, *Bulletin of the American Meteorological Society*, **80**(10), 2043–2058.

Ettrick, T. M., J. A. Mawdsley, and A. V. Metcalfe, 1987. The influence of antecedent catchment conditions on seasonal flood risk, *Water Resources Research*, **23**(3), 481–488.

Ewers, B. E., R. Oren, and J. S. Sperry, 2000. Influence of nutrient versus water supply on hydraulic architecture and water balance in *Pinus taeda*, *Plant, Cell and Environment*, **23**, 1055–1066.

Famiglietti, J. S., J. W. Rudniki, and M. Rodell, 1998. Variability in surface moisture content along a hillslope transect: Rattlesnake Hill, Texas, *Journal of Hydrology*, **210**, 259–281.

Fan, Y., and R. Bras, 1998. Analytical solutions to hillslope subsurface storm flow and saturation overland flow, *Water Resources Research*, **34**, 921–927.

Farquhar, G. D., S. von Cammerer, and J. A. Berry, 1980. A biochemical model of photosynthetic CO_2 assimilation in leaves of C_3 species, *Planta*, **149**, 78–90.

Feddes, R. A., 1971. *Water, Heat and Crop Growth*, pp. 57–72, Mededelingen, Wageningen.

Federer, C. A., 1979. A soil-plant-atmosphere model for transpiration and availability of soil water, *Water Resources Research*, **15**(3), 555–561.

Federer, C. A., 1982. Transpiration supply and demand: plant, soil, and atmospheric effects evaluated by simulation, *Water Resources Research*, **18**(2), 355–362.

Federer, C. A., and G. W. Gee, 1976. Diffusion potential and xylem resistance in stressed and unstressed northern hardwood trees, *Ecology*, **57**, 975–984.

Fenchel, T., G. M. King, and T. H. Blackburn, 1998. *Bacterial Biogeochemistry: The Ecophysiology of Mineral Cycling*, San Diego, Calif., Academic Press.

Fensham, R. J., and J. E. Holman, 1999. Temporal and spatial patterns in drought-related tree dieback in Australian savanna, *Journal of Applied Ecology*, **36**, 1035–1050.

Fernandez-Illescas, C. P., and I. Rodríguez-Iturbe, 2003. Hydrologically driven hierarchical competition colonization models: the impact of interannual rainfall fluctuations, *Ecological Monographs*, **73**(2), 207–222.

Fernandez-Illescas, C. P., and I. Rodríguez-Iturbe, 2004. On the impact of interannual rainfall variability on the spatial and temporal patterns of vegetation in a water-limited ecosystem, *Advances in Water Resources*, **27**(1), 83–95.

Fernandez-Illescas, C. P., A. Porporato, F. Laio, and I. Rodríguez-Iturbe, 2001. The ecohydrological role of soil texture in water-limited ecosystems, *Water Resources Research*, **37**(12), 2863–2872.

Foley, A. J., I. Colin Prentice, N. Ramankutty, S. Levis, D. Pollard, S. Sitch, and A. Hazeltine, 1996. An integrated biosphere model of land surface processes, terrestrial carbon balance and vegetation dynamics, *Global Biochemical Cycles*, **10**(4), 603–628.

Fowler, D., U. Skiba, and K. J. Hargreaves, 1997. Emissions of nitrous oxide from grasslands, in *Gaseous Nitrogen Emissions from Grasslands*, S. C. Jarvis and B. F. Pain, (ed.), pp. 147–164, Wallingford, UK, CAB International.

Freeze, R. A., 1980. A stochastic-conceptual analysis of rainfall-runoff processes on a hillslope, *Water Resources Research*, **16**, 391–408.

Gani, J., 1955. Some problems in the theory of provisioning and of dams, *Biometrika*, **42**, 179–200.

Gardiner, C. W., 1990. *Handbook of Stochastic Methods: for Physics, Chemistry and the Natural Sciences*, 3rd edn., Berlin, Springer-Verlag.

Gardner, W. R., 1960. Dynamic aspects of water availability to plants, *Soil Science*, **89**(2), 63–73.

Gardner, W. R., and C. F. Ehlig, 1963. The influence of soil water on transpiration by plants, *Journal of Geophysical Research*, **68**(20), 5719–5724.

Genereux, D. P., and H. F. Hemond, 1990. Naturally occurring radon 222 as a tracer for stream flow generation: steady state methodology and field example, *Water Resources Research*, **26**, 3065–3076.

Germann, P., and K. Beven, 1985. Kinematic wave approximation to infiltration into soils with sorbing macropores, *Water Resources Research*, **21**, 990–996.

Germann, P., and K. Beven, 1986. A distributional function approach to water flow in soil macropores based on kinematic wave theory, *Journal of Hydrology*, **83**, 173–183.

Gitay, H., and I. R. Noble, 1997. What are functional types and how should we seek them?, in *Plant Functional Types*, T. M. Smith, I. A. Woodward, and H. H. Shugart, (ed.), pp. 3–19, Cambridge, UK, Cambridge University Press.

Gleeson, S., and D. Tilman, 1990. Allocation and the transient dynamics of succession on poor soils, *Ecology*, **71**, 1144–1155.

Goldberger, A. L., V. Bhargava, B. J. West, and A. J. Mandell, 1985. On a mechanism of cardiac electrical stability: the fractal hypothesis, *Biophysical Journal*, **48**, 525–528.

Gollan, T., N. C. Turner, and E. D. Schulze, 1985. The responses of stomata and leaf gas exchange to vapour pressure deficit and soil water content, III. In the schlerophyllous woody species *Nerium oleander*, *Oecologia*, **65**, 356–362.

Golluscio, R. A., O. E. Sala, and W. K. Lauenroth, 1998. Differential use of large summer rainfall events by shrub and grasses: a manipulative experiment in the Patagonian steppe, *Oecologia*, **115**, 17–25.

Grayson, R. B., A. W. Western, and F. H. Chiew, 1997. Preferred states in spatial soil moisture patterns: local and nonlocal controls, *Water Resources Research*, **33**, 2897–2908.

Grime, I. P., 1979. *Plant Strategies and Vegetation Processes*, Chichester, UK, Wiley.

Grubb, P. J., 1986. Problems posed by sparse and patchily distributed species, in *Species-rich Plant Communities in Community Ecology*, J. Diamond and T. Case, (ed.), pp. 207–226, New York, Harper & Row.

Gusman, A. J., and M. A. Marino, 1999. Analytical modeling of nitrogen dynamics in soils and ground water, *Journal of Irrigation, Drainage and Engineering*, **125**(6), 330–337.

Guswa, A. J., M. A. Celia, and I. Rodríguez-Iturbe, 2002. Model of soil moisture dynamics in ecohydrology: a comparative study, *Water Resources Research*, **38**(9), 1166–1181.

Guzzetti, F., 1998. Hydrological triggers of diffused landsliding, *Environmental Geology*, **35**, 79–80.

Haas, R. H., and J. D. Dodd, 1972. Water stress patterns in honey mesquite, *Ecology*, **53**, 674–680.

Haase, P., 1995. Spatial pattern analysis in ecology based on Ripley's K-function: introduction and methods of edge correction, *Journal of Vegetation Science*, **6**, 575–582.

Hale, M. G., and D. M. Orcutt, 1987. *The Physiology of Plants Under Stress*, New York, J. Wiley & Sons.

Hanggi, P., and P. Talkner, 1985. First-passage time problems for non-Markovian processes, *Physical Review A*, **32**, 1934–1937.

Hanna, Y. A., P. W. Harlan, and D. T. Lewis, 1982. Soil available water as influenced by landscape position and aspect, *Agronomics Journal*, **74**, 999–1004.

Hansen, S., H. E. Jensen, and M. J. Shaffer, 1995. Developments in modeling nitrogen transformations in soils, in *Nitrogen Fertilization in the Environment*, P. E. Bacon, (ed.), New York, Marcel Dekker Inc.

Hanski, I., and E. Ranta, 1983. Coexistence in a patchy environment: three species of *Daphnia* in rock pools, *Journal of Animal Ecology*, **52**, 263–279.

Harr, M. E., 1962. *Groundwater and Seepage*, New York, McGraw-Hill.

Harrington, G. N., 1991. Effects of soil moisture on shrub seedling survival in a semi-arid grassland, *Ecology*, **72**(3), 1138–1149.

Hastings, A., 1980. Disturbance, coexistence, history and competition for space, *Theoretical Population Biology*, **18**, 363–373.

Hawley, M. E., T. J. Jackson, and R. H. McCuen, 1983. Surface soil moisture variation on small agricultural watershed, *Journal of Hydrology*, **62**, 179–200.

Haynes, R. J., 1986. *Mineral Nitrogen in the Plant-Soil System*, London, Academic Press.

Hedin, L., O. Chadwick, J. Schimel, and M. Torn, 2002. *Linking Ecology, Biology and Geoscience*, August Report to the National Science Foundation.

Hellerislambers, R., M. Reitkerk, F. van de Bosch, H. H. T. Prins, and H. de Kroon, 2001. Vegetation pattern formation on semi-arid grazing systems, *Ecology*, **82**(1), 50–61.

Hernandez-Garcia, E., L. Pesquera, M. A. Rodríguez, and M. San Miguel, 1987. First-passage time statistics: processes driven by Poisson noise, *Physical Review A*, **36**, 5774–5781.

Hillel, D., 1998. *Environmental Soil Physics*, San Diego, Calif., Academic Press.

Hsiao, T. C., 1973. Plant responses to water stress, *Annual Review of Plant Physiology*, **24**, 519–570.

Ingram, J., and D. Bartels, 1996. The molecular basis of dehydration tolerance in plants, *Annual Review of Plant Physiology and Plant Molecular Biology*, **47**, 377–403.

Jackson, R. B., J. Canadell, J. R. Ehleringer, H. A. Mooney, O. E. Sala, and E. D. Schulze, 1996. A global analysis of root distribution for terrestrial biomes, *Oecologia*, **108**, 389–411.

Jarvis, P. G., 1976. The interpretation of the variations in leaf water potential and stomatal conductance found in canopies in the field, *Philosophical Transactions of the Royal Society of London Series B*, **273**, 593–610.

Jarvis, P. G., T. A. Mansfield, and W. J. Davies, 1999. Stomatal behavior, photosynthesis and transpiration under rising CO_2, *Plant, Cell and Environment*, **22**, 639–648.

Jeltsch, F., S. J. Milton, W. R. J. Dean, and N. van Rooyen, 1996. Tree spacing and coexistence in semi-arid savannas, *Journal of Ecology*, **84**, 583–595.

Jeltsch, F., S. J. Milton, W. R. J. Dean, N. van Rooyen, and K. A. Moloney, 1998. Modelling the impact of small-scale heterogeneities on tree-grass coexistence in semi-arid savannas, *Journal of Ecology*, **86**(5), 780–793.

Jenkinson, D., 1990. The turnover of organic carbon and nitrogen in soil, *Philosophical Transactions of the Royal Society of London Series B*, **329**, 361–368.

Joffre, R., and S. Rambal, 1993. How tree cover influences the water balance of Mediterranean rangelands, *Ecology*, **74**(2), 570–582.

Jones, H. G., 1992. *Plants and Microclimate: A Quantitative Approach to Environmental Plant Physiology*, 2nd edn., Cambridge, UK, Cambridge University Press.

Karlin, S., 1983. *11th R.A. Fisher Memorial Lecture*, Royal Society, April 20 1983.

Katul, G., R. Leuning, and R. Oren, 2003. Relationship between plant hydraulic and biochemical properties derived from a steady state coupled water and carbon transport model. *Plant, Cell and Environment*, **26**, 339–350.

Kemp, P. R., and G. J. Williams III, 1980. A physiological basis for niche separation between *Agropyron smithii* (C_3) and *Bouteloua gracilis* (C_4), *Ecology*, **6**(4), 846–858.

Kiang, N., 2002. *Savannas and Seasonal Drought: The Landscape-Leaf Connection Through Optimal Stomatal Control*, Ph.D. thesis, Berkeley, Calif., University of California.

Kinzig, A. P., S. A. Levin, J. Dushoff, and S. Pacala, 1999. Limiting similarity, species packing and system stability for hierarchical competition-colonization models, *American Naturalist*, **153**, 371–383.

Kirkby, M., 1978. *Hillslope Hydrology*, Chichester, John Wiley.

Kirkby, M., 1988. Hillslope runoff processes and models, *Journal of Hydrology*, **100**, 315–339.

Klausmeier, C. A., 1999. Regular and irregular pattern in semiarid vegetation, *Science*, **284**, 1826–1828.

Knapp, A. K., P. A. Fay, J. M. Blair, S. L. Collins, M. D. Smith, J. D. Carlisle, C. W. Harper, B. T. Danner, M. S. Lett, and J. K. McCarron, 2002. Rainfall variability, carbon cycling, and plant species diversity in a mesic grassland, *Science*, **298**, 2202–2205.

Koopmans, L. H., 1995. *The Spectral Analysis of Time Series*, San Diego, Calif., Academic Press.

Kramer, P. J., and J. S. Boyer, 1995. *Water Relations of Plants and Soils*, San Diego, Calif., Academic Press.

Kutzbach, J., G. Bonan, J. Foley, and S. P. Harrison, 1996. Vegetation and soil feedbacks on the response of the African monsoon to orbital forcing in the early to middle Holocene, *Nature*, **384**, 623–626.

Lai, C. T., and G. Katul, 2000. The dynamics role of root-water uptake in coupling potential to actual transpiration, *Advances in Water Resources*, **23**, 427–439.

Laio, F., A. Porporato, L. Ridolfi, and I. Rodríguez-Iturbe, 2001a. Plants in water-controlled ecosystems: active role in hydrological processes and response to water stress. II. Probabilistic soil moisture dynamics, *Advances in Water Resources*, **24**(7), 707–723.

Laio, F., A. Porporato, C. P. Fernandez-Illescas, and I. Rodríguez-Iturbe, 2001b. Plants in water-controlled ecosystems: active role in hydrologic processes and response to water stress. IV. Discussion of real cases, *Advances in Water Resources*, **24**(7), 745–762.

Laio, F., A. Porporato, L. Ridolfi, and I. Rodríguez-Iturbe, 2001c. On the mean first passage times of processes driven by white shot noise, *Physical Review E*, **63**, 36 105.

Laio, F., A. Porporato, L. Ridolfi, and I. Rodríguez-Iturbe, 2002. On the unsteady dynamics of mean soil moisture. *Journal of Geophysical Research (Atm.)*, **107**(D12), 101 029.

Lambers, H., S. F. Chapin, and T. L. Pons, 1998. *Plant Physiological Ecology*, New York, Springer-Verlag.

Lange, O. L., L. Kappen, and E. D. Schulze, 1976. *Water and Plant Life: Problems and Modern Approaches*, Berlin, Springer-Verlag.

Larcher, W., 1995. *Physiological Plant Ecology*, New York, Springer-Verlag.

Lauenroth, W. K., and P. L. Sims, 1976. Evapotranspiration from a shortgrass prairie subjected to water and nitrogen treatments, *Water Resources Research*, **12**(3), 437–442.

Lauenroth, W. K., J. L. Dodd, and P. L. Sims, 1978. The effects of water and nitrogen induced stresses on plant community structure in a semi-arid grassland, *Oecologia*, **36**, 211–222.

Lauenroth, W. K., O. E. Sala, D. G. Milchunas, and R. W. Lathrop, 1987. Root dynamics of *Bouteloua gracilis* during short-term recovery from drought, *Functional Ecology*, **1**, 117–124.

Lauenroth, W. K., O. E. Sala, D. P. Coffin, and T. B. Kirchner, 1994. The importance of soil water in the recruitment of *Bouteloua gracilis* in the shortgrass steppe, *Ecological Applications*, **4**(4), 741–749.

Leetham, J. W., and D. G. Milchunas, 1985. The composition and distribution of soil microarthropods in the shortgrass steppe in relation to soil water, root biomass and grazing by cattle, *Pedobiologia*, **28**, 311–325.

Lehman, L. E., and D. Tilman, 1997. Competition in spatial habitats, in *Spatial Ecology*, D. Tilman and P. Kareiva, (ed.), pp. 185–203, Princeton, N. J., Princeton University Press.

Leuning, R., 1990. Modeling stomatal behavior and photosynthesis of *Eucalyptus grandis*, *Australian Journal of Plant Physiology*, **17**, 159–175.

Leuning, R., 1995. A critical appraisal of a combined stomatal-photosynthesis model for C$_3$ plants, *Plant, Cell and Environment*, **18**, 339–355.

Levins, R., 1969. Some demographic and genetic consequences of environmental heterogeneity for biological control, *Bulletin of the Entomological Society of America*, **15**, 237–240.

Levins, R., and D. Culver, 1971. Regional coexistence of species and competition between rare species, *Proceedings of the National Academy of Sciences USA*, **68**, 1246–1248.

Levitt, J., 1980. *Responses of Plants to Environmental Stresses*, Vol. I, 2nd edn., New York, Academic Press.

Lhomme, J. P., 1998. Formulation of root-water uptake in a multi-layer soil-plant model: does van der Honert's equation hold?, *Hydrology Earth System Sciences*, **21**(1), 31–40.

Lhomme, J. P., 2001. Stomatal control of transpiration: examination of the Jarvis-type representation of canopy resistance in relation to humidity, *Water Resources Research*, **37**(3), 689–699.

Lhomme, J. P., E. Elguero, A. Chehbouni, and G. Boulet, 1998. Stomatal control of transpiration: examination of Monteith's formulation of canopy resistance, *Water Resources Research*, **34**(9), 2301–2308.

Liang, Y. M., D. L. Hazlett, and W. K. Lauenroth, 1989. Biomass dynamics and water use efficiencies of five plant communities in the shortgrass steppe, *Oecologia*, **80**, 148–153.

Linhart, Y. B., and M. C. Grant, 1996. Evolutionary significance of local genetics differentiation in plants, *Annual Review of Ecology and Systematics*, **27**, 237–277.

Linn, D. M., and J. W. Doran, 1984. Effect of water-filled pore space on carbon dioxide and nitrous oxide production in tilled and nontilled soils, *Soil Science Society of America Journal*, **48**, 1267–1272.

Logan, J., 1983. Nitrogen oxides in the troposphere: global and regional budgets, *Journal of Geophysical Research*, **88**, 10 785–10 807.

Lohammer, T., S. Larsson, S. Linder, and S. O. Falk, 1980. FAST – simulation models of gaseous exchange in Scots Pine, *Ecology Bulletin*, **32**, 505–523.

Ludlow, M. M., 1976. Ecophysiology of C_4 grasses, in *Water and Plant Life*, O. L. Lange, L. Kappen, and E. D. Shulze, (ed.), pp. 364–380, New York, Springer-Verlag.

Luk, S. H., Q. G. Cai, and G. P. Wang, 1993. Effects of surface crusting and slope gradient on soil and water losses in the hilly loess region, North China, *Catena Supplement*, **24**, 29–45.

MacArthur, R. H., and R. Levins, 1964. Competition, habitat selection, and character displacement in a patchy environment, *Proceedings of the National Academy of Sciences USA*, **51**, 1207–1210.

MacMahon, J. A., and D. J. Schimpf, 1981. Water as a factor in the biology of North American plants, in *Water in Desert Ecosystems*, D. D. Evans and J. L. Thames, (ed.), Stroudburg, Pa., Dowden, Hutchinson & Ross.

Major, J. A., 1963. Climatic index to vascular plant activity, *Ecology*, **44**(3), 485–498.

Marone, L., M. E. Horno, and R. Gonzalez del Solar, 2000. Post-dispersal fate of seeds in the Monte desert of Argentina: patterns of germination in successive wet and dry years, *Journal of Ecology*, **88**, 940–949.

Masoliver, J., 1987. First-passage times for non-Markovian processes: shot noise, *Physical Review A*, **35**(9), 3918–3928.

Masoliver, J., K. Lindemberg, and B. J. West, 1986. First-passage times for non-Markovian processes, *Physical Review A*, **33**, 2177–2180.

Matalas, N. C., 1967. Mathematical assessment of synthetic hydrology, *Water Resources Research*, **3**, 937–945.

Matson, P. A., and P. M. Vitousek, 1990. Ecosystem approach to global nitrous oxide budget, *Biogeoscience*, **49**, 667–672.

McAuliffe, J. R., 1995. Landscape evolution, soil formation and Arizona's desert grasslands, in *The Desert Grassland*, M. P. McClaran and T. R. Van Devender, (ed.), pp. 100–129, Tucson, Ariz., University of Arizona Press.

McCord, J. T., and D. B. Stephens, 1987. Lateral moisture flow beneath a sandy hillslope without an apparent layer, *Hydrological Processes*, **1**, 225–238.

McGinnes, W. G., and J. F. Arnold, 1939. Relative water requirement of Arizona range plants, *Technical Bulletin, Number 80*, University of Arizona Experimental Station, Tucson, Ariz., University of Arizona Press.

McNaughton, K. G., and T. W. Spriggs, 1986. A mixed layer model for regional evaporation, *Boundary-Layer Meteorology*, **34**, 243–262.

Meixner, F. X., and W. Eugster, 1999. Effects of landscape pattern and topography on the emissions and transport, in *Integrating Hydrology, Ecosystem Dynamics and Biogeochemestry in Complex Landscapes*, J. D. Tenhunen and P. Kabat, (ed.), pp. 147–175, New York, Wiley & Sons.

Milly, P. C. D., 1993. An analytical solution of the stochastic storage problem applicable to soil water, *Water Resources Research*, **29**, 3755–3785.

Milly, P. C. D., 1994a. Climate, soil water storage, and average annual water balance, *Water Resources Research*, **30**(7), 2143–2156.

Milly, P. C. D., 1994b. Climate, interseasonal storage of soil water, and the annual water balance, *Advances in Water Resources*, **17**(1–2), 19–24.

Milly, P. C. D., 1997. Sensitivity of greenhouse summer dryness to changes in plant rooting characteristics, *Geophysical Research Letters*, **24**(2), 269–271.

Milly, P. C. D., 2001. A minimalistic probabilistic description of root zone soil water, *Water Resources Research*, **37**(3), 457–464.

Milly, P. C. D., and K. A. Dunne, 1993. Global water cycle to the water-holding capacity of land, *Journal of Climate*, **7**, 506–526.

Milne, B. T., 1992. Spatial aggregation and neutral models in fractal landscapes, *American Naturalist*, **139**, 32–57.

Mladenoff, D. J., M. A. White, J. Pastor, and T. R. Crow, 1993. Comparing spatial pattern in unaltered old-growth and disturbed forest landscapes, *Ecological Applications*, **3**, 284–306.

Molchanov, A. A., 1971. Cycles of atmospheric precipitation in different types of forests of natural zones in the URSS, in *Productivity of Forest Ecosystems*, P. Duvigneaud, (ed.), pp. 49–68, Paris, UNESCO.

Mooney, H. A., 1983. Carbon gaining capacity and allocation patterns of Mediterranean climate plants, in *Mediterranean-Type Ecosystems*, F. J. Kriger, D. J. Mitchell, and J. U. M. Jarvis, (ed.), pp. 103–109, Berlin, Springer-Verlag.

Moore, I. D., G. J. Burch, and D. H. Mackenzie, 1988. Topographic effects on the distribution of surface water and the location of ephemeral gullies, *Transactions of the American Society of Agricultural Engineers*, **31**, 1098–1107.

Moorhead, D. L., W. S. Currie, E. B. Rastetter, W. J. Parton, and M. E. Harmon, 1999. Climate and litter quality controls on decomposition: an analysis of modeling approaches, *Global Biogeochemical Cycles*, **13**(2), 575–589.

Mosley, M. P., 1982. Subsurface flow velocities through selected forest soils, South Island, New Zealand, *Journal of Hydrology*, **55**, 65–92.

Munro, D. S., and L. J. Huang, 1997. Rainfall, evaporation and runoff responses to hillslope aspect in Shenchong Basin, *Catena*, **29**, 131–144.

Murray, J. D., 1989. *Mathematical Biology*, Berlin, Springer-Verlag.

Naveh, Z., 1967. Mediterranean ecosystems and vegetation types in California and Israel, *Ecology*, **48**(3), 445–459.

Nee, S., and R. M. May, 1992. Dynamics of metapopulations: habitat destruction and competitive coexistence, *Journal of Animal Ecology*, **61**, 37–40.

Neilson, R. P., 1995. A model for predicting continental-scale vegetation distribution and water balance, *Ecological Applications*, **5**(2), 362–385.

Nepstad, D. C., C. R. de Carvalho, E. A. Davidson, P. H. Jipp, P. A. Levebvre, G. H. Negreiros, E. D. da Silva, T. A. Stone, S. E. Trumbore, and S. Vieira, 1994. The role of deep roots in the hydrological and carbon cycle of Amazonian forests and pastures, *Nature*, **372**, 666–669.

Newman, E. I., 1967. Response of *Aira precox* to weather conditions, I. Response to drought in spring, *Journal of Ecology*, **55**(2), 539–556.

Newton, R., 1993. *What Makes Nature Tick?* Boston, Mass., Harvard University Press.

Ng, E., and P. C. Miller, 1980. Soil moisture relations in the Southern California caparral, *Ecology*, **61**(1), 98–107.

Nicholson, S., 2000. Land surface processes and Sahel climate, *Reviews of Geophysics*, **38**(1), 117–139.

Niemann, K. O., and M. C. R. Edgell, 1993. Preliminary analysis of spatial and temporal distribution of soil moisture on a deforested slope, *Physical Geography*, **14**, 449–464.

Nilsen, E. T., and D. M. Orcutt, 1998. *Physiology of Plants under Stress: Abiotic Factors*, New York, John Wiley.

Nilsen, E. T., M. R. Sharifi, P. W. Rundel, W. M. Jarrel, and R. A. Virginia, 1983. Diurnal seasonal water relations of the desert phreatophyte *Prosopis glandulosa* (honey mesquite) in the sonoran desert of California, *Ecology*, **64**(6), 1381–1393.

Nilsen, E. T., M. R. Sharifi, and P. W. Rundel, 1984. Comparative relations of phreatophytes in the sonoran desert of California, *Ecology*, **65**(3), 767–778.

Nobel, P. S., 1999. *Physicochemical and Environmental Plant Physiology*, 2nd edn., San Diego, Calif., Academic Press.

Noguchi, S., Y. Tsuboyama, R. Sidle, and I. Hosoda, 1999. Morphological characteristics of macropores and the distribution of preferential flow pathways in a forested slope segment, *Soil Science Society of America Journal*, **63**, 1413–1423.

Noy-Meir, I., 1973. Desert ecosystems: environment and producers, *Annual Review of Ecology and Systematics*, **4**, 25–44.

Nyberg, L., 1996. Spatial variability of water content in the covered catchment Gardsjon, Sweden, *Hydrological Processes*, **10**, 89–103.

Oertli, J. J., 1976. The soil–plant–atmosphere continuum, in *Water and Plant Life*, O. L. Lange, L. Kappen, and E. D. Schulze, (ed.), pp. 34–41, New York, Springer-Verlag.

Ogden, F. L., and B. A. Watts, 2000. Saturated area formation on nonconvergent hillslope topography with shallow soils: a numerical investigation, *Water Resources Research*, **36**, 1795–1804.

O'Loughlin, E. M., 1990. Perspectives on hillslope research, in *Process Studies in Hillslope Hydrology*, M. G. Andersen, and T. P. Burt, (ed.), pp. 501–516, Chichester, Wiley.

Oren, R., J. S. Sperry, G. Katul, D. E. Pataki, B. E. Ewers, N. Phillips, and K. V. R. Schäfer, 1999. Survey and synthesis of intra- and interspecific variation in stomatal sensitivity to vapor pressure deficit, *Plant, Cell and Environment*, **22**, 1515–1526.

Otter, L. B., W. X. Yang, M. C. Scholes, and F. K. Meixner, 1999. Nitric oxide emissions from a southern African savanna. *Journal of Geophysical Research Atmospheres*, **104**(D15), 18 471–18 485.

Parsons, D. A. B., M. C. Scholes, R. J. Scholes, and J. S. Levine, 1996. Biogenic NO emissions from savanna soils as function of fire regime, soil type, soil nitrogen, and water status, *Journal of Geophysical Research*, **101**(D19), 23 683–23 688.

Parton, W. J., D. Schimel, C. Cole, and D. Ojima, 1987. Analysis of factors controlling soil organic levels of grasslands in the Great Plains, *Soil Science Society of America Journal*, **51**, 1173–1179.

Parton, W. J., J. Steward, and C. Cole, 1988. Dynamics of C, N and S in grassland soils: a model, *Biogeochemistry*, **5**, 109–131.

Parton, W. J., P. Woomer, and A. Martin, 1993. Modelling soil organic matter dynamics and plant productivity in tropical ecosystems, in *The Biological Management of Tropical Soil Fertility*, P. Woomer and M. Swift, (ed.), pp. 177–188, New York, Wiley – Sayce.

Paruelo, J. M., and O. E. Sala, 1995. Water losses in the Patagonian steppe: a modeling approach, *Ecology*, **76**(2), 510–520.

Pastor, J., and W. M. Post, 1986. Influence of climate, soil moisture, and succession on forest carbon and nitrogen cycles, *Biogeochemistry*, **2**, 3–27.

Pastor, J., J. D. Aber, C. A. McClaugherty, and J. M. Melillo, 1984. Aboveground production and N and P cycling along a nitrogen mineralization gradient on Blackhawk island, Wisconsin, *Ecology*, **65**(1), 256–268.

Peixoto, J. P., and A. H. Oort, 1992. *Physics of Climate*, New York, American Institute of Physics.

Philip, J. R., 1957. Evaporation and moisture and heat fields in the soil, *Journal of Meteorology*, **14**, 354–366.

Pielke, R. A., 2001. Influence of the spatial distribution of vegetation and soils on the prediction of cumulus convective rainfall, *Review of Geophysics*, **39**(2), 151–177.

Pielke, R. A., 2002. *Mesoscale Meteorological Modeling*, San Diego, Calif., Academic Press.

Pike, J. G., 1971. The development of the water resources of the Okavango Delta, in *Proceeding of the Conference on Sustained Production for Semi-Arid Areas*, Botswana Notes and Records, special edition, **1**, 35–40.

Pollard, D., and S. L. Thompson, 1995. Use of a land-surface-transfer scheme (LSX) in a global climate model (GENESIS): the response to doubling stomatal resistance, *Global Planetary Change*, **10**, 129–161.

Porporato, A., and I. Rodríguez-Iturbe, 2002. Ecohydrology – a challenging multidisciplinary research perspective, *Hydrological Sciences Journal*, **47**(5), 811–822.

Porporato, A., P. D'Odorico, L. Ridolfi, and I. Rodríguez-Iturbe, 2000. A spatial model for soil-atmosphere interaction: model construction and linear stability analysis, *Journal of Hydrometeorology*, **1**(1), 61–74.

Porporato, A., F. Laio, L. Ridolfi, and I. Rodríguez-Iturbe, 2001. Plants in water-controlled ecosystems: active role in hydrological processes and response to water stress. III. Vegetation water stress, *Advances in Water Resources*, **24**(7), 725–744.

Porporato, A., P. D'Odorico, F. Laio, and I. Rodríguez-Iturbe, 2003a. Hydrologic controls on soil carbon and nitrogen cycles. I. Modeling scheme. *Advances in Water Resources*, **26**(1), 45–58.

Porporato, A., F. Laio, L. Ridolfi, K. K. Caylor, and I. Rodríguez-Iturbe, 2003b. Soil moisture and plant stress dynamics along the Kalahari precipitation gradient. *Journal of Geophysical Research*, **108**(D3), 4127–4134.

Porporato, A., E. Daly, and I. Rodríguez-Iturbe, 2004. Soil water balance and ecosystem response to climate change. *American Naturalist* (in press).

Porra, J. M., and J. Masoliver, 1993. Bistability driven by white shot noise, *Physical Review E*, **47**, 1633–1641.

Post, W., A. King, and S. Wullschleger, 1996. Soil organic matter models and global estimates of soil organic carbon, in *Evaluation of Soil Organic Matter Models*, D. S. Powlson, P. Smith, and J. U. Smith, (ed.), Vol. I 38, pp. 201–222, NATO ASI, Berlin, Springer-Verlag.

Post, W. M., and J. Pastor, 1996. Linkages – an individual-based forest ecosystem model, *Climate Change*, **34**(2), 253–261.

Potter, S. C., P. Matson, P. Vitousek, and E. A. Davidson, 1996. Process modeling of controls on nitrogen and trace gas emissions from the soils worldwide, *Journal of Geophysical Research*, **101**(D1), 1361–1377.

Prudnikov, A. P., Y. A. Brychkov, and O. I. Marichev, 1986. *Integrals and Series*, Vol. 2, New York, Gordon and Breach.

Ramanathan, V., 1988. The radiative and climatic consequences of the changing atmospheric composition of trace gases, in *The Changing Atmosphere*, F. S. Rowland and I. S. A. Isaksen, (ed.), pp. 159–186, New York, J. Wiley & Sons.

Richter, H., 1976. The water status in plant – experimental evidence, in *Water and Plant Life*, O. L. Lange, L. Kappen and E. D. Schulze, (ed.), pp. 42–55, New York, Springer-Verlag.

Ridolfi, L., P. D'Odorico, A. Porporato, and I. Rodríguez-Iturbe, 2000a. Duration and frequency of water stress in vegetation: an analytical model, *Water Resources Research*, **36**(8), 2297–2307.

Ridolfi, L., P. D'Odorico, A. Porporato, and I. Rodríguez-Iturbe, 2000b. Impact of climate variability on vegetation water stress, *Journal of Geophysical Research (Atm)*, **105**(D14), 18 013–18 025.

Ridolfi, L., P. D'Odorico, A. Porporato, and I. Rodríguez-Iturbe, 2003a. Stochastic soil moisture dynamics along a hillslope, *Journal of Hydrology*, **272**, 264–275.

Ridolfi, L., P. D'Odorico, A. Porporato, and I. Rodríguez-Iturbe, 2003b. On the influence of soil moisture dynamics on gaseous nitrogen emissions: a probabilistic framework, *Hydrological Sciences Journal*, **48**(5), 781–790.

Rietkerk, M., M. C. Boerlijst, F. van Langevelde, R. Hellerislambers, J. van de Koppel, L. Kumar, H. H. Prins, and A. M. de Roos, 2002. Self organization of vegetation in arid ecosystems, *American Naturalist*, **160**(4), 524–530.

Ripley, B. D., 1976. The second-order analysis of stationary processes, *Journal of Applied Probability*, **13**, 255–266.

Ritsema, C. J., J. Stolte, K. Oostindie, and E. Van der Elsen, 1996. Measuring and modelling of soil water dynamics and runoff generation in an agriculture hillslope, *Hydrological Processes*, **10**, 1081–1089.

Robinson, M., and T. J. Dean, 1993. Measurement of near surface soil water content using capacitance probe, *Hydrological Processes*, **7**, 77–86.

Rodríguez, M. A., and L. Pesquera, 1986. First-passage times for non-Markovian processes driven by dichotomic Markov noise, *Physical Review A*, **34**, 4532–4534.

Rodríguez-Iturbe, I., 2000. Ecohydrology: a hydrologic perspective of climate-soil-vegetation dynamics, *Water Resources Research*, **36**, 3–9.

Rodríguez-Iturbe, I., and A. Rinaldo, 1997. *Fractal River Basins: Chance and Self-Organization*, Cambridge, UK, Cambridge University Press.

Rodríguez-Iturbe, I., D. Entekhabi, and R. L. Bras, 1991. Non linear dynamics of soil moisture at climate scales, 1. Stochastic analysis, *Water Resources Research*, **27**, 1899–1906.

Rodríguez-Iturbe, I., A. Porporato, L. Ridolfi, V. Isham, and D. Cox, 1999a. Probabilistic modelling of water balance at a point: the role of climate, soil and vegetation, *Proceedings of the Royal Society of London A*, **455**, 3789–3805.

Rodríguez-Iturbe, I., P. D'Odorico, A. Porporato, and L. Ridolfi, 1999b. Tree-grass coexistence in savannas: the role of spatial dynamics and climate fluctuations, *Geophysical Research Letters*, **26**(2), 247–250.

Rodríguez-Iturbe, I., P. D'Odorico, A. Porporato, and L. Ridolfi, 1999c. On the spatial and temporal links between vegetation, climate, and soil moisture, *Water Resources Research*, **35**(12), 3709–3722.

Rodríguez-Iturbe, I., A. Porporato, F. Laio, and L. Ridolfi, 2001a. Plants in water-controlled ecosystems: active role in hydrological processes and responses to water stress, I. Scope and general outline, *Advances in Water Resources*, **24**(7), 697–705.

Rodríguez-Iturbe, I., A. Porporato, F. Laio, and L. Ridolfi, 2001b. Plant strategies to cope with stochastic soil water availability, *Geophysical Research Letters*, **28**(3), 4495–4498.

Ross, S. M., 1996. *Stochastic Processes*, 2nd edn., New York, Wiley & Sons.

Rosswall, T., F. Bak, R. J. Cicerone, R. Conrad, D. H. Ehalt, M. K. Firestone, I. E. Galbally, V. F. Galchenko, P. M. Groffman, H. Papen, W. S. Reeburgh, and E. Sanhueza, 1989. What regulates production and consumption of trace gases in ecosystems: biology or physicochemistry?, in *Exchange of Trace Gases between Terrestrial Ecosystems and the Atmosphere*, M. O. Andreae and D. S. Schimel, (ed.), pp. 73–95, New York, Wiley & Sons.

Sala, O. E., and W. K. Lauenroth, 1982. Small rainfall events: an ecological role in semiarid regions, *Oecologia*, **53**, 301–304.

Sala, O. E., W. K. Lauenroth, W. J. Parton, and M. J. Trlica, 1981. Water status of soil and vegetation in a shortgrass steppe, *Oecologia*, **48**, 327–331.

Sala, O. E., W. K. Lauenroth, and W. J. Parton, 1982a. Plant recovery following prolonged drought in a shortgrass steppe, *Agricultural Meteorology*, **27**, 49–58.

Sala, O. E., W. K. Lauenroth, and C. P. P. Reid, 1982b. A new dimension for niche separation between *Bouteloua gracilis* and *Agropyron smithii* in North American semi-arid grasslands, *Journal of Applied Ecology*, **19**, 647–657.

Sala, O. E., W. K. Lauenroth, and W. J. Parton, 1992. Long-term soil water dynamics in the shortgrass steppe, *Ecology*, **73**(4), 1175–1181.

Sala, O. E., R. B. Jackson, H. A. Mooney, and R. W. Howarth, 2000. *Methods in Ecosystem Science*, New York, Springer-Verlag.

Salisbury, F. B., and C. W. Ross, 1992. *Plant Physiology*, 4th edn., Belmont, Calif., Wadsworth Publishing Company.

Salvucci, G. D., 2001. Estimating the moisture dependence of root zone water loss using conditionally averaged precipitation, *Water Resources Research*, **37**(5), 1357–1365.

Salvucci, G. D., and P. S. Eagleson, 1992. *A Test of Ecological Optimality for Semiarid Vegetation*, Report 335, Massachusetts, USA, R. M. Parsons Lab.

Salvucci, G. D., and D. Entekhabi, 1995. Hillslope and climatic controls on hydrologic fluxes, *Water Resources Research*, **31**, 1725–1739.

Sancho, J. M., 1985. External dichotomous noise: the problem of the mean-first-passage time, *Physical Review A*, **31**, 3523–3525.

Sarmiento, G., 1984. *The Ecology of Neotropical Savannas*, Cambridge, Mass., Harvard University Press.

Scanlon, T. M., and J. D. Albertson, 2003a. Inferred controls on tree/grass composition in a savanna ecosystem: combining 16-year normalized difference vegetation index data with a dynamic soil moisture model, *Water Resources Research*, **39**(8), 1224–1237.

Scanlon, T. M., and J. D. Albertson, 2003b. Water availability and the spatial complexity of CO_2, water, and energy fluxes over a heterogeneous sparse canopy, *Journal of Hydrometeorology*, **4**, 798–809.

Schenk, H. J., and R. B. Jackson, 2002. The global biogeography of roots, *Ecological Monographs*, **72**(3), 311–328.

Schimel, D. S., B. H. Braswell, R. McKeown, D. S. Ojima, W. J. Parton, and W. Pulliam, 1996. Climate and nitrogen controls on the geography and timescales of terrestrial biogeochemical cycling, *Global Biogeochemical Cycles*, **10**(4), 677–692.

Schimel, D. S., B. H. Braswell, and W. J. Parton, 1997. Equilibration of terrestrial water, nitrogen and carbon cycles, *Proceedings of the National Academy of Sciences USA*, **94**, 8280–8283.

Schlesinger, W. H., 1997. *Biogeochemistry: an Analysis of Global Change*, 2nd edn., San Diego, Calif., Academic Press.

Schlesinger, W. H., J. F. Reynolds, G. L. Cunningam, L. F. Huenneke, W. M. Jarrel, R. A. Virginia, and W. G. Whitford, 1990. Biological feedbacks in global desertification, *Science*, **247**, 1043–1048.

Scholes, R. J., 1997. Savanna, in *Vegetation of Southern Africa*, R. M. Cowling, D. M. Richardson, and S. M. Pierce, (ed.), pp. 258–277, Cambridge, Cambridge University Press.

Scholes, R. J., and S. R. Archer, 1997. Tree-grass interactions in savannas, *Annual Review of Ecology and Systematics*, **28**, 517–544.

Scholes, R. J., and D. A. B. Parsons (ed.), 1997. *The Kalahari Transect: Research on Global Change and Sustainable Development in South Africa*, IGBP Report 42, Stockholm, IGBP Secretariat.

Scholes, R. J., and B. H. Walker, 1993. *An African Savanna*, New York, Cambridge University Press.

Scholes, R. J., P. R. Dowty, K. Caylor, D. A. B. Parsons, P. G. H. Frost, and H. H. Shugart, 2002. Trends in savanna structure and composition on an aridity gradient in Kalahari, *Journal of Vegetation Science*, **13**, 419–428.

Schulze, B. R., 1972. South Africa, in *Climates of Africa. World Survey of Climatology*, J. F. Griffiths, (ed.), Vol. 10, pp. 501–586, Amsterdam, Elsevier.

Schulze, E. D., 1986. Carbon dioxide and water vapor exchange in response to drought in the atmosphere and in the soil, *Annual Review of Plant Physiology*, **37**, 247–274.

Schulze, E. D., 1993. Soil water deficits and atmospheric humidity as environmental signals, in *Water Deficits: Plant Response from Cell to Community*, J. A. C. Smith and H. Griffiths, (ed.), Oxford, UK, Bios Scientific Publishers.

Schulze, E. D., and A. E. Hall, 1982. Stomatal responses, water loss, and CO_2 assimilation rates of plants in contrasting environments, in *Physiological Plant Ecology II. Water Relations and Carbon Assimilation*, O. L. Lange, P. S. Nobel, C. B. Osmond, and H. Ziegler, (ed.), pp. 181–230, Springer-Verlag, New York.

Schulze, E. D., H. A. Mooney, O. E. Sala, E. Jobbagy, N. Buchmann, G. Bauer, J. Canadell, R. B. Jackson, J. Loreti, M. Oesterheld, and J. R. Ehleringer, 1996. Rooting depth, water availability, and vegetation cover along an aridity gradient in Patagonia, *Oecologia*, **108**, 503–511.

Scifres, C. J., and B. H. Koerth, 1987. *Climate, Soils and Vegetation of the La Copita Research Area*, Texas Agricultural Experiment Station Report MP-1626, College Station.

Seastesdt, T. R., and A. K. Knapp, 1993. Consequences of nonequilibrium resource availability across multiple time scales – the transient maxima hypothesis, *American Naturalist*, **141**(4), 621–633.

Segal, M., R. Avissar, M. C. McCumber, and R. A. Pielke, 1988. Evaluation of vegetation effects on the generation and modification of mesoscale circulations, *Journal of Atmospheric Science*, **45**, 2268–2291.

Sellers, P. J., Y. Mintz, Y. C. Sud, and A. Dalchar, 1986. A simple biosphere model (SiB) for use within general circulation models, *Journal of Atmospheric Science*, **43**, 505–531.

Sellers, P. J., R. E. Dickinson, D. A. Randall, A. K. Betts, F. G. Hall, J. A. Berry, G. J. Collatz, A. S. Denning, H. A. Mooney, C. A. Nobre, N. Sato, C. B. Field, and A. Henderson-Sellers, 1997. Modeling the exchanges of energy, water, and carbon between continents and the atmosphere, *Science*, **275**, 502–509.

Shakya, N. M., and S. Chander, 1998. Modelling of hillslope runoff processes, *Environmental Geology*, **35**, 115–123.

Shani, U., and L. M. Dudley, 1996. Modeling water uptake by roots under water and salt stress: soil-based and crop response root sink terms, in *Plant Roots: The Hidden Half*, 2nd edn., Y. Waisel, A. Eshel, and U. Kafka, (ed.), pp. 635–641, New York, Marcel Dekker Inc.

Shmida, A., and T. L. Burgess, 1988. Plant growth-form strategies and vegetation types in arid environments, in *Plant Form and Vegetation Structure*, M. J. A. Werger, P. J. M. van der Aart, H. J. During, and J. T. A. Verhoeven, (ed.), The Hague, the Netherlands, SBP Academic Publishing.

Shmida, A., M. Evenary, and I. Noy-Meir, 1985. Hot desert ecosystems: an integrated view, in *Hot Deserts and Arid Shrublands, B, Ecosystems of the World 12 B*, M. Evenary, I. Noy-Meir, and D. W. Goodall, (ed.), Amsterdam, Elsevier.

Shuttleworth, W. J., 1989. Micrometeorology of temperate and tropical forest, *Philosophical Transactions of the Royal Society of London Series B*, **324**, 299–331.

Simmons, C. S., and P. D. Meyer, 2000. A simplified model for the transient water budget of a shallow unsaturated zone, *Water Resources Research*, **36**(10), 2835–2844.

Singh, J. S., D. G. Milchunas, and W. K. Lauenroth, 1998. Soil water dynamics and vegetation patterns in a semiarid grassland, *Plant Ecology*, **134**, 77–89.

Siqueira, M., C. T. Lai, and G. Katul, 2000. Estimating scalar sources, sinks and fluxes in a forest canopy using Lagrangian, Eulerian and hybrid inverse models, *Journal of Geophysical Research*, **105**(D24), 29 475–29 488.

Siqueira, M., G. Katul, and C. T. Lai, 2002. Quantifying net ecosystem exchange by multilevel ecophysiological and turbulent transport models, *Advances in Water Resources*, **25**, 1357–1366.

Sisson, J. B., W. M. Klittich, and S. B. Salem, 1988. Comparison of two methods for summarizing hydraulic conductivities of layered soil, *Water Resources Research*, **24**(8), 1271–1276.

Skopp, J., M. D. Jawson, and J. W. Doran, 1990. Steady state aerobic microbial activity as a function of soil water content, *Soil Science Society of America Journal*, **54**, 1619–1625.

Slatyer, R. O., 1967. *Plant Water Relationships, London, Academic Press.*

Sloan, P. G., and I. D. Moore, 1984. Modelling subsurface storm flow on steeply sloping forested watershed, *Water Resources Research*, **20**, 1815–1822.

Smith, J. A. C., and H. Griffith, 1993. *Water Deficits: Plant Responses from Cell to Community*, Oxford, UK, Bios Scientific Publishers.

Sole, R. V., and B. Goodwin, 2000. *Signs of Life*, New York, Basic Books.

Sole, R. V., and S. Manrubia, 1995. Are rainforests self-organized in a critical state?, *Journal of Theoretical Biology*, **173**, 31–40.

Sole, R. V., and O. Miramontes, 1995. Information at the edge of chaos in fluid neural networks, *Physica D*, **80**, 171–180.

Sperry, J. S., F. R. Adler, G. S. Campbell, and J. P. Comstock, 1998. Limitation of plant water use by rhizosphere and xylem conductance: results from a model, *Plant, Cell and Environment*, **21**, 347–359.

Sperry, J. S., U. G. Hacke, R. Oren, and J. P. Comstock, 2002. Water deficits and hydraulic limits to leaf water supply, *Plant, Cell and Environment*, **25**, 251–263.

Spittlehouse, D. L., and T. A. Black, 1981. A growing season water balance model applied to two douglas fir stands, *Water Resources Research*, **17**(6), 1651–1656.

Stagnitti, F., J.-Y. Parlange, T. S. Steenhuis, M. P. Parlange, and C. W. Rose, 1992. A mathematical model of hillslope and watershed discharge, *Water Resources Research*, **28**, 2111–2122.

Stark, J. M., and M. K. Firestone, 1995. Mechanisms for soil moisture effects on activity of nitrifying bacteria, *Applied and Environmental Biology*, **61**, 218–221.

Stauffer, D., and A. Aharony, 1992. *Introduction to Percolation Theory*, London, Taylor & Francis.

Stephenson, N. L., 1990. Climatic control of vegetation distribution: the role of the water balance, *American Naturalist*, **135**(5), 649–670.

Stephenson, N. L., 1998. Actual evapotranspiration and deficit: biologically meaningful correlates of vegetation distribution across spatial scales, *Journal of Biogeography*, **25**, 855–870.

Steudle, E., R. Oren, and E. D. Schulze, 1987. Water transport in maize roots, *Plant Physiology*, **84**, 1220–1232.

Stevenson, F. J., 1986. *Cycles of Soil*, New York, John Wiley & Sons.

Stroh, J. C., S. R. Archer, L. P. Wilding, and J. P. Doolittle, 1996. Detection of edaphic discontinuities in ground-penetrating radar and electromagnetic induction, in *La Copita Research Area: Consolidated Progress Report*, J. W. Stuth and S. M. Dudash, (ed.), Report of Texas Agricultural Experiment Station, Texas A&M University, College Station, Texas, USA.

Swap, R., J. T. Suttles, H. Annegarn, Y. Scorgie, J. Closs, J. Prinvette, and B. Cook, 2001. Report on SAFARI 2000 outreach activities, intensive field campaign planning meeting, and data management workshop, *The Earth Observer*, **12**(3), 26–28.

Takacs, L., 1955. Investigation of waiting-time problems by reduction to Markov processes, *Acta Mathematica, Academy of Sciences of Hungary*, **6**, 101.

Tate, R. L., 1987. *Soil Organic Matter: Biological and Ecological Effects*, New York, Wiley & Sons.

Tezara, W., V. J. Mitchell, S. D. Driscoll, and D. W. Lawlor, 1999. Water stress inhibits plant photosynthesis by decreasing coupling factor and ATP, *Nature*, **401**(6756), 914–917.

References

Thomas, D. S. G., and P. A. Shaw, 1991. *The Kalahari Environment*, Cambridge, Cambridge University Press.

Thornley, J. H. M., J. Bergelson, and A. J. Parsons, 1995. Complex dynamics in a carbon-nitrogen model of a grass-legume pasture, *Annals of Botany*, **75**, 79–94.

Tilman, D., 1994. Competition and biodiversity in spatially structured habitats, *Ecology*, **75**(1), 2–16.

Tilman, D., 1996. Biodiversity: population versus ecosystem stability, *Ecology*, **77**(2), 350–363.

Tilman, D., and D. Wedin, 1991. Oscillations and chaos in the dynamics of a perennial grass, *Nature*, **353**, 653–655.

Tinley, K. L., 1982. The influence of soil moisture balance on ecosystem patterns in Southern Africa, in *Ecology of Tropical Savannas*, B. J. Huntley and B. H. Walker, (ed.), pp. 175–192, Berlin, Springer-Verlag.

Tsuboyama, Y., R. C. Sidle, S. Noguchi, and I. Hosoma, 1994. Flow and solute transport through the soil matrix and macropores of a hillslope segment, *Water Resources Research*, **30**, 879–890.

Turner, N. C., 1986. Adaptation to water deficit: a changing perspective, *Australian Journal of Plant Physiology*, **13**, 175–190.

Turner, N. C., and M. M. Jones, 1980. Turgor maintenance by osmotic adjustment: a review and evaluation, in *Adaptation of Plants to Water and High Temperature Stress*, N. C. Turner and P. J. Kramer, (ed.), pp. 87–103, New York, Wiley.

Turner, N. C., E. D. Schulze, and T. Gollan, 1985. The responses of stomata and leaf gas exchange to vapour pressure deficits and soil water content. II. In the mesophytic herbaceous species, *Helianthus annuus*, *Oecologia*, **65**, 348–355.

Tyson, P. D., 1986. *Climatic Change and Variability in Southern Africa*, Oxford, Oxford University Press.

Tyson, P. D., and S. J. Crimp, 1998. The climate of the Kalahari transect, *Transactions of the Royal Society of South Africa*, **53**, 93–112.

U.S. Department of Agriculture, 1951. Soil survey manual, *U.S. Department of Agriculture Agriculture Handbook*, **18**.

U.S. Department of Agriculture, 1979. *Soil survey of Jim Wells County, Texas*, Washington D.C., Soil Conservation Service.

Van den Honert, T. H., 1948. Water transport as a catenary process, *Discussions of the Faraday Society*, **3**, 146–153.

Van Kampen, N. G., 1992. *Stochastic Processes in Physics and Chemistry*, Amsterdam, North Holland.

Vanmarke, E., 1983. *Random Fields: Analysis and Synthesis*, Cambridge, Mass., MIT Press.

Van Wijk, M. T., and I. Rodríguez-Iturbe, 2002. Tree-grass competition in space and time: insights from a simple cellular automata model based on ecohydrological dynamics, *Water Resources Research*, **38**(9), 1179–1189.

Walker, B. H., D. Ludwig, C. S. Holling, and R. S. Peterman, 1981. Stability of semi-arid savannas grazing systems, *Journal of Ecology*, **69**, 473–498.

Walter, H., 1971. *Ecology of Tropical and Subtropical Vegetation*, Edinburgh, Oliver & Boyd.

Wan, C., and R. E. Sosebee, 1991. Water relations and transpiration of honey mesquite on 2 sites in West Texas, *Journal of Range Management*, **44**(2), 156–160.

Waring, R. H., and S. W. Running, 1998. *Forest Ecosystems: Analysis at Multiple Scales*, 2nd edn., San Diego, Calif., Academic Press.

Webb III, T., R. A. Laseski, and J. C. Bernabo, 1978. Sensing vegetational patterns with pollen data: choosing the data, *Ecology*, **59**, 1151–1163.

Weltzin, J. F., and G. R. McPherson, 1997. Spatial and temporal soil moisture resource partitioning by trees and grasses in a temperate savanna, Arizona, USA, *Oecologia*, **112**, 156–164.

Western, D. and C. van Praet, 1973. Cyclical changes in the habitat and climate of an East African ecosystem, *Nature*, **241**, 104–106.

Western, W. A., and R. B. Grayson, 1998. The Tarrawarra data set: soil moisture patterns, soil characteristics, and hydrological flux measures, *Water Resources Research*, **34**, 2765–2768.

Western, W. A., R. B. Grayson, G. Bloschl, G. R. Willgoose, and T. A. McMahon, 1999. Observed spatial organization of soil moisture and its relation to terrain indices, *Water Resources Research*, **35**, 797–810.

Whitehead, D., 1998. Regulation of stomatal conductance and transpiration in forest canopies, *Tree Physiology*, **18**, 633–644.

Wiegand, T., K. A. Moloney, and S. J. Milton, 1998. Population dynamics, disturbance, and pattern evolution: identifying the fundamental scales of organization in a model ecosystem, *American Naturalist*, **152**(3), 321–337.

Wigmosta, M. S., L. W. Vail, and D. P. Lettenmaier, 1994. A distributed hydrology-vegetation model for complex terrain, *Water Resources Research*, **30**, 1665–1679.

Wild, A. (ed.), 1988. *Russell's Soil Condition and Plant Growth*, 11th edn., Essex, UK, Longman Group.

Williams, C. A., and J. D. Albertson, 2004. Soil moisture controls on canopy scale water and carbon dioxide fluxes in an African savanna, *Water Resources Research* (in press).

Williams, E. J., M. Keller, and M. Nunez, 1992. NO_x and N_2O emissions from soil, *Global Biogeochemistry Cycles*, **6**, 351–388.

Williams, M., E. B. Rastetter, D. N. Fernandes, M. L. Goulden, S. C. Wofsy, G. R. Shaver, J. M. Melillo, J. W. Munger, S. M. Fan, and K. J. Nadelhoffer, 1996. Modelling the soil-plant-atmosphere continuum in a *Quercus-Acer* stand at Harvard forest: the regulation of stomatal conductance by light, nitrogen and soil/plant hydraulic properties, *Plant, Cell and Environment*, **19**(8), 911–927.

Yang, W. X., and F. X. Meixner, 1997. Laboratory studies on the release of nitric oxide from subtropical grassland soils: the effect of soil temperature and moisture, in *Gaseous Nitrogen Emissions from Grasslands*, S. C. Jarvis and B. F. Pain, (ed.), pp. 67–71, Wallingford, UK, CAB International.

Yeh, J. -F., and E. A. B. Eltahir, 1998. Stochastic analysis of the relationship between topography and the spatial distribution of soil moisture, *Water Resources Research*, **34**, 1251–1263.

Yonker, C. M., D. S. Schimel, E. Paroussis, and D. Heil, 1988. Patterns of organic carbon accumulation in a semiarid shortgrass steppe, Colorado, *Soil Science Society of America Journal*, **52**, 478–483.

Zeng, N., J. D. Neelin, K. M. Lau, and C. J. Tuker, 1999. Enhancement of interdecadal climate variability in the sahel by vegetation interaction, *Science*, **286**, 1537–1540.

Zhu, A. X., and D. S. Mackay, 2001. Effects of spatial detail of soil information on watershed modeling, *Journal of Hydrology*, **248**, 54–77.

Zhu, T. X., Q. G. Cai, and B. Q. Zeng, 1997. Runoff generation on a semi-arid agricultural catchment: field and experimental studies, *Journal of Hydrology*, **196**, 99–118.

Zimmermann, M. H., 1983. *Xylem Structure and the Water Ascent of Sap*, New York, Springer-Verlag.

Zimmermann, U., U. Haase, D. Langbein, and F. K. Meixner, 1993. Mechanisms of long-distance water transport in plants: a re-examination of some paradigms in the light of new evidence, *Philosophical Transactions of the Royal Society of London B*, **341**, 19–31.

Zimmermann, U., F. Meinzer, and F. W. Bentrup, 1995. How does water ascend in tall trees and other vascular plants?, *Annals of Botany*, **76**, 545–551.

Species Index

Subject Index

Abscisic acid (ABA), 99
Active soil depth, *see* Rooting depth
Albedo, 7, 308
Ammonium, 313, 327, 338
Apoplastic water, 89, 90, 95
Assimilation, 98, 179, 188, 189, 190, 201
 water stress, 207
Atmospheric boundary layer, 179, 185

Bare soil, 14, 380, 388
Bimodality, 255, 257, 265
Biodiversity, 3, 5, 389
Biomass, 3, 5, 121, 154
 allocation, 203
 microbial, *see* Microbial biomass
Blue grama, *see* Bouteloua gracilis
Bound, *see* Saturation
Budyko, 52, 55, 249

C_3 plants, *see* Photosynthetic pathways
C_4 plants, *see* Photosynthetic pathways
C/N ratio, 312, 315, 321, 323, 335
CAM, *see* Photosynthetic pathways
Carbon assimilation, *see* Assimilation
Carbon cycle, 307, 310, 318, 338, 340
Carbon dioxide, 7, 310
Carbon-to-nitrogen ratio, *see* C/N ratio
Carboxylation, 189
Cavitation, 96, 99, 183, 204
Cellular automaton, 372, 374, 406
Chaos, *see* Self-sustained oscillations
Chapman–Kolmogorov, 238
 backward equation, 61, 82, 83
 forward equation, 33, 36, 61, 82, 355
Climate change, 25, 55, 209
Climate–soil–vegetation system, 2, 6
Closure problem, 239, 240
Clusters of vegetation, 13, 366, 384, 385, 412, 413
Coherence function, 401, 403
Cohesion theory, 93
Colonization, 12, 374, 386, 390, 406
Competition, 12, 361, 363, 386, 406
 colonization model, 386, 406
Competitive-exclusion principle, 361
Conductance to water flow
 leaf, 26
 soil–root, 96, 181
 stomatal, 96, 183, 186, 188, 194
 xylem, 181
Cross-correlation, 401, 402

between relative soil moisture and nitrate, 342
between soil moisture and nitrate, 341, 342
Crossing analysis, 59, 66, 68
 frequency of crossing, 65, 68
 mean time of crossing, 62
 number of crossing, 68
 onset of stress, 70
 relation to backward equation, 84
 wilting point, 69

Decomposition, *see* Nitrogen mineralization
Deep infiltration, *see* Leakage
Diffusion, 83, 190, 314, 330, 360
Drought, 32, 86, 97, 130, 342, 374, 384
 resistance, 159
Drought deciduousness, 11, 100, 174
Dryness index, 52, 276

Ecohydrology, 1
Ecosystem structure, 12, 360
Effective rainfall, 114
Equilibrium, 12, 118, 216, 258, 348, 349, 356, 364, 387
Evaporation, *see* Evapotranspiration
Evapotranspiration, 25
 hourly dynamics, 184
 maximum, 45
 optimization, 227, 232
 stressed, 45
 temporal scaling, 200
 three stage sequence of, 46
 unstressed, 45
Evergreen species, 11, 215
Extensive users, 213

Field capacity, 31
Fire, 87, 154, 333
First-order kinetics, 312
Forests
 Amazonian, 215
 eastern European, 13
 northwestern United States, 8, 41
 transition to pastures, 215
Fractals, 381, 386
Frequency of crossings, *see* Crossing analysis

Grasslands
 Konza prairie, 210
 North American, 118
 -to-shrubland transition, 14, 372
 turnover of soil organic matter, 311